Chemistry & Technology of UV & EB Formulations for Coatings, Inks and Paints

VOLUME VII

Photoinitiation for Polymerization: UV & EB at the Millenium

VOLUME VII

Photoinitiation for Polymerization: UV & EB at the Millenium

Dr D.C. Neckers
and
Dr Wolter Jager

JOHN WILEY AND SONS

CHICHESTER ● NEW YORK ● WEINHEIM ● BRISBANE ● TORONTO ● SINGAPORE

Published in association with

SITA TECHNOLOGY LIMITED
LONDON UK

Copyright ©1998 SITA Technology Limited
Nelson House,
London SW19 7PA

Published in 1998 by
John Wiley & Sons Ltd
in association with SITA Technology Limited

All rights reserved

No part of this book may be reproduced by any means,
or transmitted, or translated into a machine language
without the written permission of the publisher

Wiley Editorial Offices
John Wiley & Sons Ltd., Baffins Lane,
Chichester, West Sussex, England. PO19 1UD

John Wiley & Sons, Inc., 605 Third Avenue,
New York, NY 10158-0012, USA

VCH Verlagsgesellschaft mvH, Pappelallee 3
D-69469 Weinheim, Germany

Jacaranda Wiley Ltd, G.P.O. Box 859, Brisbane
Queensland 4001, Australia

John Wiley & sons (Canada) Ltd, 22 Worcester Road,
Rexdale, Ontario M9W 1L1, Canada

John Wiley & Sons (SEA) Pte Ltd, 37 Jalan Pemimpin 05-04,
Block B, Union Industrial Building, Singapore 2057

A catalogue record for this book is available from the British Library

ISBN 0471 982350

CONTENTS

CHAPTER I   PRINCIPLES OF PHOTOINITIATED POLYMERIZATION

Preface iii

1. Introduction 3

2. Photoinitiation in an Historical Perspective 8

3. The Photochemistry and Photophysics of Initiator Systems: the basic Principles of Light Absorption by Organic Compounds and the Disposition of the Energy derived by the Absorption Process 12
   i.   Light Absorption 13
   ii. The Disposition of Absorbed Energy 14
   iii. Light Emission -One method of disposing of an excited state's excess energy: Case 1 - fluorescence 15
   iv. Case 2 - Phosphorescence 19

4. Transient Spectroscopy and Characterizing Excited, or Photochemically Reactive, States 23

5. Reactions from Excited States 32
   i.   Case 1: Formation of Free Radicals by Bond Dissociation 32
   ii. Reactivity and Mechanism in a known Type I Photoinitiator ... 36
   iii. Case 2: Other reactions from the excited state, manifold-oxidation/reduction 41
   iv. Generating cations by photochemical routes 47
   v.   Case 3: Generating reactive bases 51

6. Lamps, Lasers and Other Sources 69
   i.   Lasers 77

References 83

CHAPTER II UPDATE ON PHOTOPOLYMERIZED SYSTEMS 91

1. Bond dissociation Initiators: Bond Homolysis after the
   Absorption of a Photon 91
   - i. The Carbonyl Group 92
   - ii. Characterizing the Reactivity of the Carbonyl Group 95
   - iii. Electron Transfer from Tertiary Amine 98
   - iv. Reactions of Excited Carbonyl Groups: 2+2•
     Cycloaddition 100
   - v. Hydrogen Abstraction - A Route to Crosslinks with
     Carbonyl Compound Initiators 101
   - vi. Phenyl Glyoxylate Photochemistry with Shengkiu Hu 105

2. An Imaging System based on the Phenylglyoxylate Chromophore 109
   - i. Resolution of an Imaging System based on Polymeric
     Phenylglyoxylate 111
   - ii. Sensitivity and Contrast 112
   - iii. Predevelopment Images 113
   - iv. Photopolymer Analysis 115

3. The Carbonyl Group as a Target using Carbonyl Group
   Photochemistry to promote other Reactions 116
   - i. Peroxy Ester Chemistry 116
   - ii. Triphenylmethyl Radical Photochemistry - A second
     Example of Carbonyl Groups as a target with
     Alexander Nikolaitchik 119
   - iii. p-Benzoylphenyldiphenylmethyl Radical 122
   - iv. Photochemistry of RAD in Solution 127
   - v. Photochemistry and Photophysics in the Absence of
     Dimer 128
   - vi. Fluorescence of TRB in Rigid Medium 132
   - vii. Electrochemistry of TBR 134
   - viii. Transient Studies 135

4. Oxidation Reduction: Single Electron Transfer involving
   Compounds containing Carbonyl Groups 140
   - i. Camphor Quinone 140
   - ii. Electron Transfer to Benzophenone Triplets -
     Reduction with Amines 142

5. Intramolecular Electron Transfer: Carbonyl Compound Chemistry leading to Intramolecular Oxidational Reduction with Shengkui Hu ... 144

6. Charge Transfer Complex Photoinitiators ... 154
   i. Principles of Charge/Transfer Photoinitiators ... 155
   ii. Mechanism ... 166
   iii. Electron Transfer from Organoborates with Alexander Polykarpov, Ananda Sarker, Yuki Kaneko and Kesheng Feng ... 169
   iv. Tertiary as Electron Acceptors ... 170
   v. Borates as Electron Donors ... 183
   vi. Iodonium /borates with Kesheng Feng ... 184
   vii. Photoinitiating Activities - Iodonium/borates ... 187
   viii. Photochemical Reactions of Iodonium Borate Salts with Methyl Methacrylate ... 191

7. Oxidation/Reduction Reactions with Visible Initiator Systems: The Case of Organic Dyes ... 195
   i. Photopolymerization with the Fluorones ... 204
   ii. Mechanism of Photoreduction of DIBF ... 205
   iii. DIBF Radicals: Formation, Decay and Absorption Spectra ... 210

8. Oxidation/Reduction Chains Cationic Polymerization initiated by Dye/Onium Salts ... 217

9. Summary ... 219

References ... 221

CHAPTER III ANALYTIC TECHNIQUES FOR MONITORING PHOTO-POLYMERIZATION PROCESSES

1. Introduction — 235

2. General Research Techniques for Analysing Polymeric Films — 239
   i. Introduction — 239
   ii. Calorimetric Techniques; Differential Scanning Calorimetry — 239
   iii. Spectroscopic Techniques — 241
       a) Infrared Spectroscopy — 243
       b) UV/Vis Spectroscopy — 252
       c) Emission Spectroscopy — 261
           - *Polarization of emission* — 265
           - *Lifetime of emission* — 268
           - *Intensity of emission* — 271
           - *Position of the emission maximum* (λ_{max}) — 279
               Excimer forming probes — 282
               Twisted Intramolecular Charge Transfer (TICT) probes — 285
               Intramolecular charge transfer (ICT) probes 288
               Organic salt — 306
       d) NMR Spectroscopy — 311
       e) ESR Spectroscopy — 319
       f) Dielectric Spectroscopy — 320

3. Techniques for monitoring photopolymerization processes in real time — 321
   i. Introduction — 321
   ii. Photo Differential Scanning Calorimetry (Photo DSC) — 322
   iii. Real Time (FT) IR Spectroscopy — 329
   iv. Real Time UV Spectroscopy — 340
   v. Real Time Fluorescence Spectroscopy — 351
   vi. Dielectric Spectroscopy — 362

4. On Line Monitoring — 363
   i. Introduction — 363
   ii. Fluorescence Spectroscopy for On Line Monitoring — 364

References — 367

Index — 375

CHAPTER I

PRINCIPLES OF PHOTOINITIATED POLYMERIZATION

Preface:

When I was asked to write a revision of the very successful volume on photoinitiators, I accepted too quickly. I had just stepped down, after 23 years, as Chair of the Department of Chemistry at Bowling Green. I thought I would have time and anticipated taking a sabbatical at Cambridge. But almost immediately the University found another administrative duty for me, as Chair of the Board of the University's Research/Enterprise Park initiative. This has filled my spare time, and then some. Nevertheless, with a lot of patience from Gordon Birtles and Ellinor Birtles and some tremendous help from Barbara Anderson, my Ph.D. students and post-doctorals, I have finally gotten the job done.

What we have produced embodies some of the things we teach in our Ph.D. program in the Photochemical Sciences at Bowling Green. I have assumed that the typical practitioner of radiation cure has at least an undergraduate degree in chemistry or has had undergraduate courses in organic and physical chemistry. As a result, I spend some time developing a qualitative approach to photochemical reaction dynamics and spend more time discussing the experiments that lead to various theories than on the complexities of the theories themselves. We make no apologies for that. The interested reader can find much greater detail in texts on photophysics, chemical kinetics or quantum mechanics.

I have drawn liberally on the recent literature and am particularly indebted to the thesis research, and the actual dissertations, of a number of students and graduates of the Ph.D. program in the Photochemical Sciences at Bowling Green. The work on atomic force microscopy and its use in the analysis of imaging systems is entirely that of Alexander Mejiritski. Alex, with George Hammond's sage advice, introduced atomic force microscopy to my labs as well as a number of other surface analysis techniques. Some of the work has been published but we provide more details and a rational in Chapter 1. Alexander submitted his dissertation in August, 1997. The work on phenylglyoxylates is solely that of Shengkui Hu who completed his dissertation in the spring of 1998. When reading the 400 or so pages of his thesis in detail, what I discovered was that the phenylglyoxylate chromophore, almost like no other, allows one to study a group of photochemical reactions without changing the absorbing chromophore, the excited state it forms, or the analytical methods that one can use to discern reactivity. I know of no other system as versatile so I have drawn on Hu's excellent work in some detail. As in the case of Mejiritski's work, some of Hu's data has been published elsewhere and the reader is referred to the references for greater detail. The work on the photochemistry of the triphenylmethyl systems in Chapter 2 in entirely that of Sasha Nikolaichik and has not been published elsewhere except in his dissertation which was submitted in 1995. I wanted to use this work to illustrate the principle of targeting a chromophore to achieve predictable reactivity. We have used that concept often. Synthetic compounds containing a given chromophore of known and predictable reactivity may limit the mechanistic questions that need to be probed. By maintaining the excited state characteristics of the chromophore constant, one can investigate reactions of susceptible functionalities like peroxides in a systematic way.

The intramolecular chemistry of the borate systems described in detail in Chapter 2 is a group effort. Alex Polykarpov, who completed his dissertation on the borates in 1996 stimulated a lot of work in my laboratories mainly through his enthusiasm for chemistry and his prodigious intellect. He and Ananda Sarker made a series of borate ammonium salts and then they collaborated with physical chemist Salah Hassoon who studied their photophysical properties. Not satisfied alone with that, and after hearing continuously about photopolymerization in my laboratories, Polykarpov and Sarker decided to make some polymeric onium borates, and Mejiritski talked them into making imaging systems from them. The results of this teamwork are sprinkled liberally throughout Chapter 2. Since most academic programs aren't lifetime occupations, and my students/post-doctorals are forever finding gainful employment (thank goodness), when Salah left for a job at the Technion in Tel Aviv, Yuji Kaneko arrived from Tsukuba. He did the latter work on the borates.

One of the charming things about being a university professor comes from meeting persons from all over the world who love science, love chemistry and have the necessary work ethic to translate it all into useful research. Polykarpov, Dustin Martin, Tom Marino, Anita DeRaaff and I were continually arguing about just how a system iodonium salt, borate salt, dye sensitizer was ablt to accelerate the polymerization of acrylates over binary systems used as models. So Kesheng Feng studied, in detail, the reactions of iodonium salt/borate salt initiators. He isolated stable complexes, characterized their absorption spectra and then isolated all of the products from the reactions of specific iodonium, borates with methyl methacrylate. The work is being published here for the first time.

All of the comparisons of reactions with various light sources were done at Spectra Group by Tom Marino, Dustin Martin and Anita DeRaaff and I owe these fine scientists a great deal for turning some of the concepts into practical reality. They work very hard and have done yeoman's work in developing skills in the art of photopolymerization as an industrial activity.

Wolter Jager, now at the Technical University at Delft, wrote Chapter 3. Wolter helped us understand the mechanisms by means of which fluorescence probes detect changes in their immediate environment and, with Roman Popeelarz, turned this from an idle observation to a useful analytical phenomenon.

I apologize in advance for several things that I did not cover. I spent no time on the Rhone Poulenc iodonium tetra(perfluorophenyl)borate cationic initiators. They have been covered by others in other places. Work on platinum catalysts to activate the polymerization of carbosilanes that was going on in my laboratories simultaneously (Fry,

B.E., Guo, A. and Neckers, D. C. "Photoactivated Hydrosilation Curing of a Ceramic Precursor: Crosslinking and Pyrolysis of Branched Oligo[(methylsilylene)methylene]" J. Organomet, Chem., 1997, 538, 151; Fry, B. E., Neckers, D. C. "Rapid Photo-Activated Hydrosilation Polymerization of Vinyldimethylsilane," Macromolecules, 1996, 29, 5306-5312; Guo, A.; Fry, B.E.; Neckers, D. C. "Highly Active Visible-Light Photocatalysts for Curing a Ceramic Precursor," Chem. Mater., 1998, 10, 531-536} is also not discussed. Neither is work occurring simultaneously in Fouessier's lab that reports the use of platinum catalysts for epoxysilicone polymerizations. The chapter on analytical methods omits a number of relatively clever, though specific, methods which have little general use.

Again, we appreciate the long toothed patience of Gordon Birtles. Though I swear I agreed to work on this volume on July 1 and he was asking me where the manuscript was by July 4, he has been patient with me and with Wolter Jager. Unfortunately we both have day jobs.

Several have read this manuscript but all the mistakes are mine. Please let me know if I have misquoted you, or misinterpreted your work. Thanks in advance.

Douglas C. Neckers.

Chapter One

I. Principles of Photoinitiated Polymerization

1.   Introduction

Since it was first introduced by Dupont in the 40's[1], photoinitiated polymerization has become an important industrial process. The main positive attributes of photochemical processes are that they offer a rapid conversion of formulated reactive liquids to solids.

Photopolymerization reactions can be carried out in an image forming fashion, i.e., they can form a particular pattern, generally as a thin film on a surface. Alternatively they can form layers. Among the principle commodity uses of photopolymers are coatings, paints, varnishes, inks and adhesives, and the most common chemistry used for such applications are acrylates.

Though developed for proofing applications, many of the acrylates reported by Plambeck served as clear coatings and found eventual uses in that industry which began to develop sizable sales in the 60's. In the beginning, the advantage of the photoprocess was in the production of clear uncoloured shiny coats such as those found on the covers of expensive magazines.

Photospeed, clarity, processability and the formation of a non-yellow, non-tacky thin[2] coat were the most important attributes. Surface images were also often formed by processes other than chain reactions.

The polymerization of vinyl monomers, compounds with $CH_2=CH-$ functions, was first recognized by Staudinger,[3] who outlined the concept of the 'macromolecule' in 1920 and won the Nobel prize for this work in 1933. Vinyl polymers are formed in chain reactions comprised by several discrete steps: initiation, propagation and termination. A particularly useful kinetic approach involves the assumption that the concentration of the evolving reaction intermediate, a free radical, is so low that it does not change with time after the earliest onset of reaction. Since most polymerizations were developed for the production of bulk polymers, thermal reactions were originated to initiate chain polymerization processes and kinetic treatments constructed on this basis. The assumption that radical concentrations do not change with time, allowed for the steady state treatment of chain polymerization processes and this simplified their analysis. In that treatment, Scheme 1.

Scheme 1:
Mechanism in free radical polymerization

$$In \longrightarrow In\bullet \quad R_d = k_d f [In] \quad \text{initiation}$$
$$In\bullet + M \longrightarrow M\bullet \quad R_i = k_i[In\bullet][M]$$
$$M\bullet + M \longrightarrow M'\bullet \quad R_p = k_p[M\bullet][M] \quad \text{propagation}$$
$$M\bullet + M\bullet \longrightarrow products \quad R_t = k_t[M\bullet]^2 \quad \text{termination}$$

k = the rate constants
[M] = monomer concentrations
f = efficiency

Where R_j = rate of the specific steps (j = d, t, p, etc.).

Given that the change in [M•] with time is zero, (d[M•]/dt = 0) one can equate the rate of initiation (the step that produces radicals) to the rate of termination (the step that consumes radicals) and the relationships in equations 1 and 2 result.

$k_d f [In] = k_t[M\bullet]^2$; rearranging $M\bullet = \{(k_d/k_t)f[In]\}^{1/2}$ **Equation 1**

and

$R_p = k_p \{(k_d/k_t)f[In]\}^{1/2} [M]$ or $R_p = (k_p/(k_t)^{1/2}) \bullet R_I^{1/2} \bullet [M]$ **Equation 2**

Since one can measure every variable in equation 2, one can obtain a ratio of the rate constants, $k_p(k_d f/k_t)^{1/2}$ from the steady state approach. The value of k_d can be independently measured as well, so that $k_p/(k_t)^{1/2}$ is obtained using steady state techniques. Measurement of the individual rate constants k_p and k_t requires other techniques. The adherence of polymerization rate, to the steady state approximation is given for styrene, vinyl acetate and methyl methacrylate in figure 1.

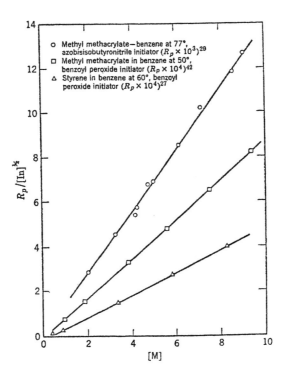

Figure 1

Methyl methacrylate initiated by azobisisobutyronitrile; initiated by benzoyl peroxide; and styrene polymerization rates over several hundred fold change in monomer concentration.

Individual values of k_p and k_t for several of the common monomers were obtained in the 50's using a rotating sector method. In this, a wheel from which a small section has been removed is placed in front of an irradiating light source so that the sample is exposed alternatively to light and dark, by rapidly rotating the wheel. The rate of polymerization is then measured viscometrically, as the rate of sector rotation is changed.[4]

The rate constants k_p and k_t are reported for a number of common monomers in Table 1. Termination, or the bimolecular reaction of radicals, is a substantially faster process, than the addition of a radical to the double bond of another monomer molecule. The values of the termination rate constants measured by the sector method are usually of the order of 1 x 10^7 litre/mol/sec, while propagation rate constants are at least 3 orders of magnitude less. In pure monomer, the monomer concentration is substantially higher than is the concentration of the chain carrying radical. The result is that, at least through the majority of the polymerization, the addition of initiator fragments to the double bond of the target monomer is accomplished and the polymer is formed by subsequent reactions of that radical, with another monomer unit, while bimolecular reactions of radicals are prevented by their low concentrations. If k_p is 1000 x less than k_t in bulk monomer every radical produced leads to the polymerization of $\approx 10^4$ monomer molecules.

Table 1:
Absolute rate constants, k_t and k_p
A Number of Common Monomers

	k_p litre/mol/sec		k_t x 10^{-7} litre/mol/sec	
Monomer	30°	60°	30°	60°
Vinyl acetate	990	2,300	2.0	2.9
Methyl methacrylate	350	705	1.5	1.8
Styrene	49	145	0.24	0.30
Methyl acrylate[1]	720	2.090	0.22	0.47
Butyl acrylate[2]	14		0.0009	
Vinyl chloride[3]	6800	12,300	1200	2300
Methacrylonitrile[4]	29	184	~1.1	2.3
4-Vinylpyridine[5]	12(25°)		0.3(25°)	
Butadiene[6i]	25	100		
Isoprene[6]	11	50		

References
1. M. S. Matheson, E. E. Auer, E. B. Bevilacqua, and E. J. Hart, *J. Am. Chem. Soc.*, **73**, 5395 (1951).
2. H. W. Melville and A. F. Bickel, *Trans. Faraday Soc.*, **45**, 1049 (1949).
3. G. M. Burnett and W. W. Wright, *Proc. Roy. Soc. London*, **A221**, 28, 37, 41 (1954).
4. N. Grassie and E. Vance, *Trans. Faraday Soc.*, **52**, 727 (1956).
5. P. F. Onyon, *ibid.*, **51**, 400 (1955).
6. M. Morton, P. P. Salatiello, and H. Landfield, *J. Polym Sci.*, **8**, 215, 279 (1952).

The number of times the chain reaction repeats itself prior to termination, R_p/R_t, is known as the kinetic chain length, γ. On the basis of the steady-state treatment of polymerization kinetics, molecular weight, a high value of which requires that the termination steps be precluded for as long as possible, should be inversely proportional to the square root of the rate of initiation which, in the steady state approximation is equal to the rate of termination. Thus the slower the initiation, the higher the molecular weight. Chain length is a theoretical parameter, which postulates that molecular weight is inversely proportional to the rate of initiation. It is shown to be correct by many experimental reports. (Note the section on 'reversal' later in this chapter.)

Acrylates constitute about 90% of the monomer market for the photoprocess business. The polymerization of acrylates developed to a relatively advanced level during World War II because clear, non-breakable sheets approaching the clarity and colour of glass were needed for many applications. Plexiglas® is one example. The polymerization of acrylics is also much faster than that of other monomers such as styrene (Table 1; Figure 1).

In the polymerization of monoacrylates, the rate of polymerization tends to increase with monomer conversion, in some cases with disastrous results. This, the so-called *gel effect,* results from a decrease in termination rate as the polymer chain becomes more and more entangled in its own polymer matrix. Smaller monomer molecules can diffuse more readily to the radicals on the growing polymer chain, but the macroradical cannot diffuse to a similar one so that propagation rates remain relatively constant. Acrylate polymerizations are also highly exothermic so that as polymerization proceeds, the reaction mixture warms up and causes an acceleration in rate.

In crosslinking systems such as those used in most photocure applications, matrix entanglement of radical chain ends results in an after effect, or dark reaction, which manifests itself in a continuing polymerization even in the absence of an obvious initiating step. Thus, even after the irradiating light source is turned off, non-terminated radical chains survive and can, for example if the temperature is increased, or if the reaction is left over a long period, continue to propagate polymerization. The result of these dark reactions can be particularly important in affecting changes in the physical properties of the finally cured polymer.

Chain transfer is generally referred to as any event in which a developing polymer radical chain undergoes termination, while generating another radical capable of reinitiating the growth of another polymer chain. With solvent, chain transfer decreases polymer molecular weight and this is often used in controlling polymer properties. When chain transfer is with another monomer unit, or with an already formed polymer chain, the result is a branched or *crosslinked* polymer. With monofunctional monomers, this results in earlier insolubilization or gelation and other changes in physical properties like hardening, embrittlement and so on.

Obviously the most critical event which differentiates radiation curing from other polymer-forming reactions, is the initiation step. In a photocuring process, light is responsible for this. In this context, an outline of some of the historical work which lead to an understanding of the effects of light on radical chain processes is useful.

2. Photoinitiation in an Historical Perspective

The formation of free radicals in photochemical reactions had been suggested for many years. Kharasch, for example, in numerous studies, aimed at developing synthetic uses for free radical reactions, found that light[5]

(i) changed the regiochemistry of the addition of hydrogen bromide to allyl bromide[6] (non-Markownikopf or Kharasch addition)

(ii) catalyzed the addition of aldehydes to olefins,[7]

(iii) accelerated the effect of peroxides on various processes.[8]

In an early classic experiment, Paneth provided experimental evidence for free methyl and ethyl radicals. This involved light to decompose tetramethyl or tetraethyl lead in the gas phase.[9]

Years ago, typical textbook annotations indicating how certain reactions, thought to be chain processes, were initiated or accelerated, included 'light' above the arrow as one potential 'initiator'. The entry from the first edition of Fieser and Fieser, "Organic Chemistry"[10] exemplifies this below.

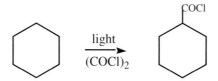

Thermal free radical chain initiators such as peroxides, for instance dibenzoyl peroxide, for instance, ($ArCO_2$-OO-CO_2Ar) and azo compounds such as azobisisobutyronitrile, $(CH_3)_2C(CN)$-N=N-$C(CN)(CH_3)_2$, however, are unsatisfactory as photoinitiators, at least in coating applications. Peroxides have limited light absorption in the regions of the spectrum, not conveniently accessible for commercial applications. Most decompose, not only to give free radicals but also the gaseous small molecules, which produce small bubbles in the coat. Benzoyl peroxide evolves CO_2 and azobisisobutyronitrile evolves nitrogen. Azo compounds do absorb light in long wavelength UV or, in some cases, in visible regions of the spectrum. Moreover, most aliphatic azo compounds decompose with good quantum yields.[11]

The photochemical decomposition of simple peroxides, such as di-tert-butyl peroxide and dibenzoyl peroxide is well known. The products of each are essentially the same as those formed in the thermal reaction; i.e., tert-butoxy radicals which produce methane, acetone, ethane and some t-butyl alcohol if there is a source of abstractable hydrogens from the former and, benzoyl radicals which decarboxylate, about 1/5 of the time, to form phenyl radicals producing benzoic acid and benzene from the latter. Di-tert-butyl peroxide has essentially no absorption except in neat solution where it has an extinction coefficient, which is just slightly greater than zero in all but deep UV. Dibenzoyl peroxide absorbs at the wavelengths of a typical non-ketone, non-aldehyde saturated carbonyl compound with an n-π* maximum (see Volume 3, Ed. 1) at about 240 nm. The quantum yield of its decomposition is reasonable; about 1/2 the molecules which absorb a photon decompose, but the accessibility of the absorption is limited to the quartz UV.[12] This is an inconvenient region of the spectrum since almost everything containing any unsaturation also absorbs at this wavelength. Furthermore ordinary glass, which is often used for the containers for the polymerization, absorbs almost everything below 290 nm.

There are ways to obviate the lack of absorption by the initiator and still develop a *photodissociable* benzoyl peroxide. One important example is p-benzoyl tert-butyl perbenzoate, equation 3.[13]

$$\text{Ph-CO-C}_6\text{H}_4\text{-COOOt-Bu} \xrightarrow{h\nu} \text{Ph-CO-C}_6\text{H}_4\text{-CO}_2{}^\bullet + {}^\bullet\text{O-tBu}$$

Equation 3

This radical initiator contains a benzophenone chromophore which provides mid-UV absorption and a dissociable peroxide which, when irradiated at the maximum absorption of the ketone, decomposes, presumably due to an intramolecular energy transfer from the excited ketone to the peroxide function and cleavage of the -O-O- bond. The products of the decomposition in hydrogen donating solvents are the corresponding acid, some benzophenone, CO_2 and t-butyl alcohol.

This system was studied briefly, as a photoinitiator by our group some years ago, and tested by scientists at Ciba-Geigy in various photoinitiator formulations. It was deemed to be thermally too unstable for their purposes. Subsequently, a deluge of patents from Japan supported the use of the similar tetra-substituted system, where decomposition was accelerated by the presence of reducing agents such as 2,4,6-triphenylthiopyrilium.[14] These patents have achieved some practical value particularly in proofing and electronics applications.

Azo compounds, such as azobisisobutyronitrile (AIBN), suffer from problems similar to those of peroxides since nitrogen gas is invariably a product. However, since they also absorb at relatively long wavelengths,[15] they have achieved greater use as photoinitiators for small molecule organic reactions which need be manipulated, such as the many different applications of radicals in synthetic processes. As obtained from most commercial suppliers, azo compounds are often yellow. As a result, their decomposition can be triggered by long wavelength UV or even visible radiation so that their usefulness is greatly increased.

Lore has it that the -N=N- bond of simple azo compounds is not susceptible to radical reactions. This is not strictly the case, but it is true that azo compounds do not undergo *induced decomposition* in a kinetic sense.

This means that most azo decompositions follow first order kinetics, since there seem to be a limited number of reactions of the formed radicals with the nitrogen-nitrogen bond of another molecule of starting azo compound. The typical aliphatic azo compound when irradiated (or thermolyzed), upon homolytic decomposition produces a radical pair in processes that are cleanly first order in initiator.

This is not the case with peroxides which are susceptible to various atom transfer and other reactions, producing dissociation of the peroxide bond. These reactions are caused by species formed in complex ways, after the dissociation of a peroxide and serve to complicate the kinetics.

Aliphatic azo compounds do suffer from certain disadvantages in that the formation of actually 'free' radicals from their decomposition is not efficient. Only a fraction of radicals formed from the decomposition of simple azo compounds like AIBN, escape the solvent cage and are able to initiate polymerization. The original evidence for this came mainly from polymer end group analysis and molecular weight studies. Somewhat later, radical traps such as chain transfer agents (e.g. thiols or molecular iodine) were employed to react with the truly 'free' free radicals formed from the decomposition of an azo compound. The notion of the *cage effect* was introduced by Hammond.[16] He discovered that the yield of the coupling product, tetramethylsuccinonitrile, isolated from the decomposition of azobisisobutyronitrile in the absence of any chain transfer in solvents like benzene was about 30% larger than the yield of products from the formed radicals that could be trapped by chain transfer. He attributed this to the fact that some percentage of radicals were trapped by the solvent cage, and unable to leave it before they recombined forming tetramethylsuccinonitrile. He referred to this as the solvent *cage effect*.

$$AIBN \longrightarrow 2\ (CH_3)_2C(\bullet)CN \longrightarrow dimer$$

$$2\ (CH_3)_2C(\bullet)CN + RSH \longrightarrow 2(CH_3)_2CHCN + RSSR \text{ - the remainder is dimer.}$$

In an early application of the use of spectroscopic techniques (the Beckman DU spectrometer) to study the mechanisms of organic reactions, the disappearance of stable radicals, whose reactions could be followed optically, was also introduced. Given that the detection of the reactions of the radical trap required different analytical procedures following the dissociation of the initiator, the two processes could be followed quantitatively, independently. Stable though these radicals were, they were sufficiently reactive to react quantitatively with other small radicals such as the radicals escaping the cage.

The first such radical trap, diphenylpicrylhydrazyl[17] above was found to give less reproducible results than did iodine or butanethiol (Hammond 1955). It was later abandoned and replaced with stable phenoxyl radicals such as galvinoxyl[18] (equation 4), which suffered fewer side reactions.[19] However, in the most definitive experiment, Hammonds group decomposed AIBN in liquid bromine and still found about 30% recombination product. That indicated to these workers that the radicals formed were of two different kinds, one that escaped to be free, while the other was solvent cage trapped.

Equation 4
Galvinoxyl Radical

The development of specific compounds where photodecomposition reactions could be controlled, and thus used beneficially for photochemical chain reaction processes, evolved in the 40's. The oldest commercial photoinitiator in general (i.e., as opposed to 'in house') use is camphor quinone, introduced by ICI and still used with tertiary amines and acrylates to form photopolymers for dental reconstructions.[20] Polybrominated (PBB's) and polychlorinated biphenyls (PCB's) were also tried as initiators and used commercially in printing applications. These however were quickly replaced, since they left residual yellows in the photochemically formed coats. Original DuPont patents taught the use of benzoin and the benzoin ethers such as α-methylbenzoin, benzoin methyl ether and α-allylbenzoin and, to an extent, introduced the concept of the Norrish Type I reaction to the field of photopolymerization.

DuPont, and specifically Catherine Chang,[21] also introduced bimolecular oxidation reduction, in the form of benzophenone/Michler's ketone as photoinitiators. This developed commercial interest in the 60's as the so-called "Hammond initiator". The chemical details of these and other initiator systems are discussed later in this volume. Here we are concerned to give essential background for concepts and techniques which form the basis for understanding photochemical processes on shorter and shorter time scales, and in greater detail. The application of such techniques to photopolymerization processes is new in the last five years, and will increase as more energy is devoted to solving important practical problems involving photopolymers.

3. The Photochemistry and photophysics of initiator systems: The basic principles of light absorption by organic compounds and the disposition of the energy derived by the absorption process

In Volume 3 of the first edition of this series, Kurt Dietliker provided an overview of the theory which forms the basic understanding of photochemical processes.[22] Photochemical processes are of commercial significance because they provide benefits not achieveable in other ways. To the first approximation, photochemical processes can be turned on and turned off. Light, or rather the absorption of specific wavelengths of light, becomes the reaction's limiting reagent. When an irradiation source emitting light at wavelengths sufficient to be absorbed by the reagent targets a sample containing the compound, reactions can occur. In contrast to thermal processes, which are non-energy specific, photochemical processes are energy specific. The reactions of initiators leading to such reactions are caused by the absorption of specific quanta of radiation. Light is absorbed and converted by processes to which specified energies can be attributed, (frequencies or wavelengths in spectroscopic terms) per atom, per molecule, per mol, etc.

The wavelength which turns on a process can be identified by the light absorption characteristics of the compound causing the reaction, its disposition of the excess energy achieved, and the characteristics of initially formed products, from that light absorption/energy disposition process. Since much of the basic research in the photochemical sciences is devoted to studying processes to achieve an understanding of the details of *absorption and the disposition of the excess energy thus obtained*. The next sections deal with some of the basic principles.

i. **Light Absorption**

Light absorption by individual compounds is studied with the techniques of spectroscopy, employing spectrometers consisting of a light source, and a method of isolating 'individual' wavelengths from that source, called a monochromator. A sample cell containing the compound, whose absorption characteristics are of interest through which the light from the source can be passed, offers a method of assessing the intensity of light emitted from the source, before and after it is absorbed by the compound. Spectrometers can cost as little as $1,000, or as much as $50,000 depending on the quality of the result sought by the experimenter. Monochromators, prisms or gratings held in moveable configurations, so that specific wavelengths can be isolated from the wavelengths emitted from the source, become more expensive as their resolution is increased. Detection devices such as photomultiplier tubes or charge coupled devices (CCD's) increase in price as they become more sensitive.

The information recorded by most spectrometers can be presented as plots of the wavelength of light (or its inverse, the wavenumber) which is absorbed, against the intensity of the radiation transmitted (reverse absorbed) by the sample. An operational relationship comes from the Beer/Lambert Law (Equation 5) which, when translated into a form useful to chemists, gives the equation

$$A = \mathrm{Log}\,(I_o/I) = \varepsilon b C: \qquad \text{Equation 5}$$

where A = the absorbance as measured on the spectrometer; I_o = the intensity of the radiation before it passes through the cell holding the sample; I = the intensity after the radiation passes through the sample; b = the pathlength of the sample cell in cm, and C = the concentration generally in mol/litre. The ε reported in units of l/mol·cm is the so-called molar extinction coefficient which is a property of the absorbing compound. The reporting of absorption spectra in the visible and UV region of the spectrum, in the literature is by molar extinction coefficient$_{(solvent)}$ at the wavelength of the maximum absorption, λ_{max}. For example, benzophenone has an extinction coefficient of 282 at its λ_{max} in ethanol (i.e., at 361 nm).[23] Though reporting, just the extinction coefficient at the wavelength of maximum absorption is a convenient shorthand for experimental sections in the technical literature, most light sources used in practical applications are not *monochromatic*. Hence an entire spectrum can be scanned the results of which give much more useful information.

Light is absorbed in bundles or packets of energy initially called "quanta" by Max Planck.[24] Planck's relationship or law relates the wavelength of a radiation to its energy

$$E = h\nu: \qquad \text{Equation 6}$$

Where E is the energy of a photon of frequency ν and h is a constant (Equation 6). In SI units Planck's constant is h = 6.626 x 10^{-34} [Joule second]. In order to convert it to [kcal /mol], one converts [J] to [kcal] and multiplies by Avogadro's number as follows: 6.626 x 10^{-34} [J s]/photon = 6.626 x 10^{-34} [J s]/photon x 6.022 10^{23} [photons/mol] /4186.8 [J/kcal] which yields h = 9.53 x 10^{-14} [kcal /mol]. The light's frequency, ν, is related to its wavelength by $\nu = c/\lambda$ where c is the speed of light and λ the wavelength. Wavelength and energy thus have an inverse relationship. The shorter the wavelength of light, the higher its energy.

To obtain energy changes resulting from an electronic transition, such as those caused by the absorption of a light quantum of specific wavelength, (ΔE), in kilocalories per mol, hc becomes 2.8 x 10^{-5} kcal m/mol when the units of λ are in meters. Thus, light of 400 nm, (4 x 10^{-7} cm) delivers by its absorption, an excess energy of about 70 kcal/mol for the compound which is absorbing it. Since bond dissociation energies of typical organic molecules range upwards from 60 kcal/mol, light absorbed at 400 nm (the long wavelength edge of the UV region of the spectrum) is often enough to cause *bond homolysis*. *This is* the equal or *homolytic* breaking of a covalent bond in the molecule[25] The result of a homolytic bond dissociation is a pair of free radicals.

$$A\text{--}B \longrightarrow A\bullet + B\bullet \quad \text{(bond homolysis)} \quad \textbf{Equation 7}$$

Two critical issues must be taken into account: --- the absorption of *and* the wavelength of the light absorbed. This data is obtained by measuring the absorption spectrum of the molecule absorbing the radiation.

ii.   The Disposition of Absorbed Energy

Compounds, after light absorption, are energetically enriched. Just exactly how much they are enriched, and how they retain their newly achieved energetic state, depends upon the rates of the various processes available to the molecule immediately after the light absorption process.

Under normal conditions, compounds exist in thermal equilibrium with their immediate surroundings in an electronic state called the *ground state*. Upon absorption of a light quantum of specific energy, the molecule is promoted to an *excited state*, the energy content of which exactly replicates the exact energy content of absorbed radiation. Just how much excess energy is distributed depends on which vibrational/rotational level of the excited state is populated by the absorption of a specific photon.

The disposition of the newly achieved energy by the molecule, begins immediately (or more accurately) in a time frame ranging upward from a few femtoseconds[26] depending on the specific molecule. Excess energy is disposed of in the form of heat transfer to the surroundings in vibrational and rotational relaxation until the molecule comes to another resting point; the so-called *zero-point vibrational level* of the first excited state. This process is referred to as *internal conversion*. Internal conversion is only important, in our context, in that it explains the relationship of light absorption processes to emission processes. The process of reaching the zero point vibrational level of the first excited state is fast on any time scale, since the energy gaps between vibrational and rotational levels (and between different excited electronic states of the same spin configuration as well) are small. The so-called Kasha Rule, accounted for by the 'energy gap law' which postulates that radiationless transitions between excited singlet states, are much faster than transitions between the first excited singlet and the ground state, is the reason that fluorescence is almost always observed as a transition from the first excited singlet state to the ground state. In addition vibrational levels of the higher excited states, always find the zero-point (or lowest) vibrational quantum level of the first excited state before emission. We can measure the separation between the different vibrational levels of excited states and of their respective ground states in a number of ways.

The most important implication of vibrational and rotational levels of various states is that molecules in solution, other than the very simplest diatomics, absorb light over regions or bands of energies rather than at single wavelengths. A typical absorption spectrum for an organic compound (benzophenone, Figure 2) gives little basic information about its various electronic energy states because of this, but it is a basic property of the molecule.

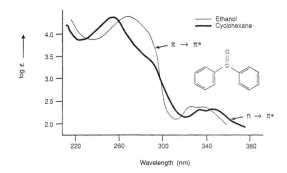

Figure 2

The absorption spectrum of benzophenone taken in ethanol and in cyclohexane.

iii.　　Light Emission - One method of disposing of an excited state's excess energy: Case 1 - fluorescence

Once a molecule resides in the zero point vibrational level of the first excited state, the structure and reactivity of that excited state as well as properties related to its transitions to other states can be probed. For example, molecules in their first excited state emit radiation which, in the ideal case, produce a spectrum which is the mirror image of the absorption spectrum but at a lower energy (longer wavelength). Emission of radiation from the first excited state is rapid in most instances, occurring prior to any significant positional reorganization of the individual atoms of the molecule. Such emission, occurring within picoseconds, is called *fluorescence*. One can record a fluorescence spectrum in virtually the same way as one can record the absorption spectrum, though there is a critical difference - the molecule must first be excited: that is it must first absorb radiation and then the emission of that radiation can be probed. The most common way for making fluorescence measurements of compounds in solution is for the excited radiation to impinge upon the sample at right angles to the observation channel for the emitted radiation.

Benzophenone, under normal circumstances does not fluoresce and cannot be used to demonstrate its fluorescence spectrum. Instead we provide spectral examples from a common organic dye, Rose Bengal to illustrate the principle, in Figure 3.

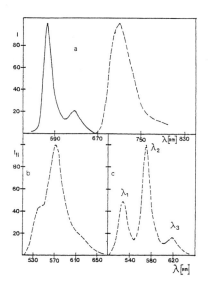 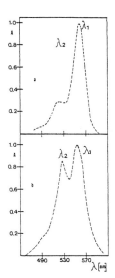

Figure 3:

Emission spectra of Rose Bengal derivatives. (a) Total emission spectra of Rose Bengal at 77 K. (b) Fluorescence emission of Rose Bengal, ditrimethylammonium salt in EtOH at ambient temperature. (c) Fluorescence emission of Rose Bengal, ditrimethylammonium salt in EtOH recorded at 77 K (λ_{ex} = 520 nm). Absorption spectra of (a) Rose Bengal in MeOH and (b) Rose Bengal, bis piperidinium salt, in CH_2Cl_2.

Several critical experiments helped photophysicists to understand fluorescence from a compound.[27] One critical issue is the *lifetime* of the fluorescing species.

Since fluorescence occurs from a spin-paired excited state (vide supra), fluorescing species generally fluoresce 'shortly' after light absorption. Thus fluorescence spectrometers are set up to observe a molecule's fluorescence at the 'instant' of its absorption of light.

A second critical issue is the *quenching* of a molecule's fluorescence. As a rule of thumb, many compounds which absorb at lower energies (longer wavelengths) may quench a molecule's fluorescence, though they do so mostly by energy transfer, rather than by the emission and then reabsorption of the emitted radiation. That means that under a certain set of conditions (concentration, solvent, temperature, etc.) the addition of such compounds to a solution of a compound known to be fluorescent when excited at a specific wavelength, will prevent (*quench*) the emission. Molecules which quench another's emission do so by *energy transfer*, that is the energy of the emitter is transferred, by some process, to the quencher. It is a corollary that the quencher will be likely to fluoresce as a result of the act of quenching. (This is not always so, but when it is, this fluorescence provides useful information about the process.) The molecule which does the quenching is known as an energy transfer *acceptor*. It receives its energy from an energy transfer *donor*. The donor is the molecule which was originally excited the fluorescence of which is quenched.

For practical reasons, the photochemical reactions of photopolymerization science require unsaturated molecules like aldehydes, ketones, aromatic compounds and olefins as light absorbers. Saturated compounds, unless they contain multiple halogen atoms, absorb little light in useful regions of the spectrum. In fact, each photoreactive compound generally has more than a single site of unsaturation; e.g. a double bond, a carbonyl group or an aromatic ring - often in conjugation. The absorption of light by such unsaturated molecules involves one or more transitions of electrons from such centres or chromophores. In a typical example, an electron is promoted, from one of the orbitals of the multiple bond, to a higher energy molecular orbital. The energy required for this electronic transition is provided by the photon, and the promotion of an electron from its ground state to an excited state, is probable, if the electronic transition occurs without change in electron spin. Thus the overall spin state of the initially formed excited molecule and that of the ground state from which it derives are the same. If the electron spins are paired in the ground state, they are also paired after the electron is promoted, even though the electrons then reside in different electronic orbitals.

A molecule with all spins paired, is said to be in a *singlet state*, and the molecule which results after the promotion of an electron stimulated by light absorption is said to be in an excited singlet state.

Fluorescence occurs when the promoted electron reestablishes its ground state equilibrium condition. The transition caused by the absorption of radiation - promotion of an electron from an orbital of the ground state to a higher energy orbital – is reversed when the attending emission occurs by fluorescence the energy of the subsequent photon is equivalent to the energy separation between the orbitals of the excited and ground states.

A simple diagram which describes this is demonstrated in Figure 4.

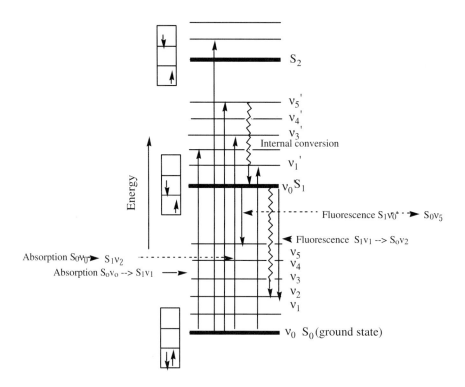

Figure 4

Absorption of the energy of a photon to promote electrons from S_0 (the ground state) to either the first excited singlet state S_1 or the second excited singlet state S_2. The transition S_0v_0 ----> S_1v_0' is that from the zero point vibrational level of the ground state, v_0 to the lowest vibrational quantum level of S_1.

Electrons promoted to higher levels of S_1 or S_2 by the absorption of photons of higher energy than those required for the S_0v_0 ----> S_1v_0' transition lose their energy very quickly by means of a radiationless transition, *internal conversion* to v_0, the zero point vibrational quantum level.

The radiative transition S_1 to S_0 (S_1v_0 ----> S_0v_2 is shown) is fluorescence. As explained, fluorescence emissions are always of the same or lower energies than the corresponding absorptions. Since vibrational levels are spaced equally in the ground state and the excited singlet states, fluorescence spectra are the mirror image of absorption spectra (see Figure 4). Non-radiative conversion of electrons in S_1 to S_0 also occurs. It is also called internal conversion.

The electrons in the bonding orbitals of the ground state, S_0 must move in opposite spins. Spin is retained even after the absorption of a photon. Thus the spin state of S_1 (and S_2) is the same as that of the ground state.

Critical Issue: Fluorescence occurs without change in electronic spin: Fluorescence is a rapid process, fluorescence generally occurs from the first excited singlet state.

iv. Case 2 - Phosphorescence

Were molecules in their excited states only to fluoresce, there would be no need for reaction photoscientists. After all, putting energy into a molecule in the form of energy absorbed from some source of radiation, and getting it back out in the form of radiation of a slightly lower energy, is an interesting, but not particularly productive, exercise.

But the first excited singlet state has several options other than light emission. One of these is, of course, chemical reaction. The short lifetime of most singlets, however, often precludes this in favor of conversions to other excited state manifolds. One of these processes is *intersystem crossing:* a transition in which the spin of one of the electrons of the first excited singlet state is reversed. The result is an excited state with two electrons in different orbitals, but now of identical spins. In this excited state, direct repopulation of the electrons into a single orbital, such as that which exists in the ground state, is improbable since one spin of one of the electrons must be reversed. An excited state molecule with two unpaired electrons of the same spin has three degenerate identical levels and is said to be in an excited *triplet state*. The diagram, (Figure 4), can now be expanded to include the triplet state. (Figure 5). Systems with two electrons of identical spin in two separate orbitals are of lower energy than identical systems, with the same two electrons of opposite spin, so that the triplet state is lower in energy than the excited singlet state.

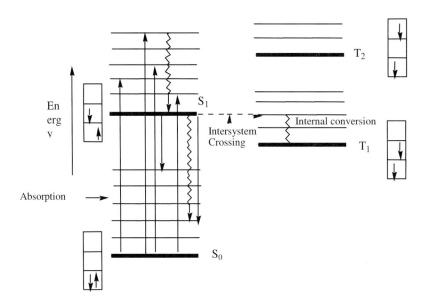

Figure 5:
The Jablonski diagram showing *intersystem crossing* from S_1v_0 to T_1v_2. The triplet state is of lower energy than the first excited singlet state because the energy of two electrons of the same spin occupying different orbitals is lower than two of opposite spin (the so-called Hund's rule - for another application of Hund's rule consider the three 2p electrons of elemental nitrogen which occupy $2p_x$, $2p_y$ and $2p_z$ atomic orbitals.)

The ground state, the excited singlet state and the excited triplet state, each have quantized vibrational and rotational levels. Fluorescence, which takes place from the zero point vibrational level of the excited singlet state can occur at any one of several such vibrational levels in the ground state. The resulting molecule may be in an upper vibrational or rotational level of the ground state, but it will equilibrate with its surroundings quickly, returning the molecule to the vibrational level, corresponding to the equilibrium vibrational state, usually the zero point vibrational level of its ground electronic state. This is the exact reverse of the excitation process resulting from the absorption of radiation which promotes the molecule from the zero point vibrational level of the ground state to any one of several vibrational levels of the first excited singlet state. Thus, light absorption occurs from zero point vibrational level to the zero point, or higher, vibrational level. The latter transitions require more energy than the 0 to 0 transition. Fluorescence occurs from the zero point vibrational level of the first excited singlet state to the zero point, or higher, vibrational levels of the ground state. Thus releasing light equivalent in energy to the 0 to 0 transition, or of lower energy. Since vibrational energy levels are spaced similarly in the ground and the excited singlet state, the absorption and fluorescence spectrum of a typical molecule have a mirror image relationship. The energy difference between the ground and first excited singlet state, the so-called *singlet energy*, is measured at the point exactly between the two - the absorption and the fluorescence (emission) spectra.

Triplet states of molecules are characterized by two electrons of identical spins. Since the spins are not paired, molecules existing as triplets are paramagnetic. The simplest example of a molecule in its triplet state is molecular dioxygen, written to illustrate the point as a diradical, •O-O•.

Molecules in their excited triplet state (as opposed to the ground state triplet state found in O_2), have a number of measurable properties. Unlike the immediately reversible transition of ground and first excited singlet state, transitions of the first excited triplet state to the ground state require a reversal of electronic spin and are therefore of limited probability. This means that in an assembly of molecules in the triplet state, transitions to the ground state occur in only a few cases. The triplet to ground state conversions are also not usually observed for molecules in fluid solution, though they can be easily observed when molecules are confined in polymeric matrices, or in glasses at low temperatures. Such transitions - triplet state to ground state - emit radiation of a lower energy (at longer wavelength) than that at which fluorescence is observed (if it occurs in the same molecule). Triplet state to ground state transitions occur after much longer time intervals (lifetimes) and are called *phosphorescence*. A compound which emits light after the excited radiation is turned off is therefore called a *phosphor*. We can now expand the Jablonski diagram to include phosphorescence, (Figure 6).

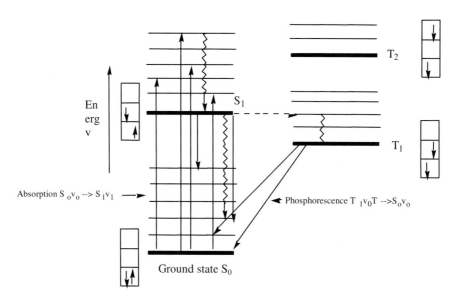

Figure 6

Jablonski diagram showing phosphorescence either from T_1v_0 to $S_1V_0^t$ ----> or of $T_1V_0^t$ to S_0V_1.

A typical phosphorescence spectrum, that of benzophenone in an inert matrix, is shown in Figure 7. As was the case with fluorescence, the phosphorescence spectrum also gives us a measure of the energy of the phosphorescing state and provides a pattern of vibrational modes of motion in the ground state. The emission of shortest wavelength in the phosphorescence manifold gives a measure of the energy of the triplet state, the so-called *triplet energy*. As we shall see later, this energy is a critical predictor of energy transfer from the triplet state.

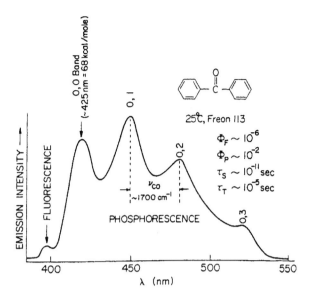

Figure 7
Total emission spectrum of benzophenone at room temperature in Freon 113 solvent. The fluorescence band (395 nm) is 'delayed fluorescence'. Benzophenone provided an almost perfect model for most of the reactive systems used in UV photoreactions. The reactivity will be demonstrated later in this chapter. Taken from *Turro Modern Molecular Photochemistry*

4. Transient Spectroscopy and Characterizing Excited or Photochemically Reactive States

Molecules in their excited states have all, or at least most, of the properties of molecules in their ground states. One can record their UV or visible absorption spectra as well as their infrared spectra. If they possess *radicaloid* character, as do most triplet states, one can also record their electron paramagnetic resonance (EPR) spectra. Unfortunately molecules in their excited states, even triplets, are very short lived. To record their spectral properties, one has to make them very quickly and in high concentrations. R. G. W. Norrish and Sir George Porter won the Nobel Prize in chemistry in 1967 for teaching us how to do just this. What Norrish and Porter developed was 'flash photolysis';[28] the act of forming a high concentration of excited states very quickly. Norrish and Porter used a huge bank of focused flash lamps which were turned on and off much more quickly than the life times of the excited states under investigation. Subsequently (in a few micro seconds) they recorded the electronic absorption spectrum of the compound in its excited state and, by measuring the intensity of its absorption as a function of time (admittedly short times), developed a kinetic profile for its decay.

With later development of the pulsed laser,[29] high powers of monochromatic radiation could be delivered to a sample in even shorter times or pulses. This meant that additional plethora of excited states were rendered characterizable, their absorption and emission spectra recorded, and their decay chemistries evaluated. Beyond this, the products of their decay could also be characterized, *their* spectra recorded, and *their* decay rates measured. Current systems, admittedly of a different design and using something called 'pump and probe', have pushed the time envelop even further. Many laboratories can now do experiments which characterize molecules, whose excited states live less than 100 femtoseconds. These time scales are not yet of interest for our applications, and - for the most part - what we need to know happens rather slowly, (nanosecond timescale), in the parlance of the kinetic spectroscopist. As additional intermediates, however, become of interest in photopolymerization science, our time frame will be pushed to shorter and shorter values.

The typical laser pulse experiment uses wavelengths of laser light absorbed by the chromophore of interest. In some instances, these pulses are provided by a nitrogen laser which can be frequency multiplied to produce light at 336 nm. In a more common setup, which is flexible enough to produce either UV or visible excitation, the exciting light comes from a neodynium/YAG laser (YAG = yttrium-aluminum-garnet). The laser is characterized, for the experimental setup, by the wavelength, the power and the time durations of the pulse. In a typical experiment, for example, Nd-YAG laser pulses at 355 nm for 7 ns with a power of 30 mJ/pulse were used. That means that the laser is delivering approximately 5×10^{16} photons for each laser pulse. A sketch of a typical kinetic apparatus is given in Figure 8.

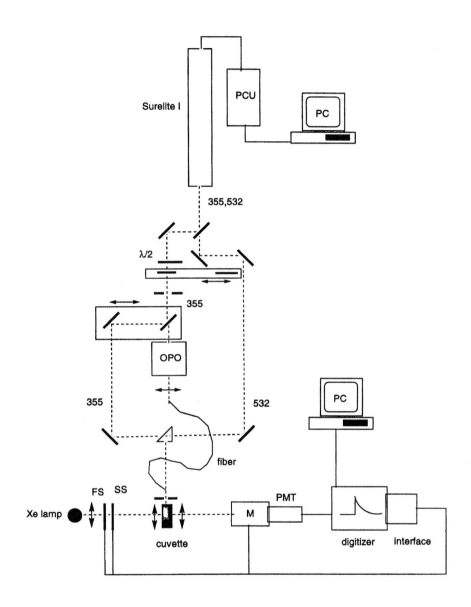

Figure 8
Sketch of nanosecond flash photolysis apparatus.

The absorption spectra of the intermediates formed from a flash of intense light can normally be detected by using an analyzing beam placed at right angles to the exciting beam and a photo-multiplier tube as a detector. Spectral analysis is provided by measuring light intensity at individual wavelengths after the pulse. There are various ways to report this, but a typical example is that shown for methyl phenyl glyoxylate (Figure 9) in Figure 10*.

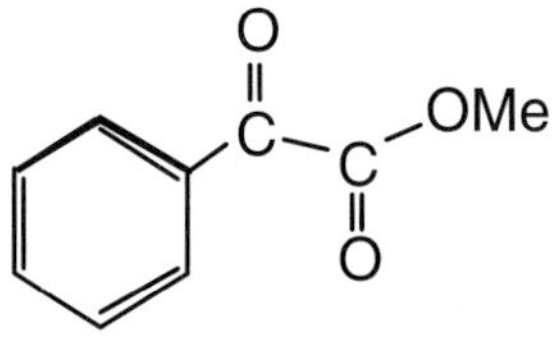

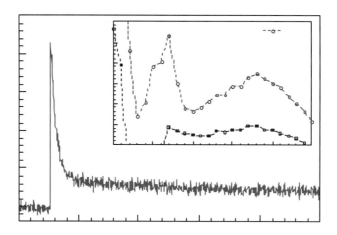

Figure 9
Methyl phenylglyoxylate.

Figure 10
Transient absorption decay trace of methyl phenylglyoxylate in benzene (0.041 M) Monitored at 440 nm. Inserted are the transient absorption spectra.

---

*This is the first of many references to the excellent work of Ph.D. students and post-doctorals, who have worked in my laboratories. The lazer flash photolysis data for methyl phenylglyoxylate was obtained by Shengkui Hu.

Methyl phenylglyoxylate is a relatively old photoinitiator which is still in use.[30] Figure 10 shows the absorption spectrum of the triplet state and the kinetic trace reporting the decay of optical density at the wavelength of maximum absorption. The triplet state is characterized by the rate constant of the first order decay of the absorption and by its lifetime $\tau = 1/k_d$. The lifetime of the methyl phenylglyoxylate triplet state is 5.6 µs, as characterized by the change in optical density at 440 nm. In this particular instance, a second intermediate having a much longer lifetime was also detected. This is the reason why the optical density trace, measured at the wavelength of maximum absorption of the transient (440 nm), does not return to the baseline during the time of the experiment. Thus, in a single flash experiment, the triplet state is formed and its absorption spectrum recorded. Its lifetime is measured in the particular solvent and the formation of the first reactive intermediate from the triplet is observed. Based on the route to its synthesis in the transient domain, it was proposed that this second species was the corresponding radical, (figure 11), produced by hydrogen abstraction. Simple structure reactivity relationships complete the story. Some of these will be discussed later.

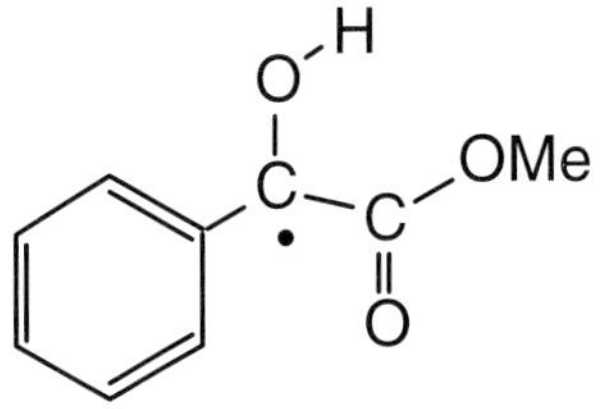

Figure 11

The light absorption characteristics of transients appear to follow the same rules as do those for ground state molecules. However, interpreting kinetic events from transient spectra is as much an art as a science. The wise investigator should do as we do, collaborate with kineticists. Their experience in dealing with transient phenomena offers great benefits.

As has been previously pointed out, benzophenone does not fluoresce. Its first excited singlet state is too short lived in that it intersystem crosses to a much more stable triplet, in just a few picoseconds. The triplet absorption decay of a benzophenone, recorded after a flash at 355 nm is recorded in Figure 12.[31]

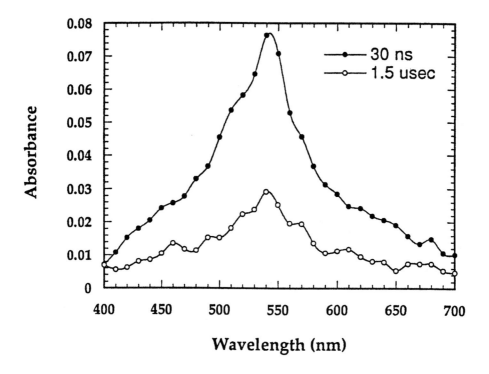

Figure 12
Transient triplet absorption spectrum of benzophenonemethyl-tri-n-butylammonium tetrafluoroborate ($<10^{-4}$ m) in benzene recorded 30ns and 1.5µs after the pulse

The kinetic characteristics of a compound's triplet state are recorded, as pointed out above, by measuring the intensity of absorption at a single wavelength, generally, the wavelength of maximum absorption for the state which is absorbing.

Reference has already been made to the concept of excited state quenching, a process where the excited state of one molecule, transfers its energy to another. This process, energy transfer, also results in a number of measurable parameters. Triplet state energy transfer, for example, obliterates the phosphorescence of the donor, or more likely decreases the concentration of triplets available for reaction, as measured either in transient or steady state studies. It also decreases triplet state lifetimes when it occurs. Thus energy transfer, or triplet state quenching, is a classical way to decide whether or not a reacting molecule is reacting from its triplet state. If the reaction is retarded by the introduction of a known triplet state quencher, it is likely to be occurring from the triplet state of the reacting species. In steady state experiments, energy transfer obliterates the property of the donor being measured. It also produces excited states of the molecule which is doing the quenching. In transient experiments, energy transfer (1) reduces the excited state lifetime of the donor and (2) produces an excited state of the acceptor, the latter is recognizable by its own absorption spectrum with its own lifetime under the conditions of the experiment. Evidence for triplet state quenching in the transient experiment is both the observation of a shorter lifetime of the donor triplet and the appearance of a new species. This new species is either the triplet state or some subsequent intermediate formed from the triplet state of the acceptor.

Triplet states, are paramagnetic, since they possess one or more unpaired electrons and are quenched by other paramagnetic species. Jablonski, in several theoretical papers published in the 30's,[32] argued that phosphorescence, the emission detected substantially after light excitation has ceased, was the result of a deactivation of a lower energy electronic configuration of the excited state of a molecule, than that responsible for fluorescence. He also postulated that all slow emission processes, (emissions occurring after the exciting light had been turned off) resulted from the original population of the same electronically excited state. In other words, the singlet state preceeded formation of the triplet state in a majority of instances. A major problem with this was confusion over emissions from organic molecules at different temperatures. In a critical landmark paper, Lewis and Kasha reported experimental evidence that the phosphorescence, or 'afterglow', was - if the emission was at longer wavelengths than fluorescence - the result of a lower energy electronic transition. Agreeing with Jablonski, they called this the 'triplet state'. They studied this in an apparatus which they called "so simple that any laboratory could set it up". In that apparatus the sample was held at the desired temperature in a quartz Dewar and illuminated through a can-like system that had an orifice on one side. The can could be rotated at varying speeds, to assess the half-life of the afterglow, which was observed by a spectrograph positioned at 180° angle to the irradiation light source. This meant that after the sample was excited, the can could be rotated and the residual emission recorded.

In experiments which considered the 'afterglow' of the dye fluorescein, Lewis, Lipkin and Kasha, used a glass of boric acid at temperatures reduced to as low as -40°C. Under those conditions the only 'afterglow' they observed was at much longer wavelength, relative to normal fluorescence. Following Jablonski, they attributed this emission to the radiative conversion of excited states of lower energy than the fluorescing state, to ground state. At higher temperatures, also in a boric acid glass, they also observed fluorescence as part of the afterglow. This emission, though of substantially longer lifetime, was attributed to a thermal repopulation of the first excited singlet state, from which subsequent emission occurred. This was called *slow fluorescence*.

Lewis and his students employed phosphorescence to study the normal vibronic levels of the excited triplet state, and identified the phosphorescent state or triplet state, as the 'biradical state.' This analysis eventually led to the conclusion that paramagnetic quenchers, such as dioxygen, would react with the triplet state causing its deactivation. When oxygen quenching of triplets was observed experimentally and the product was found to be singlet oxygen,[33] it was reasoned that since oxygen was paramagnetic that the so-called long-lived emitting state must also be paramagnetic. This also confirmed a theoretical principle put forward by Jablonski;[34] namely that the triplet state had unpaired electron spins. Oxygen quenching of a molecule in its triplet state, which has many different implications in polymer photochemistry, is evidenced in the transient spectrum of the molecule being quenched by a shorter triplet state lifetime in oxygen containing solutions. From this one can measure a bimolecular quenching constant which, for benzophenone for instance, in fluid solution is about $2.4 \times 10^9 \, M^{-1} \, sec^{-1}$. This quenching is somewhat slower than the rate of diffusion; in other words the oxygen gets to the benzophenone triplet and has time to swim around it for a few cycles, before it gets around to quenching it.

As indicated, excited state quenching is not without impact on the quencher. In the case of oxygen quenching of triplet benzophenone it results in energy transfer to oxygen, forming one of its excited states. The excited state is the first excited singlet state of oxygen; so-called *singlet oxygen*, O=O.[35] Thus, the energy transferred from a benzophenone triplet state to ground state triplet oxygen is sufficient to produce an electronic transition in the latter, forming its first excited state which is a singlet. This energy transfer has the important practical effect that one can use 365 nm radiation, where oxygen does not absorb radiation, to form oxygen's excited states. Singlet oxygen, which has a number of important properties, emits light at 1216 nm in the near infrared region of the spectrum. Thus one experiment which is done to assess energy transfer from donors to oxygen is to measure the emission of energy from the singlet oxygen formed as it reverts to its ground state triplet. This energy transfer is shown in a Jablonski diagram in Figure 13. The emission of energy from oxygen singlet to oxygen ground state triplet is shown in Figure 14, as singlet oxygen emission.

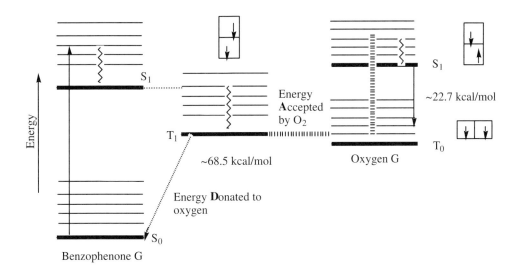

Figure 13
Energy transfer from benzophenone triplet (E_t = 68.5 kcal/mol) to ground state oxygen, T_0. Benzophenone triplets are quenched by oxygen and the latter promoted to S_1v_n.

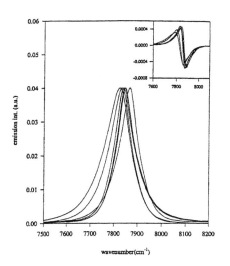

Figure 14
Emission of singlet oxygen at ≈ 1270 nm.

Energy transfer is also a source of excited states inaccessible by direct irradiation. Naphthalene, for example, fluoresces with a near unit quantum yield. Fluorescence is observed from naphthalene when it is excited by direct irradiation and this means that each photon absorbed by a molecule of ground state naphthalene is subsequently emitted as fluorescence. Thus the intersystem crossing rate of singlet to triplet naphthalene is essentially zero. Naphthalene triplet can, however, be obtained by energy transfer from benzophenone triplet to ground state naphthalene. The phosphorescence of naphthalene can be observed this way. Thus, in a classic experiment carried out by Terenin and his students in St. Petersburg in the 50's, naphthalene phosphorescence was observed when benzophenone was irradiated in a solid matrix in the presence of naphthalene under conditions where benzophenone absorbed most of the radiation.[36]

5. Reactions from Excited States

Photoscientists are interested in any reaction which produces a reactive intermediate that goes on to do useful work. The reactivity of excited molecules varies from compound to compound, but there are some general principles can be established.

In the instance of photopolymerization, by useful work we mean, any reaction that produces an intermediate which will promote or initiate a polymerization reaction. Such intermediates include free radicals, cations and anions. Their formation from excited states is discussed in this section.

i. Case I: Formation of Free Radicals by Bond Dissociation

The Norrish Type I reaction, so-named by Wagner and Hammond in the 60's[37] was first reported by R. G. W. Norrish and his students at Cambridge in the 30's.[38] Norrish was a gas phase kineticist who used photochemical processes as a source of reaction intermediates which could be studied, using the techniques available to him at the time. These were manometric pressure measurements and product isolation. Norrish focused his studies on the photochemistry of aliphatic ketones one of which was 2-pentanone, which absorbed UV light (λ_{max} = 280 nm) from one of the resonance emissions of the mercury lamp. The molecule subsequently underwent a unimolecular elimination reaction (Equation 8). Norrish also studied acetone which loses CO while forming methane, ethane and some higher molecular weight products, Equation 9.

$$CH_3CH_2CH_2CCH_3 \xrightarrow{h\nu} CH_3CCH_3 + CH_2=CH_2 \quad \text{Equation 8}$$

$$CH_3CCH_3 \xrightarrow{h\nu} CO + CH_3CH_3 + CH_4 \quad \text{Equation 9}$$

The first reaction (Equation 8) Norrish identified as general, requiring at least one hydrogen atom at the γ-carbon of the original ketone, $\underset{\text{CH}_3\text{CH}_2\text{CH}_2\text{CCH}_3}{\overset{\gamma\quad\beta\quad\alpha\quad\text{O}}{}}$. It was later shown that the initially formed product was the enol of acetone, $\text{CH}_3\overset{\text{OH}}{\text{C}}=\text{CH}_2$.

The photoreactions of acetone at low pressures (dilute conditions) in the gas phase (Equation 9) represent classic bond homolysis chemistry and were also identified as general by Norrish. The reactions are not particularly efficient with acetone, because the formation of a primary radical, in this case the methyl radical, is not a favoured process. The results are just the same and form the model for what became known as the Norrish Type I reaction, Equation 10.

$$\underset{}{\text{CH}_3\text{CCH}_3} \longrightarrow \underset{\text{methyl}}{\text{CH}_3\bullet} + \underset{\text{acetyl}}{\text{CH}_3\text{C}\bullet} \longrightarrow \text{CO} + \text{CH}_3\text{CH}_3 + \text{CH}_4$$

Equation 10

The excited state chemistries of 2-pentanone and acetone are complicated in that reaction occurs from both the singlet state and the triplet state and none of the initially formed products possesses extensive unsaturation. Dissociation of acetone triplets is faster than dissociation of acetone singlets, so that most bond dissociation in the acetone case is of acetone triplets going to a triplet radical pair. Observing such intermediates in the transient time domain is complicated, because they absorb radiation at such short wavelengths. It behoves us not to dwell on the details of the spectroscopic mechanism now. Instead we shall describe the classic studies which outlined the scope and reactivity in the Norrish Type I reaction, since that is the most important to the polymer photochemist.

Roughly translated, the quantum yield of a photochemical process is the chemical yield, wherein photons absorbed are the limiting reagent. Defined originally by Albert Einstein[39] and confirmed experimentally by Otto Warburg,[40] one measures quantum yields either, directly, using a photometer to measure the power of the incident radiation, or indirectly, by comparison with a chemical reaction for which the quantum yield is already known. One of many methods to analyze the starting materials as they disappear or the products as they appear is adapted to the specific experimental system under consideration. Avagadro's number of photons is termed an 'Einstein' and the chemical yield in mols, per Einstein of photons, is given the symbol for quantum yield, Φ. It is generally easier to measure quantum yields actinometrically, using a photochemical reaction for which the quantum yield is already known and is similar in reactivity to the reaction being studied, as the calibration standard. The latter is called an actinometer. Several precautions are necessary however, if one does not use an optical bench - a linear device in which everything is kept the same for every sample and the power emitted by the lamp is measured directly. First, both the actinometer and the reaction being studied must be irradiated under conditions in which the light from the lamp reaching the samples is averaged over time.

For this, one usually uses a device which rotates the samples and the cells containing the actinometer samples around the lamp. Second, all the samples must absorb all the light. If this is not the case, the amount of light absorbed by the sample, must be carefully calibrated from the molar absorptivities of the absorbing compound. For the best results, one should use radiation filtering to isolate the wavelengths actually absorbed by the reacting compounds.

Assuming that all precautions are followed, comparisons of quantum yields between similar reactions is one of the best ways to assess structure/reactivity relationships in photochemical processes. There is no better example which illustrates the principle uses of quantum yield in mechanistic studies, than the classic studies which led to our basic mechanistic understanding of the Norrish Type I reaction.

Structural features, facilitating the loss of CO from a ketone upon photolysis were first identified by Starr and Eastman[41], in experiments conducted in solution for synthetic purposes. The aim was to develop a route to convert a ketone to a hydrocarbon or an olefin. They pointed out that the incorporation of α-alkyl substitution, β, γ -unsaturation or a cyclopropyl carbinyl system (figure 19) facilitates decarbonylation, that is the Norrish Type I process. Such structural features stabilize most reactive intermediates; carbocations or free radicals.

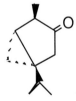

α-methyl substitution;   tert-substitution;   β, γ unsaturation

cyclopropylcarbinyl group

Yang,[42] in a subsequent study, showed that the Norrish Type I reaction for methyl t-butyl ketone had a higher quantum yield of CO elimination than the same reaction for acetone. He also demonstrated that the direction of the elimination in the former was toward the t-butyl side, rather than toward the methyl side. This was the first quantitative study, in solution, identifying that the driving force for the elimination of carbon monoxide from the excited state of the ketone, was the formation of the more stable alkyl radical.

$$\text{t-Bu-C(=O)-CH}_3 \longrightarrow \text{t-Bu}\cdot + \cdot\text{COCH}_3$$

$$\not\downarrow$$

$$\text{t-Bu-C(=O)}\cdot + \cdot\text{CH}_3$$

The n, π* triplet states of aliphatic ketones dissociate more rapidly than corresponding n, π* singlets, and the triplet is much more reactive to alpha cleavage than the singlet. Structural features which stabilize the formed radicals decrease the strength of the bond in the ketone. The rate constants for the dissociation of aliphatic ketones increase in the series of alkanones, CH_3COR, as follows:

$$R = Me, Et, iPr, t\text{-}Bu, CH_2\phi$$

$$k_d = 10^3, 10^6, 10^8, 2 \times 10^9, 10^{10} \text{ s}^{-1}$$

Cleavage rates for aliphatic ketones are about 100 times greater than they are for comparable aromatic ketones. Thus the k_d for ØCO-tBu is about 10^8 s^{-1}. In other words the phenyl group has a deactivating effect.

Other structural features which impact the dissociation rates of ketones involve the excited state configurations of the absorbing chromophore. p-MeOØCO-tBu decomposes about 100 times more slowly than ØCO-tBu, because the excited state in the methoxyphenyl case has substantial charge transfer character; that is, the excited state characteristics of the carbonyl group are substantially delocalized through the phenyl to the methoxy function. Naphthyl ketones undergo no significant n, π* photochemistry at all, since the energy resulting from light absorption ends up in the aromatic ring of the naphthalene.

The use of photodecarbonylation in synthesis takes advantage of the principles described above. Turro made good use of the latter in projecting that tetramethylcyclobutane-1,3-dione, a commercial side product of the synthesis of dimethylketene, would readily eliminate CO and form an isolable cyclopropanone.[43] The chemistry of small ring compounds[44], containing unsaturated centres such as a carbonyl group in cyclopropanone, had attracted synthetic chemists for decades, but cyclopropanones were difficult to make, since their structure was highly strained. The first successful synthesis of stable cyclopropanone involved a photodecarbonylation. Note that a substantial driving force in the elimination reaction must be the formation of two tertiary radicaloid centres on each terminus of the dione after the initial elimination of CO. More recent work has shown that the decarbonylation of racemic and diastereoisomeric cyclobutanediones produce the same cyclopropanones.

ii. Reactivity and mechanism in a known Type I photoinitiator

Transient spectroscopy provides evidence of the pathway of decomposition for many simple photoprocesses. The commercial photoinitiator (2,4,6-trimethylbenzoyl) diphenylphosphine oxide,[45] is best able to illustrate this principle

Both UV and IR detection of the transients formed were required to outline the kinetic characteristics of the formed radicals. The benzoyl radical is too nonabsorbing to be observed in the UV. The extinction coefficient of the unsubstituted benzoyl radical (λ_{max} = 368 nm; ε = 150 M^{-1}cm^{-1}) is sufficiently low to presuppose that its detection by time resolved UV spectroscopy[46] is likely to be difficult. Based on studies resulting from this work, it was determined that the phosphonyl radical is more reactive than the benzoyl radical toward olefin addition.[47] Laser flash photolysis was employed to generate and detect the phosphonyl radical, (λ_{max} = 330 nm) and to determine its bimolecular rate constants for addition to a series of alkenes (k_v 10^6 to 6 x 10^7 M^{-1} s^{-1}).[48] Though the

2,4,6-trimethylbenzoyl radical,

had been previously detected by time resolved EPR spectroscopy,[49] time resolved infrared spectroscopy is preferred in that the benzoyl radical is readily detectable by this method.[50] Bimolecular rate constants for radical reactions with BrCCl$_3$, CCl$_4$ and thiophenol were thus easily measured by the disappearance of IR transient absorption. Reactions of the diphenylphosphonyl radical, with the same hydrogen and halogen donors were also studied.

The time resolved IR spectrum obtained from the laser flash photolysis of (2,4,6-trimethylbenzoyl) diphenylphosphine oxide (355 nm excitation in deoxygenated heptane solution) is shown in Figure 15. The radical absorbs at 1805 cm^{-1} in this solvent. In the experiment, one measures the optical density point by point.

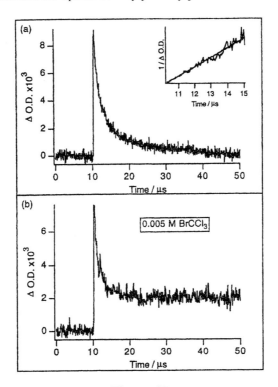

Figure 15
(a) Transient decay trace recorded at 1805 cm^{-1} from a 0.02 M solution of (2,4,6-trimethylbenzoyl)diphenylphosphine oxide in deoxygenated n-heptane. The inset shows the fit of the data according to equation 4 in reference 54. (b) Transient decay trace recorded at 1811 cm^{-1} from a 0.02 M solution of (2,4,6-trimethylbenzoyl)diphenylphosphine oxide in deoxygenated heptane containing 0.005 M bromotrichloromethane.

The addition of bromotrichloromethane, thiophenol or oxygen to the solution decreased the lifetime of the transient by a reaction which followed pseudo first order kinetics in each case. The kinetic data for the decay of the radical, in the absence of quencher, can be obtained from a decrease in the intensity of the absorption at 1805 cm^{-1} and follows pseudo first order kinetics. The values obtained for the rate constants are slightly solvent dependent, but approximately $5 \pm 2 \times 10^7$ s^{-1} at [BrCCl$_3$] = 0.005M. Decay in the presence of donors such as bromotrichloromethane is accelerated by reaction forming the corresponding benzoyl bromide (λ_{max} = 1802 cm^{-1});

Typical decay traces are reported in Figure 16.

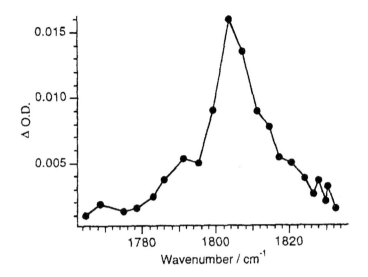

Figure 16

Time resolved IR spectrum recorded 1μs following laser flash photolysis (355 nm excitation) of a deoxygenated 0.02 M n-heptane solution of (2,4,6-trimethylbenzoyl)diphenylphosphine oxide at 23 ± 2 °C.

The quenching constant, k_q is approximately 1.7×10^8 $M^{-1}s^{-1}$. The I.R. spectrum recorded 1μs and 40μs after the flash is shown in Figure 17.

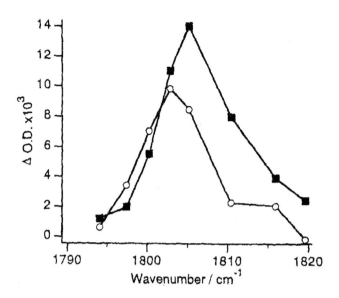

Figure 17

Time resolved IR spectra, recorded (solid squares) 1 μs and (circles) 40 μs following the laser pulse, from flash photolysis (355 nm excitation) of a 0.02 M solution of (2,4,6-trimethylbenzoyl)diphenylphosphine oxide in deoxygenated n-heptane containing .005 M bromotrichloromethane.

The reactivity of the phosphonyl radical with bromotrichloromethane and thiophenol was measured using transient UV spectroscopy. The transient absorption due to this radical is shown in Figure 18. Bimolecular rate constants for reaction with BrCCl$_3$ and thiophenol are of the order of 8 x 10^{6}M^{-1}•s^{-1} for BrCCl$_3$ and 1 x 10^{6}M^{-1}•s^{-1} for thiophenol.

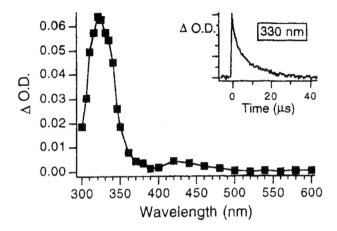

Figure 18:

Transient UV spectrum recorded 0.1 to 1 μs after laser flash photolysis (355 nm) of a 0.001 M solution of (2,4,6-trimethylbenzoyl)diphenylphosphine oxide in deoxygenated heptane. The inset shows the decay of the transient measured at a monitoring wavelength of 330 nm.

The reactivity of the partner radicals, and toward reactive atom donors, were similar with the phosphonyl radical being slightly more reactive. This reactivity is attributed to the fact that, though both radicals have a substantial degree of s character, as can be demonstrated from the hyperfine coupling constants measured by electron paramagnetic resonance spectroscopy (EPR),, and though most of the spin density is located on the carbon and phosphorus atoms respectively.[51] The hyperfine on phosphenyl radical proves the radical to be almost entirely localised in an s orbital. This similarity of reactivity towards atom abstraction is not like the addition reactions to olefins of the two radicals where the phosphonyl radical is 10 to 100 times more reactive.

iii. Case II: Other reactions from the excited state, manifold oxidation/reduction

Bond dissociation reactions of the Norrish Type I type are the predominant reactions used to initiate polymerization in the UV and short wavelength visible regions of the spectrum.

The bond dissociation energies of most organic compounds however range upward from 70 kcal/mol, and visible wavelengths are insufficient to cause dissociations even in appropriately substituted molecules. Excited states, however, are also able to oxidize and reduce partner molecules of appropriate reduction or oxidation potentials. The thermodynamics of such excited state oxidations or reductions, can be predicted by the Rehm/Weller equation.[52] This equation predicts that excited state electron transfer will be spontaneous, if the difference between the oxidation potential of the donor, the reduction potential of the acceptor and the energy of the excited state is negative; a final term represents electrostatic interaction between the cation and anion in the solvent cage. If the ions are separated this term is generally neglected; Equation 11.

The Rehm-Weller equation is

$$\Delta G_{ET} = F(E_d^{ox} - E_a^{red}) - E^{oo} - kN_A° e^2/Ed$$

Equation 11

Where ΔG_{ET} = Gibbs free energy of electron transfer [J/mol]

F = Faraday constant (86,989 C/mol)

E_d^{ox} = oxidation potential of the donor molecule (volts)

E_a^{red} = reduction potential of acceptor molecule (volts)

E^{oo} = zero zero transition energy of the excited state (J/mol)

The last term, the electrostatic interaction energy term, represents the interaction between radical ions formed by the electron transfer where

k = conversion factor between the electrostatic system and the SI system
($k=8.99 \times 10^9 [NM^2/c^2]$)

$N_A°$ = Avagadro's number (6.022×10^{23} l/mol)

e = elementary charge (1.602×10^{-18} [e])

d = distance between the radical ions formed [m]

E = dielectric constant of the solvent

In water at room temperature (E=80). If the radical ion pair is separated by 0.5mm the electro static term accounts for only 3.5 kJ/mol (0.83 kcal/mol).

Examples of the application of oxidation/reduction to generate radical chain processes are the cyanine/borates, the so-called 'Mead initiators.' The cyanine borates, a general example of which is shown in equation 12, were developed by Gottschalk in consultation

Equation 12

with Schuster and Neckers to serve to initiate polymerization in microcapsules for a color copy application called Cycolor. Known examples can initiate polymerization throughout the visible region of the spectrum.[53] Wavelength flexibility comes from the fact that the methylene chain connecting the two benzthiazole rings, can be either shorter or longer, depending on whether one wishes to have short wavelength or long wavelength light absorption. In early studies, it was shown by flash photolysis that the lifetime of the excited singlet state of the cyanine, as measured by transient absorption spectroscopy, is reduced in non-polar solvents such as ethyl acetate in the presence of triarylalkyl borate counter ion, but unaffected in more polar solvents such as acetonitrile.[54] Non-reducing counter ions such as BF_4^- had no effect on the singlet lifetime in any solvent. Fluorescence was also quenched in non-polar solvents in the presence of the reducing borate anion. The flash results shown for the carbocyanine, $Ø_3(Bu)B^-$ are consistent with a reaction in which the organic borate, $Ar_3(Bu)B^-$, but not the inorganic borate, BF_4^- quenches the excited singlet. Observation of Ar_3B among the products points to electron transfer with the elimination of butyl radical. The quenching reaction is solvent dependent and depends on the ions being paired and in tight proximity. This is more due to the properties of the cyanine than the borate. Cyanine singlet lifetimes are very short since the molecules undergo in ersatz at reversible isomerization in the excited state. This process uses excited state energy but serves no useful purpose.

The transient absorption spectra, for a simple case, are shown in Figure 19.

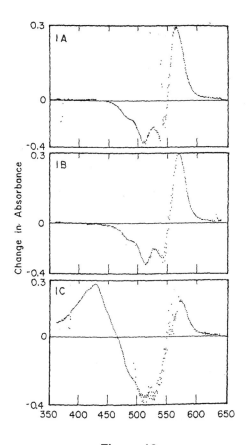

Figure 19

Transient absorption spectra recorded 1 μs after irradiation of cyanine dyes at 532 nm: (1A) irradiation of the cyanine borate in acetonitrile solution; (1B) irradiation of cyanine hexafluorophosphoate in ethyl acetate; (1C) irradation of cyanine borate in ethyl acetate solution.

Fluorescence quantum yields are not generally affected by the counter ion in any solvent and lifetimes for cyanine cation borate counter ions pairs are given in Table 2.

Table 2
Fluorescence efficiencies and lifetimes, various cyanines
Fluorescence Efficiencies and Lifetimes

Cy$^+$X$^-$	solvent	ϕ_f^a	τ (ps)
[n-C$_4$H$_9$B(Ph)$_3$]$^-$	EtOAc^b	0.13 ± 0.02	300 ± 35
PF$_6^-$	EtOAc	0.16 ± 0.03	303 ± 40
Br$^-$	EtOAc	0.14 ± 0.03	360 ± 20
[n-C$_4$H$_9$B(Ph)$_3$]$^-$	DBS	0.53 ± 0.03	1040 ± 100
PF$_6^-$	DBS	0.55 ± 0.04	915 ± 100
Br$^-$	DBS	0.55 ± 0.01	1070 ± 100
[n-C$_4$H$_9$B(Ph)$_3$]$^-$	DES	0.42 ± 0.014	690
Br$^-$	DES	0.45 ± 0.012	670
[n-C$_4$H$_9$B(Ph)$_3$]$^-$	CH$_3$CN	0.03 ± 0.01	120 ± 30
PF$_6^-$	CH$_3$CN	0.03 ± 0.01	115 ± 30
Br$^-$	CH$_3$CN	0.04 ± 0.01	130 ± 40

^{a}Determined by comparison with rhodamine B (ϕ_f = 0.5 in C$_2$H$_5$OH).
^{b}EtOAc is ethyl acetate; DBS is dibutyl suberate; DES is diethyl succinate.

The critical step which triggers the initiation of radical chains is the dissociation of the oxidized borate to triphenylboron and an alkyl radical. Until recently the evidence for this was inferential. This is because the oxidation step produces an intermediate so short-lived, that it also has not been directly observed. Since alkyl radicals have no significant absorption in regions of the spectrum routinely accessible in UV, and their adducts with acrylates are hard to detect, direct evidence for the radical addend of the acrylate was not available. The only direct evidence for radical formation from the borate is in a related case in which an α-naphthyl radical was eliminated and thus could be observed directly after the flash.[55] Reduction potentials of the borates can be estimated from kinetic arguments. Since borate salts have become extremely important coinitiators in systems other than those using cyanine counter ions, full discussion of this is included in Chapter 2.

Excited state reducing agents include the above mentioned borates, where the initiation step involves oxidation of the borate and cleavage to trisubstituted boron and a radical (equation 13) as well as amines where the initiation step involves oxidation at the nitrogen, followed by the loss of an α-proton (equation 14).

$$Ar_3\bar{B}\text{-}R \xrightarrow{-1\ e} Ar_3B\ +\ \cdot R$$

Equation 13

$$Ar\bar{N}(CHR_2)R' \xrightarrow{-1\ e} Ar\overset{+\bullet}{N}(CHR_2)R' \longrightarrow ArN(\overset{\bullet}{C}R_2)R'\ +\ H^+$$

Equation 14

Miscellanous processes for the oxidation of arsines, phosphines, tetrasubstituted tin[56] and enolates are also reported. Most of the latter have limited commercial use at this point. Details are provided in the previous Volume.[46]

Two critical issues in every case control the effectiveness of the reducing coinitiator. First, the electron transfer reaction, forming the critical first stage of the reaction, cannot be rapidly reversible, i.e. back electron transfer must be prohibited. Second, the radical formed from the initially formed intermediate must react rapidly with olefin. In the cases of borates and trisubstituted amines, back electron transfer is essentially prohibited by the rapid dissociation of the intermediate formed initially. Borates lose an alkyl radical and form the trisubstituted boron compound in just a few picoseconds. Evidence for this will be reported in Chapter 2. In the case of tertiary amines, deprotonation and radical formation is also very rapid. If back electron transfer occurs at all, it does so in a limited number of cases. Evidence for this comes from the studies of Yubai Bi and this will be reported shortly.

The oxidation of excited states is also a possibility and an important issue. Onium salts, for example, iodonium salts and pyrillium salts, are strong enough oxidizing agents to capture excited states by electron transfer from the excited state to the salt. The important reactions are shown in equations 15 & 16. The reduction of iodonium salts produces

$$Ar_2I^+, SbF_6^- \xrightarrow{+1\ e} ArI\ +\ Ar\cdot$$

Equation 15

Equation 16

aryl radicals which can carry a radical chain. Reduction with pyrillium ion is more frequently used in conjunction with a photodissociable oxidizing agent like a peroxide which we will discuss in a later chapter.

iv.   **Generating cations by photochemical routes**

In the first edition, Crivello discussed, in some detail, the photoreactions of onium salts in which the objective was to produce cationic chain reactions. During the last several years, considerable new information about the mechanism of such reactions in the presence and the absence of sensitizers, and in the presence and absence of polymerizable monomers, has appeared.

The earliest report of the photoreactions of aryl iodonium salts came from McEwen's group.[57] Crivello and Lam later reported that diaryliodonium salts with non-nucleophilic counter ions (BF_4^-, PF_6^-) were efficient photoinitiators for the polymerization of cyclic ethers.[58] These reactions received immediate and intense industrial interest because epoxides have very important properties. Previous literature had supported the conclusion that the photoreactions of iodonium salts were messy and tenuously perched. Ionic and radical processes were in competition with one another. Kampmeier and Nalli[59] offered definition of the mechanism in a classical mechanistic study of the photolysis of diaryliodonium salts in the polymerizable ethers tetrahydrofuran (THF) and 1,3-dioxolane. Products from iodonium salts irradiated in these solvents are the arenes and iodoarenes, and convincing evidence for a radical chain reaction as the initial step in the photodecomposition is reported. Earlier Crivello had argued that iodoarene radical cations ($ArI^{+\bullet}$) and protons were the key initiating species. His mechanism was based, in part, on the products of the reaction, as well as a quantum yield of product formation of less than 1.0 in acetonitrile suggesting a non-chain process. Since this mechanism has basically been disproven by more recent results, we shall not comment further on it. An alternative is outlined in Scheme 2.[60] This is the mechanism recent results have favoured in the iodonium case.

$$Ar_2I^+ \xrightarrow{h\nu} Ar_2I^{+*} \dashrightarrow ArI + Ar\bullet \quad \text{Initiation step}$$

$$Ar\bullet + S-H \dashrightarrow ArH + S\bullet$$
$$S\bullet + Ar_2I^* \dashrightarrow S^+ + ArI + Ar\bullet \quad \bigg| \text{Chain reaction}$$

$$S^+ + SH \dashrightarrow \text{Polymer}$$

Scheme 2
Mechanism for cation initiated photopolymerization

The role of light in a photochemical process involving an iodonium salt is to generate a source of radicals. The quantum yield for the formation of 4-tert-butyliodobenzene from the iodonium salt in less polar 1,3-dioxolane is 14±2, sharply in contrast with the value of 0.22 at 313 nm previously reported in more polar acetonitrile.[61] The role of the solvent in this process is to furnish a source of hydrogen atoms and both THF and 1,3-dioxolane are well suited for this role. The solvent derived radical so formed then serves as a reductant for another molecule of iodonium salt. Acetonitrile, on the other hand, is a relatively poor hydrogen atom donor. Thus it seems reasonable that the quantum yield measured in this solvent, approximates more closely to the quantum yield of iodonium salt photolysis.

The problem with previous mechanistic studies was their failure to recognize reaction temperature as a critical variable. In Kampmeier's report, the thermal reaction of 4,4'-di-tert-butyldiphenyliodonium chloride was studied at 75°, in polymerizable solvents like THF, while the photoreaction of the same iodonium salt was studied at 25°. At 75° 4,4'-di-tert-butyldiphenyliodonium chloride gives 4-tert-butyliodobenzene and 4-tert-butylchlorobenzene in an apparent S_NAr reaction. At room temperature upon photolysis, tert-butylbenzene, a product of a radical reaction, was formed and there is a clear difference in the products formed from the two reaction routes. The same radical chain reaction with 4,4'-di-tert-butyldiphenyliodonium chloride can also be initiated thermally by tert-butoxy radicals formed from the low temperature thermal initiator, di-tert-butyl peroxyoxalate (DBPO);[62]

$$\text{t-BuOOC--COO-tBu} \xrightarrow{30°} \text{t-BuO}\bullet + 2\,CO_2 + \bullet\text{O-tBu}$$

Thus at low temperatures in the presence of DBPO, the iodonium salt also gives the radical product. Visible light irradiation of phenylazoisobutyronitrile, $PhN=NC(CH_3)_2CN$ (λ_{max} = 395 nm) in the presence of the iodonium salt also gives the radical product. The addition of azo(bisisobutyronitrile) to the iodonium salt at 75° led to the formation of the radical product in competition with the previously observed S_NAr products.

The relative reactivity of the two THF derived radicals, at the 2-position and at the 3-position, (see equation 19) and kinetic analysis, suggest that the termination step for the chain reaction of Ar_2I^+, **Scheme 2,** is unimolecular, as opposed to bimolecular, in radicals. Steady state treatment of the mechanism with the termination step b equation 19 below predicts the observed kinetic behavior (zero order relative to Ar_2I^+ throughout the course of the reaction). If bimolecular termination were important, the reaction should be first order in monomer. This also means that ratios of rates of the two processes would be large.

$$k_{SET}[THF\bullet][Ar_2I^+]/2k_T[THF\bullet]^2 > 10$$

Equation 17

Since the rates of formation of Ar• and S• radicals during the initiation step are equal to the rate of their disappearance, $R_i = R_t$ according to the steady state assumption, Equation 17 becomes Equation 18:

$$k_{SET} [Ar_2I^+]/(2k_t Io\Phi_i)^{1/2} > 10$$

Equation 18

The quantum yield of iodonium salt decomposition is approximately 0.22, the iodonium salt concentration under which the experiments were carried out ≈ 0.015M, and the light intensity about 10^{-5} [Einstein/dm^{3}s]. If the rate constant of radical coupling is diffusion controlled and taken as 1.5 x 10^9 M^{-1}s^{-1}, then a lower limit for the rate constant for SET (Single Electron Transfer) can be estimated to be at least 10^5 M^{-1}s^{-1}. Including equation 19 as the termination step leads to the kinetic expression in Equation 20.

Equation 19

$$d[ArI]/dt = (k_\alpha + k_\beta)/k_\beta \cdot I_0\Phi$$

where I_0 is in [Einstein/dm•s]

and Φ = the initial photochemical quantum yield

Equation 20

When $k_\alpha \gg k_\beta$, Equation 20 becomes $d[ArI]/d_t = I_0\Phi k_\alpha k_\beta$. This predicts zero order behaviour, and a kinetic chain length for the iodonium salt chain reaction equivalent to the relative rates of hydrogen abstraction from the alpha and beta positions of THF. Since this is the observed result, the chain termination anticipated is the abstraction of hydrogen from the 3 position rather than from the chain carrying 2 position of THF.

In short it seems clear from Kampmeier's analysis that cationic polymerization of the cyclic ether, THF, and likely other monomers said to form polymer by cationic chains, is initiated by a radical chain process, in which the role of the iodonium salt is two fold: to form the initiating radicals and to oxidize solvent radicals to cations. It is the latter that carry the polymer chain growth.

Evidence for this mechanism in a polymer forming system was provided independently by the detailed study of the polymerization of cyclohexene oxide by Yubai Bi, reported in detail in his Ph.D. thesis at Bowling Green in 1994.[63] Bi sought a visible light initiator for cationic polymerization and discovered essentially the same chain process as Kampmeier suggested, without knowing about Kampmeier's work. Bi found that only visible light sensitizers, whose structures contained amino groups, were effective in catalyzing cationic chains.[64] Early Crivello results had led to the idea that amines terminated cationic polymerization and that it did not occur in the presence of amines. In the absence of a radical forming step this might be true, but not when the first step in which the amine is involved leads, eventually, to a radical product. Bi constructed a system from a visible dye, a tertiary aromatic amine, and an onium salt. To polymerize cyclohexene oxide he found that ethyl erythrosin, dimethylaniline, and p,p'-dimethyldiphenyliodonium hexafluoroantimonate, gave satisfactory rates of polymerization. He proceeded to analyze all the products in the absence of monomers from the various partners in these reactions.[65] Based on electron transfer rates measured by flash photolysis, Bi projected that the primary reaction of the excited triplet state of the dye was reduction by the amine, followed by proton transfer to form a radical and that the iodonium salt served mainly to oxidize the formed α-amino radical.

$$\text{Dye}^* + \text{ArN(CHR)R'} \longrightarrow \longrightarrow \text{ArN}(\overset{\bullet}{\text{C}}\text{R})\text{R'}$$

Equation 21

$$\text{ArN}(\overset{\bullet}{\text{C}}\text{R})\text{R'} + \text{Ar}_2\text{I}^+ \longrightarrow \text{ArN}(\overset{+}{\text{C}}\text{R})\text{R'} + \text{ArI} + \text{Ar}\bullet$$

Equation 22

An important observation made by Bi, was that the carbocation from the ammonium ion could be detected as an end group in the poly(cyclohexene oxide). This thus provided capstone evidence for the chain suggested.[66]

In summary, the mechanism for cationic polymerization using iodonium salt in the presence and absence of sensitizer is given in Scheme 3. The role of the iodonium salt in such photolyses is to initiate the formation of a radical and to subsequently take part in the radical chain process of radiationless salt decomposition.

$$\text{Ar}_2\text{I}^+ \xrightarrow{h\nu} \text{Ar}\bullet + \text{ArI}$$

$$\text{Ar}\bullet + \text{R-H} \longrightarrow \text{Ar-H} + \text{R}\bullet$$

$$\text{R}\bullet + \text{Ar}_2\text{I}^+ \longrightarrow \text{R}^+ + \text{ArI} + \text{Ar}\bullet$$

Scheme 3

Mechanism of initiation of polymerization of cyclic ethers by iodonium salts

v. Case III: Generating reactive bases

Though amines had been generated by photochemical processes, in a number of earlier literature examples, the yields were generally low.[67] The first report that one could generate bases photochemically in a potential imaging system was that of Kutal and Willson who showed that classical metal-amine complexes such as $[Co(NH_3)_5Cl]^{+2}$ which were known to release ammonia when irradiated, could be used to initiate the crosslinking of epoxides.[68] Cameron and Frechet[69] later reported a non-quantitative study of photoreactive dimethoxyphenyl carbamates, the reactions of which were triggered by short wavelength UV radiation, to produce alkyl amines, Equation 23.

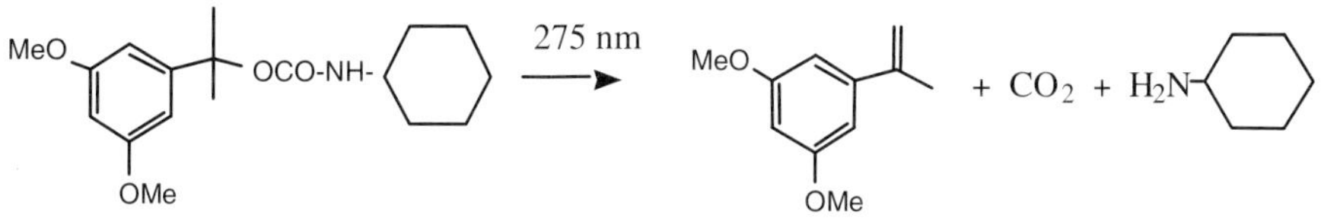

Scheme 23
Photo-induced Reaction of Dimethoxy phenyl carbamates

Ortho-nitrobenzyl carbamates, in a reaction modelled after one first discovered in the 20's by Romanian scientists,[70] but exploited by Patchornik,[71] were also reported to be sources of bases from a photodeblocking reaction (Equation 24).[72] These systems, Frechet reports, are more sensitive at longer wavelengths than are the 3,5-dimethoxy analogs which require radiation below 300 nm. However the quantum yields, reported at 254 nm, depended on the alkyl group and ranged from a high of 62% to a low of 11%. Both methods were successfully used to generate bases to crosslink polymers with pendent epoxy groups,[73] in image tone reversal of known negative-tone photoresists, in polyimidization,[74] and in several other reactions catalyzed by bases.[75]

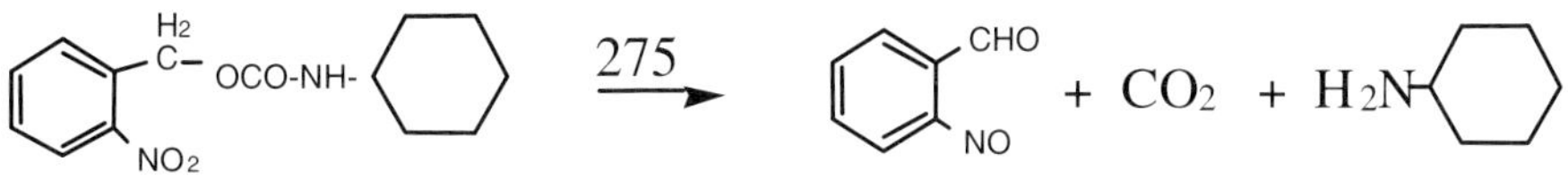

Equation 24
Photo-induced Reaction of Orthonitrobenzyl carbonates

Cameron, Willson and Frechet[76] were the first to report 3',5'-dimethoxybenzoinyloxycarbonyl masked amines using the absorption of the benzoin chromophore to trigger decomposition of the carbamate. This reaction,

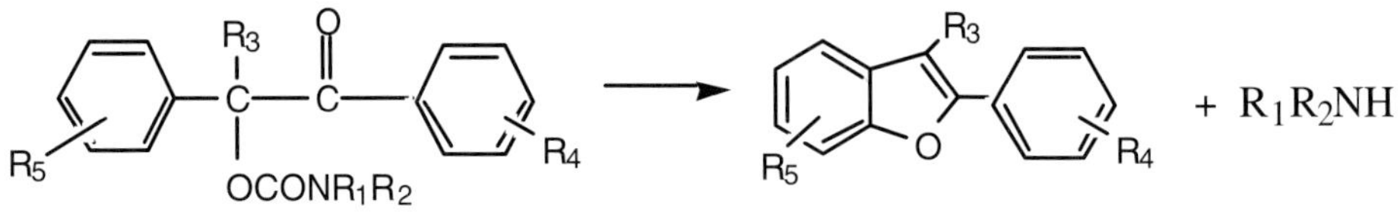

Equation 25

and similar ones, have received some patent attention and will be discussed in detail later.

In recent studies Hassoon, et al have found that the benzophenone-like triplet state of N,N,N-trialkyl-N-(p-benzoylbenzyl)ammonium triphenylbutylborate[77] is rapidly quenched ($\tau \approx 300$ ps in benzene/acetonitrile (1%)) presumably by single electron transfer (SET) from the borate anion to the carbonyl portion of the molecule followed by methylene carbon-nitrogen bond cleavage and reduction of the formed radical cation of the amine, by the enolate of benzophenone which leads to the formation of the free amine, Scheme 4.

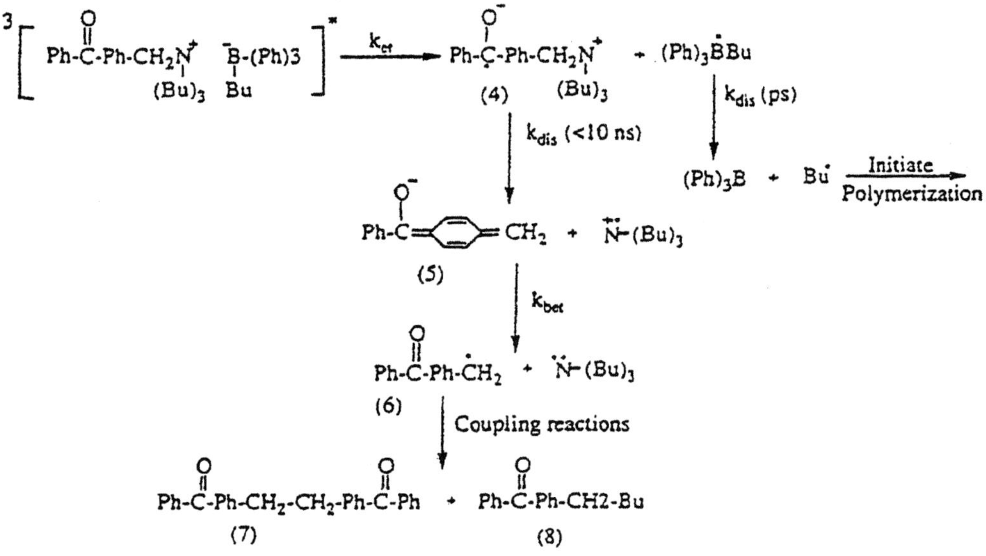

Scheme 4

Photodecomposition of N,N,N-trialkyl-N-(p-benzoyl)benzylammonium triphenylbutylborate

It was postulated that application of the same process, to suitably constituted polymeric quarternary ammonium borates, would lead to the photochemical generation of polymeric amine. Since amines are soluble in acids, this could result in a new class of positive-tone photoimageable polymers which, after irradiation, could be developed by dissolution of the formed polymeric amines in aqueous acid. A photochemical route was established towards the generation of polymeric amine from poly 2-[N,N-dimethyl,N-(p-benzoyl)benzylammonium (triphenyl-n-butylborate)] ethyl methacrylate (polymer **1**). Upon UV irradiation of polymeric quarternary ammonium borate **1** with 365 nm light, rapid quenching of the excited state via SET from the triphenyl-n-butylborate anion and subsequent reactions, led to the formation of polymeric amine **2**. The liberation of the target polymer with its pendant dimethylamino functionalities is accompanied by the formation of biphenyl (from triphenylboron) and products of the coupling of p-benzoylbenzyl radical **3** (with itself and with n-butyl radical). The total reaction is presented in **Scheme 5.**[*]

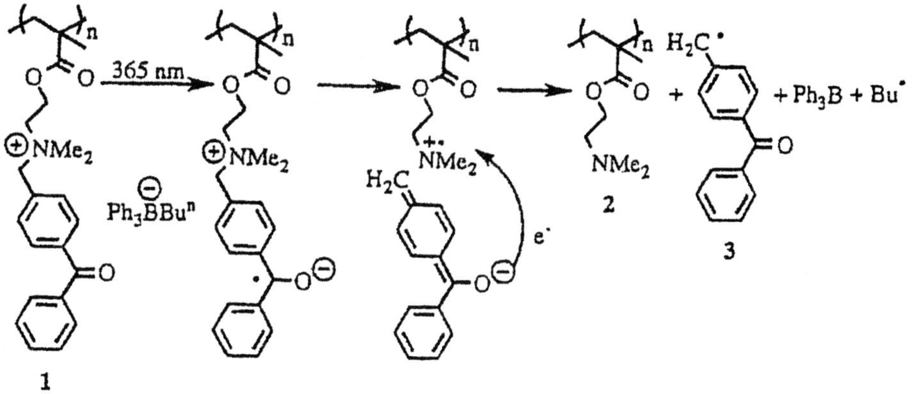

Scheme 5

Photolysis of 2-[N,N-dimethyl,N-(p-benzoyl)benzylammonium(triphenyl-n-butylborate)] ethyl methacrylate (polymer 1)

---

[*] Work of Alexander Mejiritski, Ananda Sarker and Alexander Polykarpov. Taken in part, from the Ph.D. dissertation of Alexander Mejiritski, 1997

Polymer **1** was synthesized according to **Scheme 6** with 70% polymerization yield. Solutions (20 wt%) of polymer **1** in N,N-dimethylformamide where used to yield uniform, almost defect free, glassy films of ca. 0.5 μm thickness on silicon wafers (13 x 13 mm) (non dust free conditions) by spincoating at 6000 rpm for 30 sec with a Headway Research, Inc. spincoater. The films were subsequently prebaked in an oven at 80°C for 30 min in order to remove residual solvent.

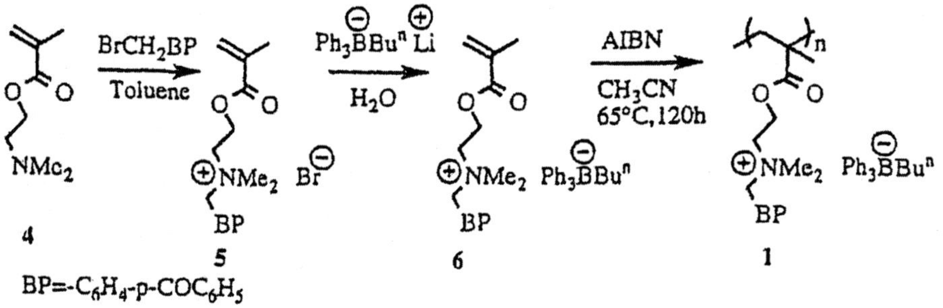

Scheme 6

Synthesis of 2-[N,N-dimethyl,N-(p-benzoyl)benzylammonium(triphenyl-n-butylborate)] ethyl methacrylate (polymer 1)

Film thickness was determined by measuring the intensity of fluorescence at 500nm (λ_{ex}=350nm) from a probe, 5-N,N-dimethylaminonaphthalene-1-(N',N'-di-n-butyl) sulphonamide, which had been incorporated in the films.[78] The intensity of the fluorescence is linearly proportional to film thickness, provided that concentration of a probe is small and remains constant throughout the film. To ensure uniformity of dispersion, the probe was added to the original polymer solution at 0.5-1% level, based on the overall polymer weight. Additionally, several substrates with films of varying thickness (all containing the fluorescent probe) were broken and the resulting film cross-sections investigated by scanning electron microscopy (SEM). Using film thickness determined by SEM a linear calibration graph was established (probe fluorescence intensity vs. measured thickness) to ensure that the fluorescence probe technique provides a quick and nondestructive method for measuring film thickness.

A number of substrates bearing films of the same thickness (0.60 ± 0.03μm) were produced for resolution and sensitivity measurements. In order to measure sensitivity, films were exposed at room temperature to different doses of defocused UV irradiation from a 200W high pressure mercury arc lamp filtered through a 365 nm filter (bandwidth ~40 nm). The distance from film to lamp was fixed at 11cm. Flux was measured by a PMT powermeter appropriately placed in an identical geometrical arrangement. Irradiated films were developed by immersing the substrates sequentially in aqueous HCl (~1.0 M)

for
15 sec, in distilled water for 30sec and in 100% ethanol for 15sec at room temperature. Developed films were air dried.

Areas exposed to the largest doses were completely cleared of polymer film upon development. Since an acrylate backbone is the carrier for the pendant dimethylamino functionalities in polymer **2**, it was assumed that the disappearance of a spectroscopic feature which could be unequivocally attributed to the acrylate network would be utilized as a measure of a relief image formation. Hence, film thinning was evaluated by obtaining the ratio between integrated areas of FT-IR peaks corresponding to the methacrylate carbonyl (1730-1734 cm-1) measured for the given film before irradiation and after development. The system shows rather high contrast of γ_p=1.82.[79] The sensitivity value, D_c (the minimum exposure dose required completely to remove the film of positively functioning polymer of given thickness, under specified processing and development conditions), for the aforementioned system is estimated at 4.5 J/cm^2. The D_o value, with the minimum exposure dose required to start observing changes in film thickness as a result of a photochemically induced process, with subsequent development under the specified processing and development conditions) being close to 1.3 J/cm^2. This sensitivity lies significantly below the 25-200 mJ/cm^2 region of sensitivities for currently employed photoresists.[80]

We would not expect sensitivity to equal that obtained with systems in which the base, liberated by irradiation, serves as a catalyst for film destruction or crosslinking providing chemical amplification of the photochemical reaction. However, materials such as polymer **1** may be useful in a variety of photoimaging processes requiring less speed than microlithography.

Methods for amplifying images formed by polymeric amines are also being explored. In order to increase absorption by a chromophore in the film, a number of them are being investigated as a replacement for benzophenone.

To evaluate the attainable resolution TEM grids G200-Cu and T2000-Cu, modelling the soft contact type masks with the smallest detail size of 38 and 3µm, respectively, were employed. Substrates received the exposure dose of 5J/cm^2 and were developed in the manner described above. A pattern with resolution of the order of 3µm was achieved. The slight "footing" of the detail profile is the result of an imperfect exposure setup. However, our measurements suggest that resolution of less than 3µm is achievable. Investigation with smaller masks using collimated light are currently under way in order that the resolution limit may be determined more accurately.

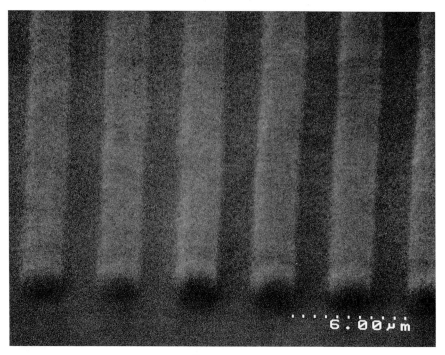

Figure 20
SEM micrograph with 2μm resolution

Polymer (**1**) has a melting endotherm T_m at 115°C as determined by differential scanning calorimetry (DSC) at a heating rate of 20°C/min. In order to determine thermal stability (T_d) the polymer was subjected to thermogravimetric analysis (TGA) in nitrogen. A loss of 5 wt% was recorded at 235 °C when the sample was heated with a heating rate of 20°C/min.

Although the original polymer **1** is soluble in concentrated acids (48% HF and 70% HNO_3) it is insoluble and stable upon exposure to aqueous bases (25% tetramethylammonium hydroxide). This allows for pattern transfer onto the underlying substrates which are destroyed by wet etching with basic solutions.

Amines participate in the thermocrosslinking reactions of epoxy materials and polyurethane oligomers, to produce polymers used as adhesives, coatings and thermosets.[81] Amines photogenerated from thermally stable precursors are especially important, since they alleviate storage stability problems incurred in curable formulations, which include epoxy resins and polyurethane oligomers. Generally photogenerated organic bases used as curing agents are produced from the photolysis of low molecular weight aliphatic amines which induce the crosslinking of polymer matrices, containing appropriate reactive functionalities.[82]

The previously mentioned polymeric amines produced in photoprocess have been assessed as catalysts for thermally driven epoxy crosslinking. If this process was to yield a matrix which was insoluble in organic solvents, it would form the basis for a negative-tone imaging system. Thus this system could act in both tones. There are very few such dual tone systems.[83] Since polymerization and crosslinking of polyepoxides is catalytic in nature, we expected chemical amplification in the negative-tone mode thus increasing the speed of photoimage formation.[*]

Tertiary amines (R_3N:) are not normally employed as sole curing agents for epoxides since they are also good chain terminators. They can be used as curing agents in a two stage process - nucleophilic attack on carbon leading to epoxide ring opening, and a chain reaction leading to a crosslinked product in the case of the polyfunctional epoxides. Most often the presence of a hydroxyl group, in the form of a polyol, is required for the first step. Transfer of the proton to the oxygen of the epoxy substrate produces an alkoxide and quarternary ammonium ions. Chain propagation then involves repeated attack of alkoxide on the available epoxides, Figure 21.

---

[*] The work which follows is a collaborative effort of Alexander Mejiritski, Ananda Sarker and Adrian Lungu

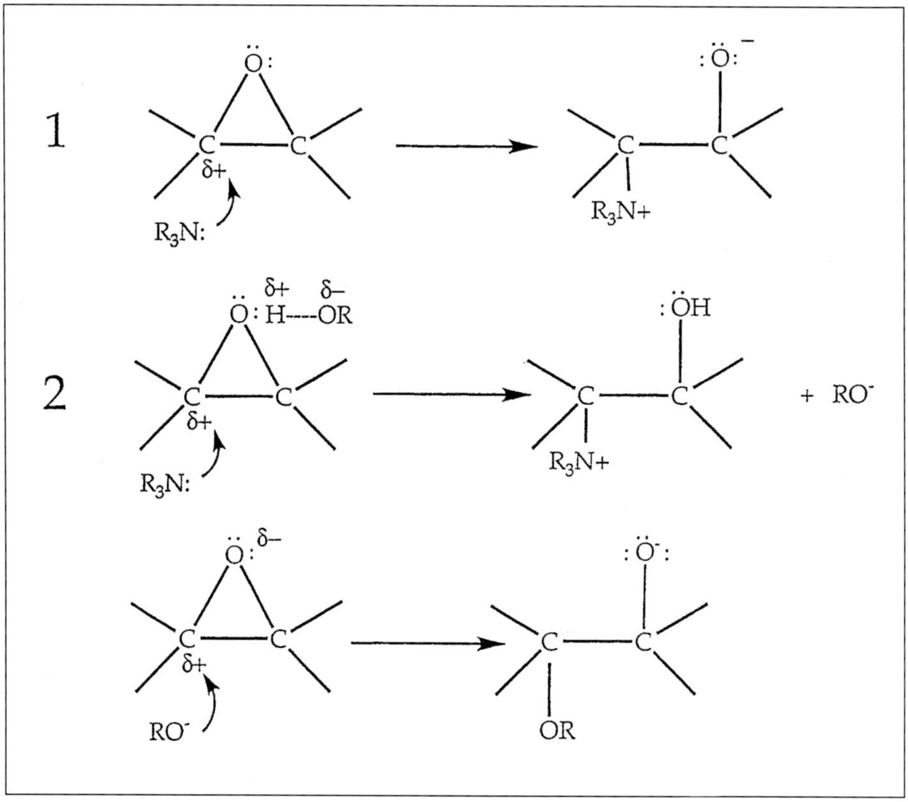

Figure 21

Mechanism of anionic polymerization of epoxides with tertiary amines as initiators.

An important implication of this mechanism is that nucleophilic attack on the epoxide in the initiation step should be sensitive to steric effects. As a result only substrates with terminal epoxy groups are subject to catalytic polymerization by tertiary amines, since the relatively unhindered CH_2 group is available for nucleophilic attack in this case.

N,N-dimethyl-n-butylamine (DMBA), a close analog of the amino groups in polymer **2**, Scheme 5, was used as a model compound to determine which epoxides could be polymerized with such a catalyst. Portions of mixtures of DMBA with commercial epoxides UVR 6110, EPON 828 and DEN 438 (Figure 22) (molar ratio epoxy/amine= 8:1) were spread between NaCl rectangular plates and heated to 80°C. Curing was measured by

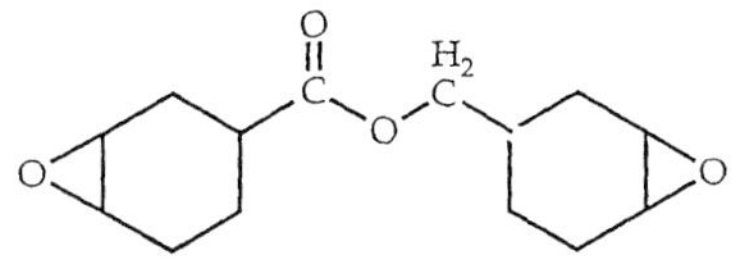

3,4-epoxycyclohexylmethyl-3',4'-epoxycyclohexane carboxylate, Union Carbide resin UVR 6110

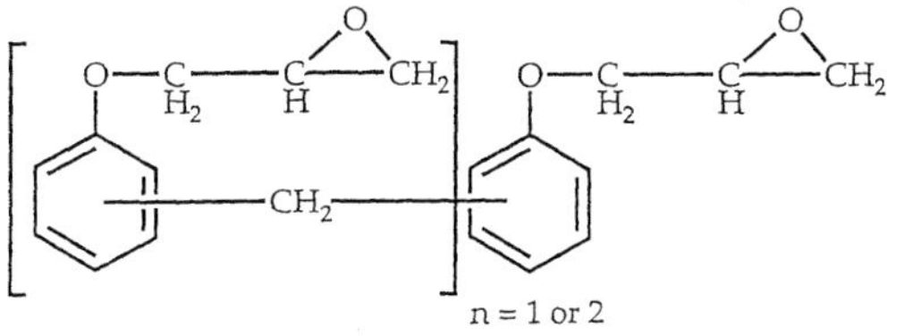

Epoxy novolac resin from Dow, DEN 438

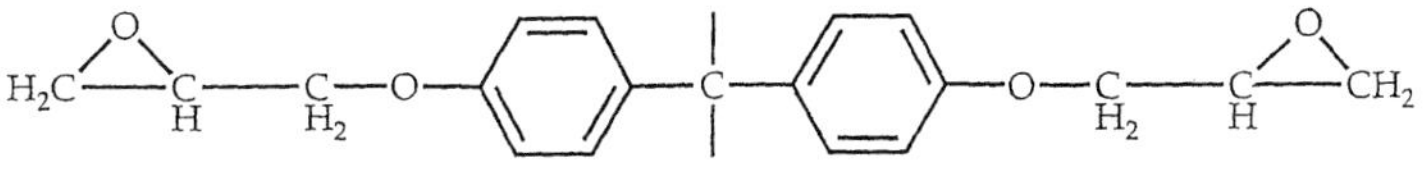

Diglycidyl ether of bisphenol A, Shell resin EPON 828

Figure 22
Epoxy resins tested.

recording transmission IR spectra for all of the mixtures. DMBA cured EPON 828 and DEN 438 in 30 minutes to the point that the NaCl plates could not be separated. There was also a significant decrease in intensity in an IR band assigned to the epoxide rings at 916 cm^{-1} and a significant increase in the intensity of an IR band at 1120-1145 cm^{-1} assigned to the ether linkages formed after ring opening, Figure 23.

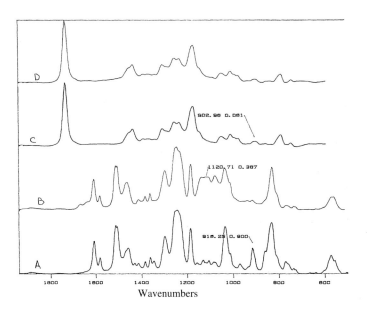

Figure 23

Transmission IR spectra of mixtures between NaCl plates for: A) EPON 838 exposure and DMBA at room temperature; (B) EPON 838 and DMBA after thermal exposure at 80 °C for 30 minutes; C) UVR 6110 and DMBA at room temperature; D) UVR 6110 and DMBA after thermal exposure at 80 °C for 30 minutes.

Further thermal exposure of these mixtures led to no changes in the IR spectrum. At the same time, no curing was observed in the mixture with UVR 6110, a cycloaliphatic epoxide. The addition of glycerol (a polyol) had no influence on the curing behavior and caused no changes in the IR spectrum, with any of the three mixtures. This clearly demonstrates that amines of the type outlined above are suitable for the thermal curing of polyfunctional epoxides with terminal epoxide groups of the types normally used in UV curing.

Single crystalline Si wafers are transparent in the IR fingerprint region[84] and offer an ideal substrate for the monitoring of polymerization of epoxies by infrared. For spin coated films containing the polymeric quarternary ammonium salt and epoxides of various compositions, Table 3, the molar reactivity ratio of the photopendants to epoxy moieties was chosen to be 1:2 (1:2.5 in the case of DEN 438.) This is a lower ratio than that used by Beecher et al.[85] in which the mole ratio was biased in favour of the amine precursor.

Table 3
Polymeric sources of amines and their reactions with epoxides
Sensitivity parameters for mixtures containing the polymeric amine photoprecursor and epoxides in different compositions

Mixture	Molar ratio, amine/epoxy/OH (where present)	Sensitivity parameters, (mJ/cm-2)
(A) UVR 6110 + 3	1 : 2	no image
(B) EPON 828 + 3	1 : 2	$D_O = 1600$ $D_C = 3800$
(C) DEN 438 + 3	1 : 2.5	$D_O = 600$ $D_C = 3200$
(D) DEN 438 + 3 + glycerol (prebake)	1 : 2.5 : 5	$D_O = 600$ $D_C = 3200$
(E) DEN 438 + 3 + glycerol (no prebake)	1 : 2.5 : 5	$D_O < 200$ $D_C = 1050$

Prebaking the films, containing both epoxides and quarternary polymer at 80 °C for 30 minutes, induced no curing. No changes in the IR spectra were observed (Figure 31) and films of all compositions remained soluble in acetonitrile.

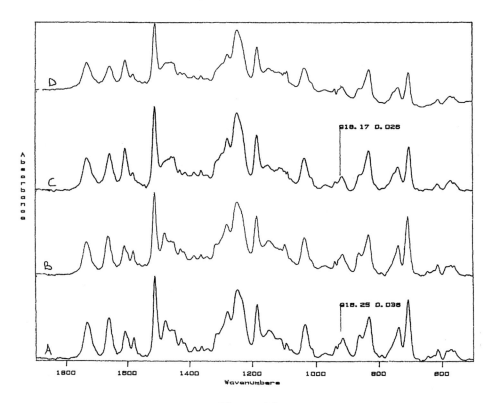

Figure 24
Transmission IR spectra of spincoated film including polymeric amine precursor 1 (Ph$_4$B$^-$ gegen ion) and EPON 838 resin: A) before prebake; (B) after prebake at 80 °C for 30 minutes; C) after prebake and UV irradiation for an accumulated dose of 4500 mJ/cm^2; D) after prebake, irradiation and postbake at 80 °C for 30 minutes.

Parts of the films were then exposed through a 365nm cutoff filter (bandwidth π 40 nm) to different doses of defocused UV irradiation. No changes were observed in the epoxy region (902-916cm^{-1}) for the irradiated areas of any composition, even though light doses yielded ≈ 75% conversion to polymeric amine.[79] However, the changes in the carbonyl region of the spectrum noted previously,[86] during irradiations of polymer **1** (gegen ion, Ph$_4$B$^-$) alone were again observed. It can be inferred, therefore, that photochemical processes leading to the formation of polymeric amine, proceed uninhibited in the epoxy matrix. Films of all compositions (Table 3) were then exposed to the doses (4500mJ/cm^2) known to be adequate to render the films completely removable, by our acid development procedure (see above)[86]. UV exposure was followed by a subsequent heating step chosen to be below the Tg for polymer **1**. After this treatment the decrease in intensity for the IR band at 916cm^{-1} in the irradiated portions of the films of compositions **B** and **C** is indicative of epoxy ring opening, Spectrum D, Figure 24. However, no changes in this region of the IR spectrum were observed for films of composition **A**, Table 3. Subsequent immersion of irradiated films containing **B** and **C** in acetonitrile, leads to the immediate dissolution of non-irradiated areas with film retention in the irradiated portions of the wafer surface. Apparently, sufficient epoxide polymerization occurred to insolubilize the network. It is likely that the complete consumption of the epoxides was inhibited by the low mobility of the polymers, both those containing amino groups and those containing epoxides while at room temperature, in the crosslinked matrix. Films of composition **A** were completely dissolved upon immersion even though polymeric amine was formed. These results are entirely consistent with the view that photolysis produces polymeric tertiary amines, which can initiate the polymerization of epoxides. However, only those epoxides which contain terminal amino groups are crosslinked.

Since the irradiated and the unexposed areas evidence differential solubility in the developing solvent (exposed areas are less soluble) the formation of negative photoimages is possible.[87]

A number of films of all compositions, Table 3, were exposed to differing doses of UV light imagewise through soft contact Cu foil grids. After the postbake and development procedures outlined above, the three-dimensional relief images of the mask (ca 7 x 7 µm square mounds) were formed in all of compositions except composition **A** as exemplified in Figure 25.

Figure 25

In the AFM Micrograph of negative-tone relief image obtained in the film polymeric amine precursor 1 (Ph$_4$B$^-$ gegen ion), epoxy resin EBON 838 after prebake at 80 °C for 30 minutes, UV irradiation for an accumulated dose of 3800 mJ/cm^2, postbake at 80 °C for 30 minutes and development in acetonitrile; 3-dimensional view.

Atomic force microscopy (AFM) was used to determine the thickness of the remaining film portions as well as its surface structure.

*Atomic force microscopy (AFM) is one of the scanning probe microscopies (SPM) which allows one to visualize surface topography with a resolution theoretically close to atomic separations. In AFM a tip, which is an integral part of a spring cantilever, is brought within a distance from the surface, so that there is an interatomic interaction between the end of the tip and the surface of interest (Figure 26). As the tip is scanned across the surface, it follows the contours of the surface bouncing up and down. The displacement of the tip (laser beam deflection) monitored by an optical detector serves as a measure of surface topography. Furthermore, the detector is an important part of the negative feedback loop which brings the sample to its original position after deflection by means of contraction/extension of the piezoelectronic tube (PZT).

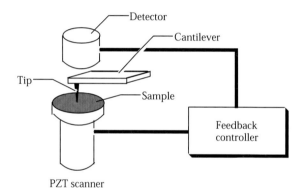

Figure 26
AFM operating principle

---

* This section was contributed by Alexander Mejiritski. Taken in part, from the Ph.D dissertation submitted to the Center for Photochemical Sciences in 1997.

It is an ideal technique by means of which to measure the thickness, roughness and topology of an imaged surface.

Si tips with a spring constant of K = 0.1 Newt/meter were utilised. A twenty five micron by twenty five micron scan module was used. The overall scan time was 2 min. Where higher resolution was desired, and in order to avoid substrate damaging by the tip, the overall scan time was increased and reference force was lowered (3.0 V).

In the experiments reported the thickness of the film remaining after development (dZ parameter in Figure 27) was plotted versus the exposure dose in order to obtain a sensitivity curve for each composition. Sensitivity parameters D_c and D_o can be determined from the corresponding sensitivity curve and are shown in Table 3. D_o is defined as the minimum exposure dose after which the thickness of the remaining film reaches saturation. For compositions **B** and **C**, D_c is equal to about 3200-3800 mJ/cm^2 which corresponds to about 70% conversion of the polymeric amine precursor. This value quantitatively corresponds to the conversion necessary for positive-tone photoimaging as shown previously for system **1**. Thus, no significant chemical amplification has been achieved from the expected catalytic chain reaction of an epoxide curing by a polymeric tertiary amine.

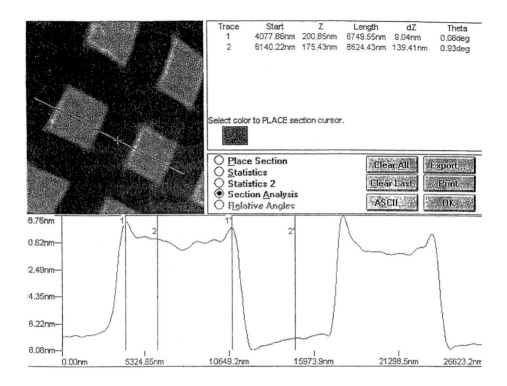

Figure 27
AFM Micrograph of negative-tone relief image obtained in the film polymeric amine precursor 1 (Ph$_4$B$^-$ gegen ion), epoxy resin EBON 838 after prebake at 80 °C for 30 minutes, UV irradation for an accumulated dose of 3800 mJ/cm^2, postbake at 80 °C for 30 minutes and development in acetonitrile; cross-sectional view and film thickness measurements.

The appearance of the surface of the developed cross-linked matrix undergoes significant transformation as the irradiation dose, followed by post-bake, is gradually increased. While the film appears to be an ensemble of polymer islands only slightly fused together at a dose equal to D_o, it becomes a unified solid with island tops comprising 'waves' at doses equal to D_c, Figure 28.

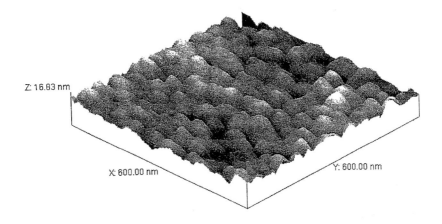

Figure 28

AFM micrograph showing surface morphology of film of epoxidized resin crosslinked with polymeric amine generated by UV photolysis of polymer precursor 1 (Ph$_4$B$^-$ gegen ion).

Highly sensitive photoresists are sought in modern lithography because they may be effectively used with low-brightness sources. Since quantum yields of most non-chain photoprocesses are less than 1.0, thermal catalytic processes have been investigated to enhance the photochemical effect by chemical amplification.[88] We expected to achieve amplification by adding a polyol, glycerol, to the film **C**, Table 3, with an OH/epoxy ratio chosen to be 2:1. A series of spincoated films of this composition (**D**) on Si wafers were prebaked, exposed imagewise, postbaked and developed according to the procedure above. This resulted in very little amplification still requiring ca. 70% of the precursor conversion to reach D_c. It had been previously noted that the intense glycerol IR stretch at 3450cm^{-1} corresponding to the OH stretching frequency of the polyol disappeared upon baking. This step was subsequently omitted for the composition **E** (Table 3). After exposure, post-bake and development, the remaining film thickness as assessed by AFM yielded sensitive parameters D_o and D_c equal to 200 and 1050mJ/cm^2 respectively, Figure 29.

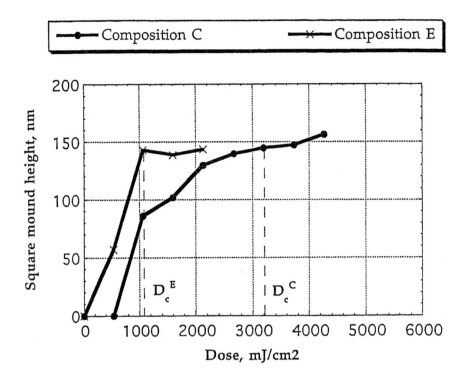

Figure 29
Sensitivity graphs for negative-tone image formations for compositions C and E; See Table 3.

This corresponds to only 20% conversion to polymeric amine in order that D_c be reached. In accordance with the above mechanism of epoxide crosslinking, the presence of hydroxyl function significantly enhances the process. The hydroxyl function thus leads to a more than 3 fold increase in activity.

The conclusion is that a single system produces a polymeric amine, a Lewis acid and a radical in one photochemical step. The result of the photoprocess itself is degradation of the film in an imagewise fashion. The catalysts formed by the process, the radical and the base, can be used to initiate subsequent reactions. In the case of the base, these subsequent reactions can be used in imaging applications via polymerization reactions of terminal epoxides.

6. Lamps, Lasers and Other Sources

This introductory section concludes with a short discussion of artificial light sources. There is no intention to review the field. The reader is referred to the first edition for that, and to the book being written by Professor R Mehnert of Leipzig IOM. Here, some of the issues which are important in developing visible initiators to their present state are discussed. The light source used for a curing event is an extremely important matter and even experienced practitioners find difficult to predict it.

Broadly speaking, artificial light sources divide into lamps and lasers. Either may be continuous wave (CW) or pulsed. Scientists in this field have become adept in the evaluations of new formulations for specific radiation cure applications. Formulations are assessed under conditions close to those employed in practice. Hence a number of lamp types ranging from the tungsten/halogen lamp of the overhead projector to lasers are involved. The range incorporates the broad wavelength output dental lamps, pulsed sources and various phosphor doped mercury lamps used in commercial photocure apparatus. The outputs of (1) a dental lamp from a commercial supply house and (2) a simple office overhead projector are given in Figures 30 and 31. One can do a lot worse for simple testing than using these radiation sources. The critical parameters of any radiation source are power and wavelength output. All of the graphs show plots of intensity vs. wavelength. These can be obtained from a simple powermeter if one is available in your laboratories. Otherwise they are readily and inexpensively available.

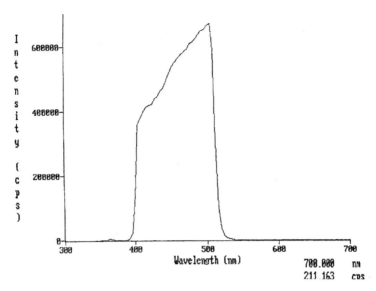

Figure 30a
Dentist's lamp with filter

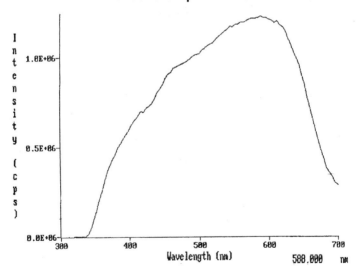

Figure 30b
Output of a typical dentist's lamp (Dentsply)

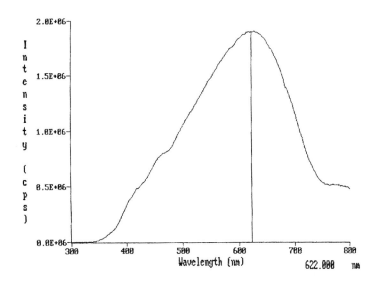

Figure 31
Output of typical overhead projector (tungsten/halogen) bulb.

Commercial mercury lamps can be coated with phosphors and various areas of the spectrum may be isolated to enhance performance. Other lamp configurations also work, so that there are now a number of sources which allow a user to optimize photoperformance. The outputs of the Fusion Systems D, H, Q and V bulbs are compared in Figure 32. The 'D' bulb outputs mainly in the long wavelength UV and short wavelength visible regions of the spectrum. 'Q' bulb radiations are more in the visible. Emissions of the 'V' bulb are centered mainly around 420nm. The 'H' bulb is an uncoated mercury arc the output of which is essentially balanced between the visible and the UV. Principle wavelengths are 313nm, 366nm, 410, 441nm, and so on. Radiations below 300nm emitted from this bulb are only important in instances in which the sample being radiated is not covered by, or contained in, a glass vessel. Therein lies a trick used by lampmakers and practitioners in the field. One can filter the emissions of the mercury arc merely by using a glass envelop to contain the sample emitting the radiation.

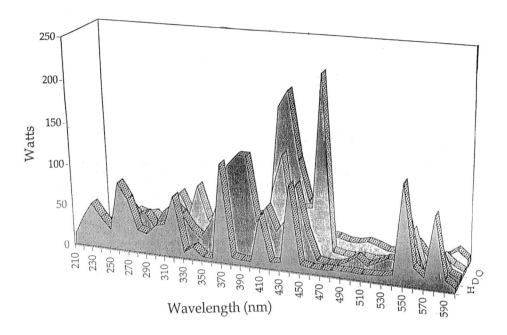

Figure 32
Output of Fusion Systems H, D, Q, and V bulbs.

A typical comparison of performance based on the irradiation source for a highly pigmented white coating is given in figure 33. The formulations contain a combination of UV and visible initiators. UV initiators which are available from specialist chemical companies. DIBF is H-Nu 470 available from the Spectra Group. Depth of cure controls the thickness of a sample that can be photohardened, and the adhesion of that cured formulation to a substrate.

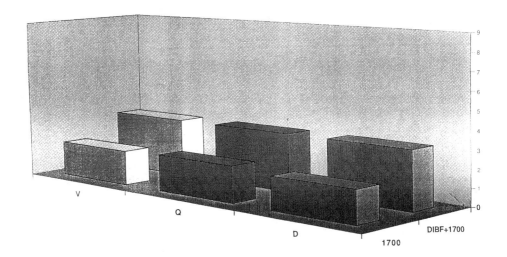

Figure 33
Depth of Cure for 2 Formulations Each Containing 40% Titanium Dioxide: Formulation 1700 Contains UV Initiator Only; Formulation 1700 + DIBF Contains Both a UV and Visible Initiator System:

Solvent resistance is a measure of hardness and resistance to environmental factors. Fixed thickness samples of the similar formulations cured at various belt speeds with Fusion Systems D, Q and V bulbs are compared in Figure 34.

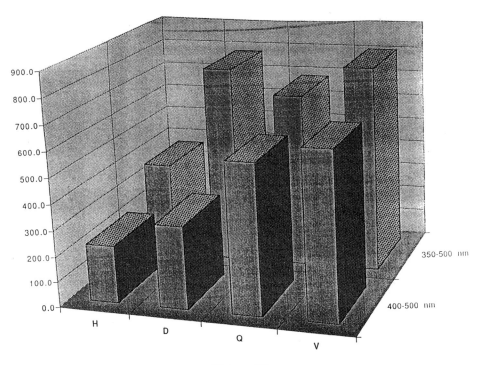

Figure 34
Surface Resistance to Solvent: 2 Formulations Each Containing 40% Titanium Dioxide: Formulation 1700 Contains UV Initiator Only; Formulation 1700 + DIBF Contains Both a UV and Visible Initiator System (a) D bulb (b) Q Bulb (c) H Bulb.

Xenon emits more evenly than mercury throughout the UV and visible regions. Its spectral output is given in Figure 35.

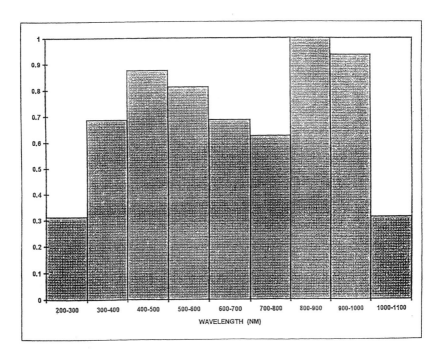

Figure 35
Emission spectrum of xenon lamps.

Pulsed sources deliver a much higher density of photons in a shorter time. This can help cure up to a point but it is fallacious to think that this alone will improve cure speed in anything approaching a linear way.

The fallacy is best illustrated by kinetic arguments. In the steady state approximation, one assumes that the efficiency of radical formation, or escape of initiating radicals from the cage in which they are formed, is always the same, no matter the light intensity. However the rate of bimolecular reactions of the radicals formed is intensity dependent in that the rate of initiation is accordingly decreased by radical consuming reactions. The larger the amount of recombination is, the lower the rate of initiation of the subsequent chain reaction.

One thus needs to subtract the recombination rate, $k_c[/R\bullet/]^2$ from the rate of initiator decomposition $k_dI_0[In] - k_c[/R\bullet/]^2$ to obtain the overall rate of initiation.

Moreover at higher radical concentrations, the kinetic chain length becomes shorter. The average molecular weights of the polymer formed (or its crosslink density) is lowered and this, in turn, generates a softer or more tacky polymer.

Pulsed sources are the current day flash bulb or lamp. They deliver much higher photon densities in a short period of time presenting exactly the situation described kinetically above. The higher the concentration of radicals generated, the more radicals wasted in recombination and the tackier the polymer formed. This is sometimes called 'reversal', in that higher photon density leads to a lesser photo-response. The relative difference between a pulsed source and a CW source is illustrated for a mercury lamp in Figure 36. A pulsed lamp can deliver, at peak power, as much as 10^6 watts with a pulse width (duration of the pulse) of a few microseconds.

Pulsed sources have advantages and disadvantages over CW sources. The former are safer in that there are none of the heat transfer problems which arise with either lasers or CW sources. They can also be more easily directed, aligned, or 'shot' at the target, and they warm up instantaneously. However the higher radical concentration which they may generate has disadvantages deriving from reversal. One of these is poor surface cure. The same TiO_2 formulations cured with D, Q and V bulbs (above) were cured with a pulsed xenon source and with a projector bulb, Table 4. Samples cured with xenon pulses were tacky no matter how many repetitions of the flash lamp were used. Those cured with a projector bulb cured more slowly but with a tack free surface.

Table 4
Curing TiO$_2$ filled compositions with CW and pulsed xenon sources

TiO$_2$ Filled Formulations Cured with Pulsed Xenon and a Tungsten/Halogen Bulb - Formulation 1700 (UV Initiator only) Formulation 1700 + DIBF (UV Initiator + Visible Initiator).			
PULSED XENON (time/result)		TUNGSTEN/HALOGEN BULB (t/r)	
1700	1700 + DIBF	1700	1700 + DIBF
5 sec tacky	tacky		
2 min ------	------------	tacky	tacky
3 min ------	------------	tacky	tack free
4 min ------	------------	tacky	tack free
6 min tacky	tacky	-------	----------

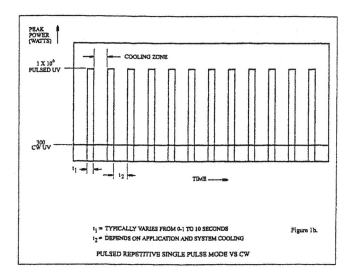

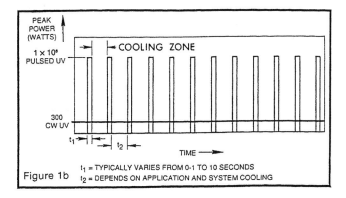

Figure 36
Pulsed continuous mercury emission vs. xenon emission

i. **Lasers**

With the introduction of the ruby laser in the late 1950's, commercial single wavelength, bright light sources became laboratory realities. Though the red emission of synthetic ruby is not yet useful in photopolymerization science, since it is too low power and wavelength too long, the introduction of rare gas lasers in the 60's and tunable dye and pulsed lasers in the early 70's both increased the available power and greatly increased the wavelengths that lasers could offer. Today the lasers used in photopolymerizations are mainly rare gas lasers and include the helium/cadmium laser (324nm) which delivers wavelengths in the UV targeted to many of the older UV initiators, and the argon ion lasers which can be tuned in the UV (351; 355nm) and at several visible wavelengths (488 and 514nm). Fast laser direct write systems for stereolithography such as the 3D Systems SLA 500 use argon ion lasers tuned to UV lines.

The objectives for lasing action UV must be that all excited state deactivation processes of the emitter, save for fluorescence, are prevented, when excitation from the source is triggered by very high energies. Under such conditions, with certain materials, the zero point vibrational level of S_1 becomes overpopulated with respect to S_0. The transition S_1v_1 ---> S_0v_0 is subsequently stimulated by interaction of photons in vibrational energy levels higher than S_1n. A simplified energy profile diagram is given in Figure 37.

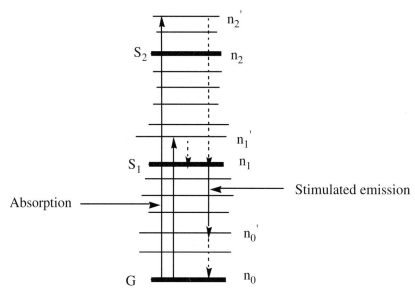

Figure 37
Lasing action: Jablonski diagram: Stimulated emission from $S_1v_0^1$ to S_0v_2.

All rare gas lasers have high power requirements for the lasing condition to be achieved and a great deal of heat must be dissipated while they are in use. Lower wattage argon ion lasers (< 250 mwatt peak power) are air cooled. Bigger lasers require water cooling. This is a major disadvantage. Another disadvantage is effective laser life. Many lasers, particularly those of higher power, offer limited hours of service at peak power. Beyond this replacing the tubes is expensive. The alternative commercial solid state lasers are no alternative. Most are of too low power and too long wavelength for the available photopolymerization chemistries, presenting an interesting chemical challenge for future photopolymerization scientists.

Rare gas lasers use various pumps but offer emissions at the various spectral lines of rare gas emissions. These are the lines in He, Ar, Ne, etc. of the common lasers: He/Cd (helium emission), Argon ion (Argon emission), He/Ne (neon emission) etc. One tunes the laser to the line of interest.

Table 5
Wavelengths emitted by common lasers used in photochemistry

Laser	λ/nm
F_2	157
ArF, excimer	193
KrCl, excimer	223
Ruby (tripled)	231.4
KrF, excimer	248
Nd: YAG (quadrupled)	266
XeCl, excimer	308
Nitrogen	337.1
Ruby (doubled)	347.2
Krypton ion	350.7
XeF, excimer	351
Nd: YAG (tripled)	355
Nitrogen	428
Helium-Cadmium	441.6
Argon ion	488
Argon ion	514.5
Nd: YAG (doubled)	532
Krypton ion	568.2
Helium-Neon	632.8
Krypton ion	647.1
Ruby (fundamental)	694.3
Gallium arsenide	904
Nd: glass	1060
Nd: YAG (fundamental)	1064
CO_2	10600

The pulsed laser is made to emit by the absorption of radiation often from another laser.

Lasers, even CW lasers, can be highly focused and their beams interrupted by shuttering to deliver bursts of photons. An interesting feature of photopolymer formation particularly in acrylate formulations is the effect of intensity differences caused by the beam profile. A typical example is the result of the sinusoidal beam profile. Since the intensity is higher in the middle of the beam than at the edges under these conditions, the photopolymer forms as a spike. The refractive index of the polymer is larger than is that of the monomer as well, causing a focusing of the beam as it proceeds into the monomer mix. An A̲tomic F̲orce M̲icrograph (AFM) of the surface profile of a C̲ompact D̲isc (CD), the original of which was probably printed this way illustrates the profile produced during a laser initiated photopolymerization, Figure 38.

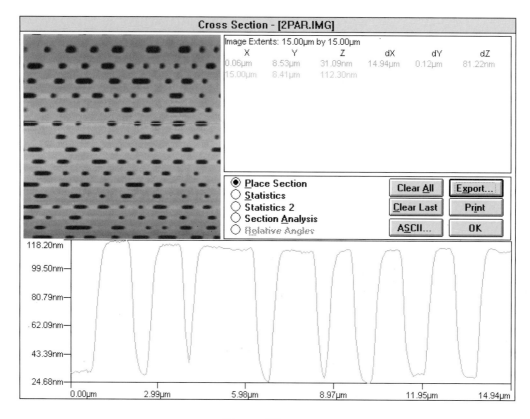

Figure 38
An AFM micrograph of the surface of a CD.

We compare results for the photocuring TiO_2 filled systems containing both a UV and a visible initiator with various light sources, (see Table 4) including two lasers, below.

1. Ar ion laser (514nm) with a power of 145 mW, scanning speed of 4 mm/s

- Irgacure 1700 - no cure;

- Irgacure 1700 + DIBF - cured (tacky surface).

2. He-Cd laser (325nm) with 44 mW power, scanning speed of 4 mm/s

- Irgacure 1700 - no cure;

- Irgacure 1700 + DIBF - no cure;

3. UV bath (300-400nm) 2 x 400 W lamps

- Irgacure 1700 - cured in 30 s;

- Irgacure 1700 + DIBF - cured in 10 s

The major cure mechanism in the highly filled system is mainly via the visible initiator (Irgacure 1700 does not absorb 514 nm radiation). An auxiliary effect may be provided by the 1700 when the source contains long wavelength UV radiation in addition to some short wavelength visible (UV bath 300 - 400 nm). The laser is an intense source and suffers the same reversal phenomena as does a flash lamp. Surface cures are incomplete. The results of reversal and from the effects of O_2 are ameliorated with slow, surface cure assisting initiators.

References

[1] Plambeck, L. F., U. S. Patent #2,760,863 August 28, 1956

[2] For the purposes of this book we'll use metric units of depth. Arbitrary, 10 µ is thin.

[3] Staudinger, H. *Ber*, **12920**, *53*, 1073.

[4] The mathematical analysis enabling the conversion of rates measured as a function of rotating sector speed is unnecessary to the point I wish to make here. Please see Walling, C. *Free Radicals in Solution* Wiley, NYC, 1957 for additional details.

[5] Light, as we shall see later, comes in many different forms. An important section of this book will be devoted to radiation sources, and their effect on photochemical processes.

[6] Kharasch, M. S.; Mayo. F. R.. *J. Am. Chem. Soc.*, **1933**,*55*, 2468; Kharash, M. S.; Walling, C.; Mayo, F. R. *J. Am. Chem. Soc.* **1939** *61*, 1559. Walling, C.; Kharash, M. S.; Mayo, F. R. *J. Am. Chem. Soc.* **1939** *61*, 2693.

[7] Kharasch, M. S.; Urry, W. H.; Kuderna, B. M. *J. Org. Chem.* **1949**, *14*, 248-253.

[8] Kharasch, M. S., Reinmuth, O.; Urry, W. H. *J. Am. Chem. Soc.* **1947**, *69* 1105.

[9] Paneth, F.; Hofeditz, W. *Berichte* **1929** *62* 1335-1347

[10] *Organic Chemistry*, Fieser, L. F.; Fieser, M. Reinhold, NYC, 1950

[11] The concept of the quantum yield will be discussed later.

[12] Common laboratory glassware absorbs most radiation at wavelengths shorter than about 290nm. For samples that must be irradiated at shorter wavelengths, quartz is required. Synthetic quartz of which Vycor® is an example, is transparent to about 215 nm.

[13] Thijs, L.; Gupta, S.N.; Neckers, D.C. *J. Org. Chem.* , **1979**,*44*, 4123.; Gupta, S.N.; Gupta, I.; Neckers, D.C. *J. Poly. Sci., Polymer Chem. Ed.* , **1981**, *19*, 103.

[14] Nippon Oil and Fat

[15] Curran, D. P.; Porter, N. P.; Giese, B.

[16] Hammond, G. S.; Sen, J. N.; Boozer, C. E. *J. Am. Chem. Soc.* **1955**, *77* 3244.

[17] Goldschmidt, S.; Renn, K. *Ber.* **1922**, *55*, 628.

[18] Named after the discoverer, Galvin Coppinger, then employed by Shell Development Corporation in Emeryville, CA.

[19] Bartlett, P. D.; Funchushi, T., *J Amer. Chem. Soc.*, **1962**, *84*, 2596

[20] camphor quinone, Volume 3, Ed. 1, pp 266-68.

[21] Hammond, G. S.; Wamser, C.C., Chang, C. T.; Baylor, C. *J. Am. Chem. Soc.* **1970** *92* 6362

[22] Volume 3, Edition 1, this series.

[23] It's an obvious and simple exercise to calculate the projected absorbance (A as measured by the spectrometer) for a solution of benzophenone (0.1 molar) at its wavelength of maximum absorption.

[24] Planck, M. *V. Deutsch. Phys. Ges.* **1900**, *2* 237-245.

[25] This is in contrast to bond heterolysis which produces a non-identical pair of partners from bond cleavage. Bond heterolysis would produce a pair of ions, for example.

[26] One femtosecond is 10^{-15} seconds; other units of time which become important in the spectroscopies that follow excited states are picoseconds (10^{-12}); nanoseconds (10^{-9}); microseconds (10^{-6}); and milliseconds (10^{-3}).

[27] The difference between MOLECULE and COMPOUND is a conceptually crucial one. Molecule is mainly a theoretical concept. One can only observe a collection of molecules, or a compound. One puts compounds in bottles; one writes molecules on blackboards.

[28] Norrish, R. G. W.; Porter, G. *Nature* **1949**, *164* 658.

[29] Gould, G. U. S. Patent # 3,586,998, June 22, 1971.

[30] Hu, S.; Neckers, D. C. *J. Org Chem.*

[31] Reporting the solvent is critical. Reactive solvents reduce the lifetime by their reaction.

[32] Jablonski, A. *Nature*, **1933**, *131* 839.

[33] Khan, A. U.; Kasha, M. *J. Chem. Phys.* **1963**, *39* 2105

[34] Jablonski, A. *Zeit. f. Phys.* **1935**, *94* 38-46

[35] Written this way for simplest and to contrast it with the ground state of the same molecule.

[36] Terninin, A.; Ermolaev,V. *Trans. Faraday Soc.*, **1956**, *52* 1042-1052.

[37] Wagner, P.; Hammond, G. S. *Advances in Photochemistry, Volume 5, 1965.*

[38] Norrish, R. G. W.; Appleyard, M. E. S. **1934**, 874-880.

[39] Einstein, A. *Ann. d Physik*, **1912**, *37*, 832-838.

[40] Warburg, O. *Z. f. Electrochem.* **1920**, *26* 54-59.

[41] Starr, J. E.; Eastman, R. H. *J. Org. Chem.* **1966**, *31* 1393-1402.

[42] Yang, N. C.; Feit, E. *J. Am. Chem. Soc.* **1968**, *90*, 504.

[43] Turro, N. J.; Hammond, W. B.; Leermakers, P. A. *J. Am. Chem. Soc.* **1965**, *87*, 2774.

[44] Turro, N. J.; Hammond, W. B. *J. Am. Chem. Soc.* **1966**, *88*, 3672.

[45] Sluggett, G. W.; Turro, C.; George, M. W.; Koptyug, I. V.; Turro, N. J. *J. Am. Chem. Soc.* **1995**, *1177*, 5148.

[46] Huggenberger, C.; Lipscher, J.; Fischer, H. *J. Phys. Chem.* **1980**, *84*, 3467.

[47] eg. Sumiyoshi, T.; Schnabel, W.; Henne, A.; Lechtken, P. *Polymer* **1985**, *26*, 141.

[48] Sumiyoshi, T.; Schnabel. W. *Maacromol. Chem.* **1985** *186*, 1811.

[49] Baxter, J. E.; Davidson, R. S.; Hageman, H. J.; McLauchlan, K. A.; Stevens, D. G. *J. Chem. Soc., Chem. Comm.* **1987**, 73.

[50] Neville, A. G.; Brown, C. E.; Rayner, D. M.; Lusztyk, J.; Ingold, K.U. *J. Am. Chem. Soc.*,**1991**, *113* 6433.

[51] Geoffroy, M.; Lucken, E. A. *Mol. Phys.* **1971** *22* 257.

[52] Rehm, D.; Weller, A.; *J. Phys. Chem. NF* **1970**, *69,* 183; *Israel J. Chem.*, **1974**, *8,* 259.

[53] Gottschalk, P; Neckers, D. C., Schuster, G. B. U. S. Patent # 4,842,980 June 27, 1989; U. S. Patent # 5, 151, 520. Sept. 29, 1992;

[54] Chatterjee, S.; Gottschalk, P. Davis, P. D.; Schuster, G. B. *J. Am. Chem. Soc.* **1988**, *110*, 2326.

[55] Lan, J.Y.; Schuster, G. B., *J. Am. Chem. Soc.,* **1985**, *107,* 6710.

[56] Eaton, D. F. *Photogr. Sci. Eng.* **1979**, *23* 190.

[57] Knapczyk J. W.; Lubinkowski, J. J.; McEwen, W. E. *Tetrahedron Lett.* **1972**, 3739.

[58] Crivello, J. V.; Lam, J. H. W. *J. Polym. Sci., Polym Symp.* **1976** *56* 383.

[59] Kampmeier, J. A.; Nalli, T. W. *J. Org. Chem.* **1994**, *59* 1381.

[60] Ledwith, A. *Makromol. Chem. Supply* **1979** *3*, 348.

[61] Crivello, J. W.; Lam, J. H. W. *Macromolecules*, **1977**, *10* 1037.

[62] Studies of this initiator are very old. It is still used commercially in the synthesis of poly(bis-allylcarbonate), the clear polymer used in plastic lenses.

[63] Bi.Y.; Neckers, D. C. *Macromolecules* **1994** *27* 3683.

[64] Crivello, J. V.; Lam, J. H. W. *J. Polym. Sci. Polym. Chem. Ed.* **1978**, *16* 2441.

[65] Bi, Y.; Neckers, D. C. *J. Photochem. Photobio. A: Chem.* **1993** *47* 221; Bi, Y.; Neckers, D. C. *Tetrahedron Lett.* **1992**, *33*, 1139.

[66] Bi, Y. Ph. D. Dissertation, Center for Photochemical Sciences, **1994**.

[67] See, e. g. Inoue, H.; Hiroshima, Y.; Miyazaki, K. *Bull. Chem. Soc. Jpan.* **1979**, 52 *664*.

[68] Kutal, C.; Willson, C. G. *J. Electrochem. Soc.* **1987**, *134*, 2280.

[69] Cameron, J. F.; Frechet, J. M. J. *J. Org. Chem.* **1990**, *55* 5919.

[70] Tanaseiscu, I. *Bull. Soc. Chim.*, **1926**, *39*, 1443.

[71] Amit, B.; Zehavi, U.; Patchornik, A. *J. Org. Chem.* **1974**, *39* 192.

[72] Cameron, J. F.; Frechet, J. M. J. *J. Am. Chem. Soc.,* **1991**, *113*, 4303.

[73] Frechet, J. M. J. *Pure Appl. Chem.* **1992** *64*, 1239.

[74] McKean, D. R.; Briffaud, T.; Volksen, W.; Hacker, N. P.; Labadie, J. W. *Polym. Prepr.* **1994**, *35*, 387.

[75] Uranker, E. J. Freschet, J. M. J. *Polym. Prep.* **1994**, *35*, 933.

[76] Cameron, J. F.; Willson, C. G.; Frechet, J. M. J. *Polymer Prep.* **1994**, *35,* 323.

[77] Hassoon, S.; Sarker, A; Neckers, D. C.; Rodgers, M.A.J.*R. Amer. Chem. Soc.* **1995**, 117, 11369.

[78] Hassoon, S.; Sarker, A.; Polykarpov, A. Y.; Rodgers, M.A.J.; Neckers, D. C. *J. Phys. Chem.*, **1996**, 100(30), 12386-12393.

[79] Mejiritski, A., Polykarpov, A.Y., Sarker, A. M., Neckers, D. C. *Chem. Mat.*, **1996**, *8*, 1360-1362.

[80] Mejiritski, A., Polykarpov, A.Y., Sarker, A. M., Neckers, D. C. *Chem. Mat.* submitted.

[81] Bauer, R. S. *Epoxy Resin Chemistry II*. ACS Symposium Series 221, American Chemical Society, Washington, D. C., 1983; Buist, J. M.; Gudgeon, H. *Polyurethane Technology*; Maclaren and Sons, Ltd: London, 1968.

[82] Kutal, C.; Willson, C. G. *J. Electrochem. Soc.,* **1987**, *134* 2280.

[83] MacDonald, S. A.; Willson, C. G.; Frechet, J. M. J. *Acc. Chem. Res.* **1994**, *27*, 151.

[84] Lee, E. J.; Bitner, T. W.; Ha, J. S.; Shane, M. J., Sailor, M. J. *J. Am. Chem. Soc.* **1996**, *118*, 5375.

[85] Beecher, J. W.; Cameron, J. F.; Frechet, J. M. J. *J. Mater. Chem.* **1992**, *2* 811.

[86] Mejiritski, A.; Polykarpov, A.; Sarker, A. M.; Neckers, D. C. *J. Photochem. Photobio.* **1997** *xx* yyy.

[87] Thompson, L. F.; Willson, C. G.; Bowden, M. J. *Introduction to Microlithography*, 2nd ed.; ACS Professional Reference Book, American Chemical Society, Washington, D. C., **1994**.

[88] Reichmanis, E.; Houlihan, F. M.; Nalamusu, O.; Neenan, T. X., *Chem. Marter.* **1991**, *3* 394.

CHAPTER II

UPDATE ON PHOTOPOLYMERIZATION SYSTEMS

D. C. Neckers

Chapter Two

II. Update on photopolymerization systems

In this chapter, we will review the known methods of initiating polymerization using photochemical routes paying particular attention to those systems which are new. Since a preponderance of the new material involves electron transfer reactions, much of the chapter's contents will be devoted to that. However we will also review older methods of generating radical intermediates, in particular bond dissociation reactions, outlining some of the ways in which we prefer to think about such systems, and focusing discussion on compounds whose chemistry is particularly instructive. We also plan an extensive review of the photochemical reactions of aromatic carbonyl compounds. Though most bond dissociation photoinitiators are from this class of compounds, reactions other than dissocation affect initiator performance both positively or negatively. We concentrate on some new chemistry of certain carbonyl systems hoping to share certain insights gained from studying these compounds in the hope that it will trigger creative approaches to new photopolymerization chemistry on application.

1.   Bond Dissociation Initiators: Bond Homolysis after the Absorption of a Photon

In his book, Kurt Dietliker identified two routes to free radical formation and the subsequent initiation of polymerization by formed radical products. Type I photoinitiators, or bond homolysis initiators, are compounds that absorb a photon and dissociate with some measureable efficiency into a pair of free radicals either or both of which is capable of starting a radical chain reaction. Recent results for one such photoinitiator system, the phosphine oxides, were observed in Chapter 1.

Type II initiators form radical pairs by oxidation/reduction reactions known as single electron transfer (SET) processes. Electron transfer genesis of radicals needs lower photon energy and can thus be achieved with the absorption of much longer wavelengths than can bond dissociation. Moreover, common UV initiator systems such as benzophenone/Michler's ketone and camphor quinone/tertiary amine also operate by such a mechanism. The spontaneity of electron transfer can be predicted by the Rehm-Weller equation, quoted in Chapter 1. Several other examples of electron transfer systems covered in this chapter.

Since aldehydes and ketones form the basis of many UV photopolymerization initiator systems, the first section of Chapter 2 is devoted to them.

i.   **The Carbonyl Group**

The structure of the compounds containing carbonyl groups which are not bonded to conjugated groups, is determined by a sigma (σ) skeleton, a π bond, and two non-bonded pairs of electrons localized on the oxygen atom. Valence bond descriptions of carbonyl compounds indicate them as resonance hybrids of two limiting resonance contributors, the higher energy of which is a charge separated structure

Resonance Structures of Carbonyl Group

The resonance hybrid accounts for the fact that aldehydes and ketones have dipole moments ranging upward from 2.0 D with the negative end directed toward the oxygen atom. The resonance hybrid also rationalizes the many, many reactions of aldehydes and ketones in which negative, or negatively polarized reagents add to carbon, and positive or positively polarized reagents react first at oxygen. It also accounts for the acidity of protons alpha to the carbonyl group in that the conjugate base is stabilized by the dispersal of the charge to the negative oxygen atom

Acidity of proton alpha to a carbonyl group

A slightly more detailed picture is given by considering the entire molecular orbital description of the carbonyl group which includes the sigma skeleton, a pair of pi (π) orbitals and two distinct non-bonded electron pairs. The lower energy π orbital is occupied by a pair of electrons indicated as a double bond in the valence bond picture, the other π orbital is unoccupied in the ground state. The non-bonded pairs localized entirely on the oxygen atom, one of which has more s character than the other.

The bonding π orbital, the higher energy of the two orbitals resulting from the linear combination of atomic p orbitals on oxygen and carbon (the so-called pi star (π*) orbital), as well as the non-bonded electron pairs on oxygen, are responsible for the UV absorption spectra of the molecules which are of most interest to us. The so-called π-π* transition results from the promotion of an electron from the bonding π to the anti-bonding π* orbital. This transition is symmetry allowed, since the symmetry of the originating orbital (the π orbital), and the final orbital (the π* orbital), is the same. Both orbitals have a nodal plane (where there is no probability of electron density) along the x axis connecting the bonding atoms. The statement the π-π* transition is highly probable means that it occurs for a large number of individual molecules in an assembly of such molecules, so that π-π* transitions generally have large extinction coefficients. The energy of the transition measures the separation of the energy levels, π and π*. The π-π* transition for any molecule possessing a π bond is shown in Figure 1.

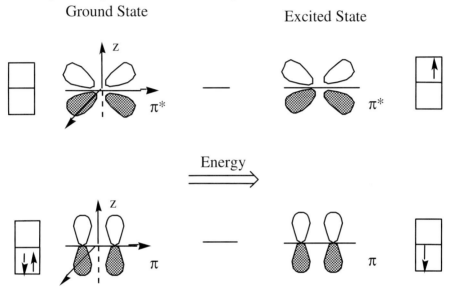

Figure 1
The π-π* transition for any molecule containing a π system.

One of the results of the absorption of a photon by a molecule containing a π system is that in the π-π* transition which ensues, the bonding π orbital becomes depopulated. Since the π bond is directionally specific, accounting for the stereochemistry of atoms and groups bonded to atoms at its terminus, the π-π* transition in compounds for which stereochemistry is possible (olefins, for example) produces an excited state capable of free rotation about the terminae. A subsequent result of this transition may be rotation about the π bond and formation of the other stereoisomer.

In aldehydes and ketones, of course, since no stereoisomerization is possible, the π-π* transition is evidenced by a large extinction coefficient for the transition, with π-π* absorptions in non conjugated systems generally occurring in the mid-to-deep UV region of the spectrum.

The n-π* transition of ketones results from the promotion of one of the non-bonded electrons on oxygen into the anti-bonding π* level. The n-π* transition is a symmetry forbidden transition, in that the symmetries of the originating orbital and the orbital to which the electron is promoted differ. The experimental observation is that n-π* transitions have small extinction coefficients, though they occur at lower energy than does the accompanying π-π* transition for the same molecule. The orbital picture of an n-π* transition for an aldehyde or ketone is shown in Figure 2.

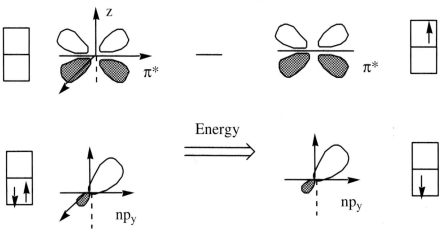

Figure 2

The n-π* transition for any aldehyde, ketone or other carbonyl containing molecule.

The n-π* transition results in a depletion of electron density on oxygen and an augmenting of electron density associated with the π system, some of which is shared by carbon. One manifestation of this is that the dipole moment of the excited state resulting from the transition, is smaller than is the dipole moment of the ground state in the same direction. Another manifestation is that n-π* transitions occur at shorter wavelengths in more polar solvents. This is because the excited state is less stablized by the polar solvent since it is less polarized. Another important result is that the oxygen atom assumes more free radical character. The n-π* excited state reactivity of aldehydes and ketones often resembles the reactivity of free radicals centered on oxygen both quantitatively and qualitatively, an example of which is the tert-butoxy radical, t-Bu-O•.

Though photopolymerization science concentrates on those reactions that form reactive intermediates, which subsequently initiate chain reactions, other processes are often faster, even in the context of a polymerization system. They can cause lower than anticipated reactivity in specific cases. As this chapter proceeds, we'll see some examples of this, so it makes sense to consider some of the general reactions of excited carbonyl groups in this introductory section.

The chemical reactivity of a carbonyl compound in its n-π* excited state (either singlet or triplet - the differences are important but subtle) is the factor which allows us to characterize the properties of the excited state as we have above.

ii.   Characterizing the Reactivity of the Carbonyl Group

On a rudimentary level, the dissociation of the n-π* excited state can be pictured as in Equation 1. As we have seen, the stability of the group R• controls the ease of dissociation in typical ketones. The important observation is that n-π* excited carbonyl functions are electron deficient on oxygen relative to the ground state, and somewhat electron rich on carbon relative to the ground state. The radical R•, if it's an alkyl radical, can be characterized as reacting as though it were a planer, as opposed to a tetrahedral, intermediate.   It does not retain its stereochemistry, for example, in the additional products which it forms. Its radicaloid character is largely localized in a sigma orbital.

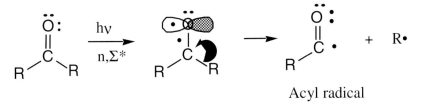

Acyl radical

Equation 1
Formation of acyl radical

The acyl radical, which we have shown as bent in Equation 1, may also be linear, and for each geometry the radical can be either primarily σ or primarily π, Equation 2. Bent radicals have lower energy and the degree of sigma character or π character depends on the other substituents. The hybridization can be predicted as is shown for the benzoyl radical below, from ^{13}C Electron Spin Resonance (ESR) hyperfine coupling constants[*]. In the general case, the dissociation to an acyl radical and a radical partner, leads to a radical pair for which there are three singlet states and one triplet state.

---

[*] An introductory discussion of this can be found in J. W. Leffler, Introductions to Free Radicals, Wiley, 1995

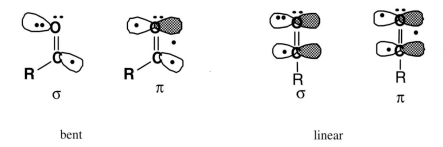

Equation 2
Proposed structures for an acyl radical

Since in the carbon radical, R•, the odd electron must always reside in a sigma hybridized orbital with some s character, the four states of the radical pair are $(\sigma,\sigma)^1$, $(\sigma,\sigma)^3$, $(\sigma,\pi)^3$ & $(\sigma,\pi)^1$. The singlet/triplet configuration of the radical pair plays a critical role in the rate of recombination reactions, particularly in the solvent cage. Singlet pairs combine much more rapidly than triplet pairs, since one of the two radicals in the latter case must undergo a spin inversion before recombination is possible. In the case of the benzoyl radical, which could be either sigma or pi, it has been shown by ESR hyperfine coupling constant C^{13} measurements, that the free electron is located almost entirely in a pure sigma orbital. In specific terms, both the degree of s character of the radical and its electrophilic nature predict reactivity in initiating radical chains.

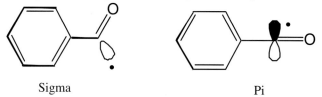

Possibilities for Electron Distribution in a Benzoyl radical

In polar terms, the acyl radical which forms from the Norrish Type I dissociation may be thought of as either a donor radical or electrophilic. Given the choice, it would prefer to donate an electron and become a carbenium ion rather than accept an electron producing the corresponding carbanion. The carbenium ion, called an acylium ion, is much more likely than its negative counterpart, Equation 3. Many reactions of carbonyl compounds, which go through this intermediate, are known in strongly acidic solution.

Equation 3
Oxidation/Reduction of an acyl radical

In transition states, in which the acyl radical participates, it forms as the electropositive end of a developing, though imaginary, charge separated species. In terms of its polar contribution to the transition state, it is an electron donor radical, whereas the developing end adjacent to the carbonyl function in the reaction monomer, is an electron acceptor radical, Equation 4.

Equation 4
Electron accepting radical

In the latter, the electron withdrawing carbonyl group supports the formation of this particular transition state, as an acceptor of developing charge in the same imaginary dipolar species.

If one considers chain initiating species formed by radical reactions in which light is used to initiate the polymerization of acrylates, virtually all are electron donor radicals and the best have very high degrees of sigma character. The reactions of other such species are considered in detail later, but they include the alpha amino alkyl radical formed by electron transfer from tertiary amines, as well as the alkyl radicals generally thought to initiate chains, when tetrasubstituted borates are used as the chain initiating species, Equation 5.

$$Ar_2\ddot{N}CH_3 \longrightarrow Ar_2\overset{\bullet+}{\ddot{N}}CH_3 \overset{-H+}{\longrightarrow} Ar_2\ddot{N}\overset{\bullet}{C}H_2$$

Equation 5
Further examples of transition state formation

iii.   Electron Transfer From Tertiary Amines

Of all the coinitiators available for use with oxidizing excited states (benzophenone, camphor quinone, various dyes) none are more ubiquitous or important than are those containing tertiary amines. The mechanism of amine reduction of excited states is well known at this time, and has been outlined previously. What has not been systematically studied is the effect of the structure of the amine electron donor on the rate of polymerization. Tertiary aromatic amines provide opportunity to carefully study the effect of substituent, on the rate of polymerization, when such amines are used as coinitiators. This has been done in a series of most important papers by Paczkowski and his coworkers.[1]

Paczkowski studied the rate of photopolymerization of TMPTA in 1-methyl-2-pyrrolidinone and in N-vinylpyrrolidinone as catalyzed by the RBAX/tertiary aromatic amine. RBAX is decarboxylated Rose Bengal which has been esterified and its structure is shown on page 99. It undergoes one electron reduction by both tertiary amines and borates in a reaction studied in some detail by Valdes-Aguilera.[2]

RBAX $X^1 = Cl;\ X^2 = I$

Structure of RBAX

It had previously been reported that rates of quenching of the triplets of Rose Bengal derivatives by tertiary amines ranged from 10^9 to 10^{10} M^{-1}s^{-1} thus being relatively close to the diffusion controlled.[3] Paczkowski therefore concluded that the rate of electron transfer to RBAX from tertiary amines of various structures was not the limiting step for the overall polymerization reaction when the product, the radical deriving from the amine radical cation, was the initiating species.

The tertiary amines studied in detail by Paczkowski were substituted N-phenylglycines,

wherein R_1 was hydrogen, except in three cases where it was methyl, acetyl and phenyl respectively. In those cases R_2 was hydrogen. In all other cases R_2 was hydrogen and R_1 = CN, NO$_2$, acetyl, benzoyl, ethoxycarbonyl, Cl, CH$_3$, t-Bu, PhO and MeO.

Assuming that intermolecular electron transfer is diffusion controlled and is therefore ΔG^0 independent, the relationship of the rate of polymerization to the substituent on the N-phenylglycine is predicted, on the basis of radical reactivity arguments. Thus variation in polymerization rates can be explained as a substituent effect on radical stability/reactivity and subsequently on the polymerization reaction. Since it was already known that the reactivity of radicals, (derived from very similar substituted aromatic amines) toward initiating the polymerization of butyl acrylate increased as the σ_p value increased (that is, electron withdrawing groups in the para position accelerated the reaction),[4] Paczkowski found that he could relate the ΔG^0 for electron transfer from substituted phenylglycines to the substituent constant in a linear fashion. The relationship between the substituent constant and polymerization rate is also linear. Thus rate is controlled by the reactivity of N-phenylglycine radical and not by the electron transfer rate. In other words, k_p is rate determining. Everything before that step is fast. A similar result was obtained by Polykarpov in assessing the reactivity of substituted borates as electron transfer initiators in a very different system.[40]

The overall mechanism proposed by Paczkowski is

$$A \xrightarrow{h\nu} A^* + HD \underset{k_{-diff}}{\overset{k_{diff}}{\rightleftarrows}} [A^* \text{----} HD] \underset{k_{-el}}{\overset{k_{el}}{\rightleftarrows}} [A^{-*} \text{----} HD^{+*}] \xrightarrow{k_{sep}} A^{-*} + HD^{+*}$$

with k_H branch leading to $AH^{\bullet} + D^*$, and k_{ret} branch leading to $A + HD$.

From $AH^{\bullet} + D^*$: k_{bl} → bleached dye; k_p → polymer + reduced dye.

Paczkowski reports this by saying that "under his best conditions, the rate of polymerization may be better described by Rp = $\rho/\rho^*(-\Delta G^0)$; where r^* is defined by the relationship between ΔG^0 and the substituent constant, s; - $\Delta G^0 = r^*s$.

On the basis of many other studies relating polymerization rates to substituents in the coinitiator part of the system, these results are extremely important. They allow one systematically to predict that anything which stabilizes the transition state, below, will accelerate the reaction.

$$R_1\text{-C}_6H_4\text{-N(H)}\overset{\bullet}{C}H(COOH) + CH_2=CHCOOR' \longrightarrow R_1\text{-C}_6H_4\text{-N(H)CH(COOH)}\cdots CH_2\overset{\bullet}{=}CHCOOR'$$

Groups that stabilize the radical are electron withdrawers, rather than electron donors. If the substituent constants are known, groups with positive values of s show faster polymerization rates.

iv.   **Reactions of Excited Carbonyl Groups: 2 + 2π Cycloaddition**

Any reaction which does not form a useful reaction intermediate, retards useful chain processes. The 2 + 2π addition of an excited carbonyl compound to an olefin or an alkyne is called the Paterno-Büchi reaction and is most commonly observed with electron rich olefins. Paterno-Büchi processes impact photopolymerization processes, only in the sense that they may interfere with them. In most instances, the regiochemistry of the Paterno-Büchi reaction indicates a transition state in which the oxygen atom of the carbonyl group selects the least hindered carbon atom of the double bond to which the addition is occurring. If one thinks of the reaction as a two step addition which first goes through a biradical adduct, (much in the manner of the Kharasch, or non-Markovnikopf addition of radicals to unsymmetrical olefins), the major product of the 2 + 2π cycloaddition can be rationalized based on the stabilities of the two radicals which might form by addition of the oxygen of the carbonyl to the cell hindered, and offered to the more hindered, of the two carbon atoms of the olefin, Equation 6. In simple reactions of the Paterno-Büchi type, for example the photoaddition of acetone to isobutylene (R = CH_3), the major product oxetane predominates by about 9:1 over the minor product oxetane.[6]

Equation 6

The reaction with alkynes shows the same regiochemistry, though the product oxetenes are not generally stable and undergo immediate ring opening to unsaturated aldehydes or ketones. As in the addition of excited carbonyl groups to olefins, the resulting product is derived from the addition of the carbonyl oxygen to the terminus of the double bond with the smaller substituent, Equation 7.

Equation 7
Formation of a biradical by attack on an alkyne

v.   **Hydrogen Abstraction - A Route to Crosslinks with Carbonyl Compound Initiators**

A model treating the excited carbonyl group as a free radical centre on the oxygen atom, works as an explanation for other processes as well. Benzophenone, when irradiated in solvents with reactive hydrogen atoms, abstracts a hydrogen from the solvent, producing a reduction product from the ketone and an oxidized or coupled product from the hydrogen source. The photoreduction of benzophenone by isopropyl alcohol is a reaction which can easily be carried out in sunlight, Equation 8.

$$Ph_2C=O + (CH_3)_2CHOH \xrightarrow{h\nu} HO-C(Ph)_2-C(Ph)_2-OH + (CH_3)_2C=O$$

Benzpinacol   Acetone

Equation 8
Reduction of benzophenone by isopropyl alcohol

Transient spectral studies among other evidences, have made it clear that the reaction is that of the benzophenone triplet state. The quantum yield of benzophenone disappearance in photoreduction reactions in alcohols approaches 2 under ideal conditions (i.e. two benzophenones are consumed per photon absorbed). This demonstrates that only one of the two benzophenone units in the product benzpinacol, comes from the reaction of an excited benzophenone molecule. In the presence of optically active alcohols such as d(+) or l(-) 2-butanol, no racemization of alcohol is recovered after reaction. Since radicals, in general, are planar reaction intermediates, this latter result strongly suggests that one molecule of optically active 2-butanol is completely consumed in the reduction reaction. There is no formation of radicals from the 2-butanol moiety, which disproportionate forming the ketone and reforming the starting alcohol, along the reaction pathway. This is totally consistent with the accompanying observations; i.e. the quantum yield of benzophenone disappearance is >> 1.0.

Consumption of optically active 2 butanol in reduction

Experimental results suggest a reaction mechanism in which a benzophenone triplet is reduced to a radical while a second molecule of benzophenone is reduced in the ground state by hydrogen atom transfer from the α-hydroxyalkyl radical formed in the first step. The driving force for the latter reaction is the substantially greater resonance stablity of the diphenylhydroxymethyl radical, relative to the α-hydroxyalkyl radical. The mechanism is shown in Equation 9.

Equation 9
Mechanism for reduction of benzophenone by isopropyl alcohol

Hydrocarbons also serve as hydrogen donors to benzophenone triplet states. One set of experiments comparing the reactivity of alkoxy radicals towards benzophenone triplets involved the irradiation of benzophenone in the presence of a large excess of two hydrocarbons. Relative reactivity is based on the relative rate of the disappearance of the two competing hydrogen donors. The same experiment can be carried out with the thermolysis of, for example, di-tert-butyl peroxide producing the reactive intermediate tert-butoxy radicals, Equation 10.[1]

Equation 10
Reactions of aromatic hydrocarbons

When such experiments are carried out, one finds that cumene, which forms the more stable tertiary cumyl radical, is about 10 times as reactive as toluene (which forms the benzyl radical) toward both benzophenone triplet states and tert-butoxy radicals on a per hydrogen atom basis. Since the structure of the transition state is similar in the two cases, the selectivity/reactivity of benzophenone triplet states, formed by benzophenone absorption of a light quantum at 350 nm, and tert-butoxy radicals formed by the thermolysis of di-tert-butyl peroxide are about the same.

Most polymeric structures of consequence in commerce contain hydrocarbon residues in some form, so that hydrogen abstraction, such as that known to occur with benzophenones, is a primary route to polymer crosslinking, photodegradation, and backbone destruction. For example in the photodegradation of styrene, introduction of a single carbonyl group in the backbone, which is equivalent to introducing an acetophenone-like residue, is tantamount to introducing a catalytic centre to enhance photodegradation. The introduction of a carbonyl function into an hydrocarbon polymer backbone can occur by means of autooxidation chains in which a radical chain oxidation process is introduced by the non-propitious introduction of an initiator fragment. This is followed by degradation of the formed hydroperoxide by -O-O- homolysis and cleavage, which leads to an acetophenone like ketone in the backbone. A following Norrish Type II reaction further cleaves the chain. This mechanism has been identified as a primary route to the uncatalyzed light catalyzed degradation of styrene and other aromatic polymers, Equation 11.[7]

Equation 11
Degradation of polystyrene

vi. Phenylglyoxylate Photochemistry with Shengkui Hu[*]

Aromatic ketones and their analogs undergo a plethora of reactions from their excited states. The relative contribution of each of these, under conditions where competitions are set up, depends on the specific system involved.

Methyl phenylglyoxalate has some use as a commercial photoinitiator, particularly for amine acrylates, and is convenient to use because it is a liquid.[8] We've shown by isolating 1:1 addition products to methyl methacrylate in dilute solution under controlled conditions that the initiating radical formed from methyl phenylglyoxylate is the acyl radical. Reactions of this chromophore are rich and varied, and worthy of extensive consideration in the context of helping to understand the reactions of aromatic carbonyl triplets.

The Norrish Type II reaction of phenylglyoxylate esters, producing an aldehyde or a ketone from the alcohol residue of the ester and the hydroxyketene, Equation 12, is an old reaction having been discovered in 1964, and it has a rate constant of about 10^6 s^{-1}.[9]

---

[*] From the Ph.D. Dissertation of Shengkui Hu; Center for Photochemical Sciences, 1998. Consult also the many references to this work.

Equation 12
Light initiated dissociation of phenylglyoxylate

The lifetime of the biradical indicated as the intermediate in the Norrish Type II pathway to the subsequently important, though unstable α–hydroxyketene, can be estimated from a ring opening rearrangement. That of the cyclopropylcarbinyl radical, for which the rate is known to depend on substituents, occurs with rate constants ranging from 10^7 to 10^9 s^{-1}. If ring opened products are observed, as they are, the lifetime of the biradical must be longer than the time required for the ring opening process.[10] The products, with two different cyclopropylcarbinyl esters of phenylglyoxylic acid (also called benzoyl formic acid), are shown in Equation 13.

R_1 = H; Φ = .88: Norrish Type II product = 45%
R_1 = Ph; Ring opening product = 98%

Equation 13
Competitive reaction of the excited state of phenylglyoxylate

Bimolecular intermolecular hydrogen abstraction as, for example, in the reaction of the keto ester triplet with another molecule like itself, Equation 14, is a little faster than the intramolecular counterpart in the phenylglyoxylate series (k = 3 x 10^6). Since the reaction is 2nd order it depends on the concentration of the starting ester, but below 10^{-3} M in non-polar solvents, the reaction becomes a non-factor. Above that concentration bimolecular hydrogen abstraction becomes an important process particularly in non-polar, non-reactive solvents. On the simplest level, dimerization is generally indicated by the formation of a precipitate, the precipitate being a dimer similar in general structure to the dimer formed by the photoreduction of benzophenone.[8]

Equation 14: see also Equation 16

Given a competition between three processes of triplet reactivity in the phenylglyoxylate system, the Paterno-Büchi reaction, the Norrish Type II reaction and intermolecular hydrogen abstraction, the Paterno-Büchi reaction is fast enough to supercede other reactions at just about any relative concentration of olefin/keto ester in non-polar solvents, though the rate of the Paterno-Büchi reaction does depend on the structure of the olefin. The Paterno-Büchi reaction is approximately 2 orders of magnitude faster with electron rich olefins, than the competitive Norrish Type II process in the phenylglyoxylate series. The Norrish Type II reaction is the slowest of the three reactions in phenylglyoxylate esters, and only occurs at low concentrations (0.001M) in nonpolar solvents.[11]

Intramolecular 2 +2 π cycloaddition can lead to some unusual products, Equation 15.

Equation 15

In the case of the intramolecular Paterno-Büchi reaction of phenylglyoxalate/olefin esters, the stereospecificity of the reaction differs from that predicted by the general rule that such an addition produces the more stable of the two possible biradicals as the first formed reaction intermediate. This can be the result either of steric constraints on the ring closure of the biradical intermediate, or caused by the formation of a product determining exciplex (see below). One would anticipate, for instance, that the tertiary radical (BR2 for biradical 2 in Equation 15) would be more stable than the secondary radical (biradical 1, BR1). The primary formation of a product derived from the less stable biradical of the pair, suggests that a slightly bound complex in the excited state may be responsible for controlling the regiochemistry of the addition rather than radical stability effects.[12] Excited state complexes are generally called *exciplexes;* that is they are complexes of an excited state and a ground state. As such they have measureable properties.

As will be demonstrated later in this chapter, the exciplexes suggested are important in directing the regiochemistry of these additions. They may subsequently be consumed by a single electron transfer reaction involving a reduction of the ketone and an oxidation of the olefin as the first step leading to products.

2. An Imaging System Based on the Phenylglyoxylate Chromophore

It has long been realized that polymer insolubilization can be achieved by crosslinking and a great majority of negative resists developed until now, have been based on the formation of such crosslinks.[13,14,15,16] Two types of reactions have been utilized to generate crosslinks: those in which reaction of excited chromophores causes crosslinks as in the (2 + 2) π addition of cinnamate esters; and those where photogenerated reactive species act on ground state materials.

The section that follows, illustrates a number of important principles, applications and analytical techniques as they are applied in an imaging system based on the phenylglyoxylate chromophore.

Biomolecular photoreactivity of a ketone chromophore such as phenylglyoxylate provides a crosslinking process which insolubilizes photoreacted glyoxylate residues at points at which they have been irradiated. The result is an imaging system based on phenylglyoxylate chemistry. The surface on which the soluble glyoxylate polymer is laid down is that of an amorphous silicon wafer.

In a comprehensive study of the photochemical reactions of alkyl phenylglyoxylates,[9] photoreduction in benzene was found, and an intermolecular hydrogen abstraction initiated radical chain process was proposed to account for this observation.

Equation 16

2'-(Meth)acrylate ethyl phenylglyoxylates (**1**) were synthesized in good yields by the (DCC) esterification of benzoylformic acid and polymerized in refluxing benzene, in which the polymer formed, called **Poly-1**, is soluble, Scheme 1. The irradiation of a benzene solution of **Poly-1**, at the λ_{max} (350nm) of the glyoxylate chromophore, led to the appearance of a white precipitate. No Norrish Type II products were detected.

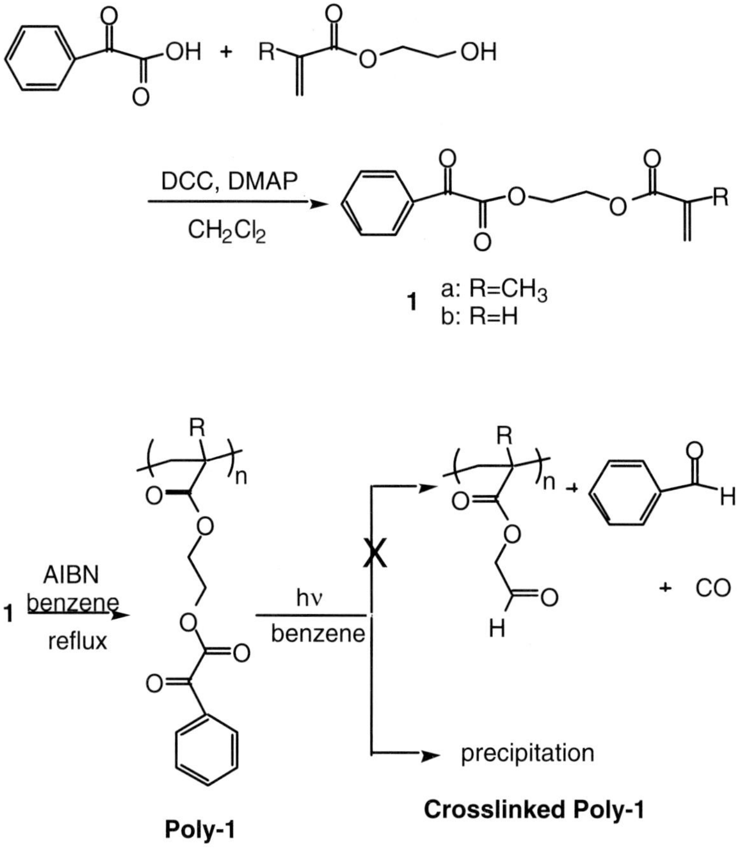

Scheme 1

As we have pointed out above, alkyl phenylglyoxylates, possessing γ-hydrogens, generally undergo Norrish Type II photolysis in dilute benzene solution. Intermolecular reactions only become competitive when the concentration of the ketoester is higher.[9] The absence of the Norrish Type II reactions in **Poly-1** was initially surprising.[17] It is surmised that dominating intermolecular reactions are assertive in the polymeric systems because the phenylglyoxylate chromophores in **Poly-1** are linked by a polymeric chain. Even though the concentration of such chromophores in solution is relatively low, individual phenylglyoxylate chromophores are close to each other and their freedom of motion restricted by the acrylic backbone. As a result, the actual concentration of a chromophore, found in relative proximity to another identical chromophore, is high.

We suspected that **Poly-1** could be used as a photoresist material, because of the difference in its solubility to that of crosslinked polymer, formed by irradiation. Imaging was tested by irradiating with 365nm light through masks. A **Poly-1a** film[18] was spin-coated on silicon wafers and crosslinking effected in the irradiated areas. These areas were rendered insoluble in benzene. A single wash with solvent served as the developing step, and the non-irradiated areas were completely cleared of polymeric material.

i.   Resolution of an Imaging System Based on Polymeric Phenylglyoxylate

The maximum resolution possible with the mask system used for these experiments is appromately 5μm. The image produced after exposure dose of 1000 mJ/cm^2 through a TEM grid and subsequent benzene wash is shown in Figure 3. The non-irradiated portion of the film reveals bare silicon wafer as shown by the AFM micrograph in Figure 4. The resolution of the mask is achieved.

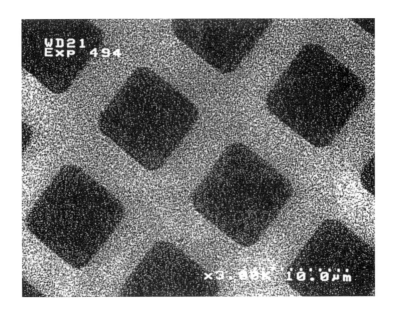

Figure 3
SEM micrograph of negative image produced.

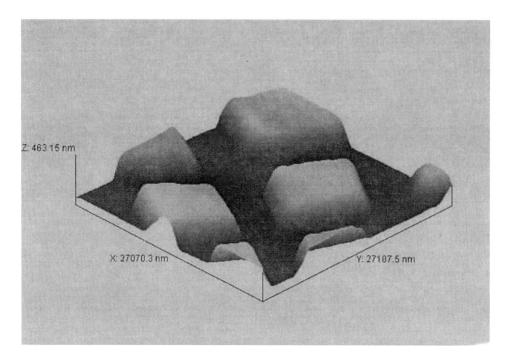

Figure 4
AFM micrograph of negative image produced.

ii. Sensitivity and Contrast

To measure the sensitivity of the imaging system, spin coated samples were irradiated at various doses. After irradiation, the samples were washed with a dry benzene solution for 30 seconds. The thickness of the crosslinked film, if any remained, was measured by AFM and a sensitivity curve, such as that of Figure 5, was constructed. The gel dose, D_g is that at which the gel first appears and is about 30mJ/cm^2.[19] The insolubilization dose, D_i is the minimum dose required to produce an insolubilized film of thickness equal to that of the initial film. In this case it is 663mJ/cm^2. The sensitivity of the resist, D_s defined as the dose required to retain 50% of the original film, is measured as 110mJ/cm^2.[20] The contrast $\gamma = 1/\log(D_i/D_g)$ is calculated as 0.75.[21] The molecular weight of **Poly-1a** as measured by gel permeation chromatography is M_n, 9,314; M_w, 27,476. The theoretical resist contrast, $\gamma = 4.606 M_n/M_w$ is 1.5, higher than the value measured.

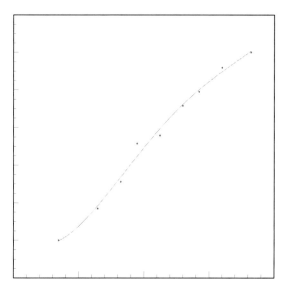

Figure 5
Sensitivity curve for Poly-1a photoresist.

iii.   Predevelopment Images

Photochemical reactions carried out on polymeric surfaces resulting in changes in surface morphology induced by the photocrosslinking process may also be detected prior to development. To do so, one uses Atomic Force Microscopy in the lateral force mode, as opposed to the direct contact mode. (Atomic Force Microscopy was described in Chapter 1, pp 63-68). A resist film which had been irradiated through a TEM mask was thus observed by AFM directly after irradiation, Figure 6. It shows that the irradiated areas are about 3nm higher than the non-exposed areas. Upon crosslinking the polymer expands rather than, as is generally the case, contracts. The image collected in the lateral force mode also shows that a distinguishable difference in the physical properties of the polymer exists in the exposed, as opposed to the non-exposed areas.

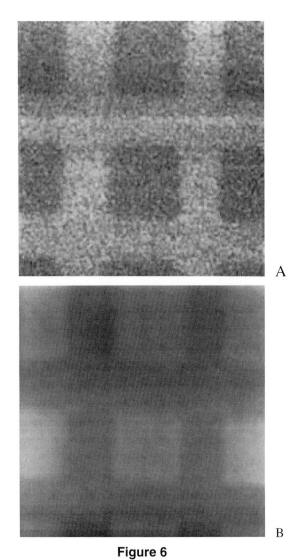

Figure 6
AFM images of a photoresist film that was mask-controlled irradiated, undeveloped. A, the topological graph. B, the lateral force image.

iv.   Photopolymer Analysis

The precipitate resulting from irradiating a benzene solution of Poly-1a has the same elemental composition as the original unirradiated polymer. Since no low molecular weight compounds, such as those expected from Norrish Type II photolysis were detected after photolysis, a crosslinking mechanism involving the mandelate radical (or benzoyl radical) is suggested and it results in dimerization of the pendent phenylglyoxylate chromophore, Scheme 2. [12]

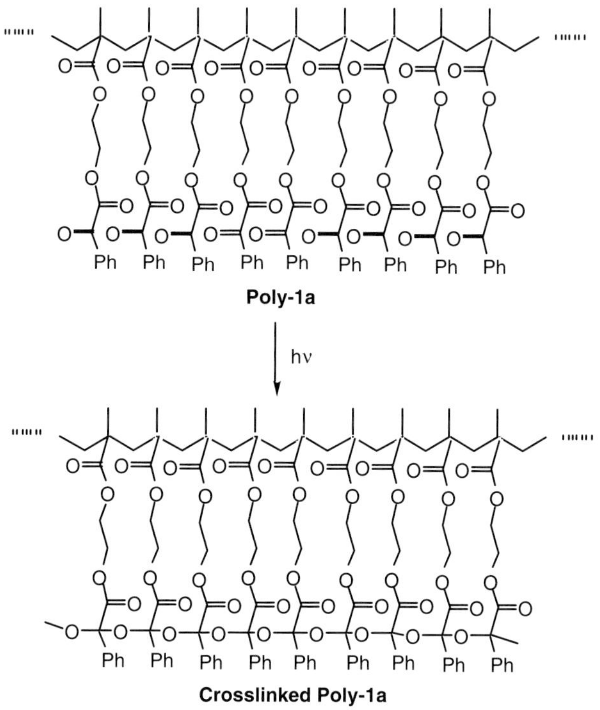

Scheme 2

The degree of polymer crosslinking after irradiation can be estimated from the amount of the residual phenylglyoxylate chromophores existing in the crosslinked film. This can be monitored using the UV absorption of a film cast on Pyrex glass before and after irradiation, (Figure 7). The absorbance at 320nm decreased to 1/2 of its original intensity with this dose, indicating that nearly half of the phenylglyoxylate chromophores crosslinked after irradiation. This point also clearly illustrates another simple experimental measure, the UV spectrum, can be used to great advantage in characterizing the reactivity of a photopolymer system.

Poly 1 is a rationally synthesized polymer the chemistry of which can be easily predicted from that of the small molecule glyoxylate analog on which it was modelled. As a photoresist it has no application. But the methods used in accomplishing the analysis of the photoproducts in the photopolymer system represent important advances for the consideration of every polymer photochemist.

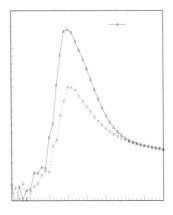

Figure 7
The UV absorbance of a resist film before and after irradiation.

3. The Carbonyl Group as a Target using Carbonyl Group Photochemistry to Promote Other Reactions

One important characteristic of the aromatic carbonyl function is that it forms, after excitation, long lived triplet states. These have radical like activity in cases where they are not dissociative producing radical pairs. The next section shows that carbonyl triplets can be oxidized or reduced. But another important characteristic is that they can be used as a messenger - a route to reactivity - for generally non-absorbing functional groups. A good example of the latter concept is p-benzoyl tert-butyl perbenzoate. This deserves further attention in the context of the information it provides about carbonyl group reactivity in polymerization systems. It will be discussed, together with the carbonyl substituted triphenylmethyl radical, in this section.

i. Peroxy Ester Chemistry

As we pointed out in the previous chapter, peroxides are generally unsatisfactory for use as photoinitiators at least in forming thin films because most, even those containing aromatic groups, absorb light in the mid to deep UV region of the spectrum. Another important problem is that peroxides produce small molecules that are gases, and these produce bubbles particularly in coatings applications. p-Benzoyl tert-butyl perbenzoate (BPE) was originally synthesized by Thijs to circumvent the lack of specific photoreactivity of the peroxide moiety[22] (BPE). It was reasoned that energy absorbed by the benzophenone carbonyl group would translate into a decomposition reaction at the para positioned -O-O- function, the result of which would be the formation of a free radical pair. This would be stimulated by absorption at the wavelengths absorbed by the benzophenone carbonyl group, ≈350nm where the non-functionalized perester is non-absorbent and virtually unreactive.

The notion was essentially a correct one, and BPE decomposed smoothly when irradiated at 350nm to the expected products; t-BuOH and acetone derived from the tert-butoxy radical and p-benzoylbenzoic acid/benzophenone formed from the p-benzoylbenzoyloxy radical. In a later paper, it was shown that the triplet in the perester had a much shorter lifetime than did benzophenone itself, suggesting that 'transfer' of energy to the dissociable function occurred within picoseconds.[23] The benzophenone function in this series of compounds is serving to harvest radiation, or be a target, to cause dissociation of the perester.

When tested as photoinitiator for acrylate polymerization, it could be shown that the p-benzoylbenzoyl radical was the main initiating species, though it underwent decarboxylation to the appropriate aryl radical nearly 20% of the time.[23] As expected, the rate of polymerization of methyl methacrylate as initiated by p-benzoyl tert-butyl perbenzoate under steady state conditions increased in a manner that was approximately first order in monomer.

As expected (Chapter 1) the rate of polymerization of styrene initiated with p-benzoyl tert-butyl perbenzoate was about an order of magnitude less than was the rate for methyl methacrylate. As the concentration of styrene was increased, the rate of polymerization decreased when initiated with p-benzoyl tert-butyl perbenzoate to the point that no polymer was formed at all in pure styrene. This was the result of the quenching of the excited state of the perester in a diffusion controlled energy transfer to styrene monomer.[24]

Peresters with substituents on the aromatic ring were also studied,[25] and all but two behaved as predicted. The two that were anomalous were the diperester,

Perester of BPE

and the p-bromomethyl perester,

p-bromomethyl perester

The quantum yield for the decomposition, even in the absence of induced decomposition, in the case of bis-perester, was slightly greater than unity. At the time, it was suggested that the pair of peroxy ester units were decomposing as a result of the absorption of a single photon. This, however, seems quite unlikely unless the dissociation reactions of the two perester units are not simultaneous. The p-bromomethyl perester proved to be the poorest photoinitiator in the series. Whereas the rate of polymerization of methyl methacrylate (bulk) under identical conditions was 1.14×10^{-3} $lmol^{-1}s^{-1}$ when the bisperoxy ester was used as the initiator, and 7.0×10^{-3} $lmol^{-1}s^{-1}$ when BPE was the initiator, the rate of polymerization under the same conditions was only 9×10^{-5} $lmol^{-1}s^{-1}$ when the p-bromomethyl perester was used to initiate methyl methacrylate polymerization. The rate of the former reaction was attributed to the biperoxy ester dissociation suggested above. The sluggishness of the latter was suggested to be the result of a bond clevage competition between the CH_2Br function and the CO_3t-Bu function. In view of the expected lessened reactivity of benzyl type radical

                    Benzyl radical   **Benzoyloxy radical**

the benzoyloxy radical, we attributed the lower rate of polymerization observed with this inititator to the fact that the CH_2Br function was more frequently dissociated than the CO_3t-Bu function. Thus the less reactive benzyl radical was the predominant chain initiating species.

Though all of these points bear reinvestigation using contemporary flash photolysis equipment, each observation illustrates clearly some important points. First, the benzophenone moiety provides a convenient, mid-UV photochemical target where some normally unexpected chemistry can be facilitated. Secondly, dissociation of the peroxy ester function is in competition with a number of other reactions. Much like the competitive reactions of phenylglyoxylate competitions outlined in Section 2.ii, bimolecular reactions often interfere with intramolecular processes, even though the latter might be expected to be rapid. Third, intramolecular competition for the energy of the excited state can be either beneficial and complementary, or decrease reactivity of the initiator system depending upon the radical products.

As was pointed out in the previous chapter, the practical application of such peresters for coatings applications at least, is limited by their thermal stability and the fact that CO_2 is an important product. The systems had a substantial amount of interest at the time they were first reported and several patents on their composition and use were filed.[26] Applications in electronics are clearly implicated.

ii. **Triphenylmethyl radical photochemistry. A second example of carbonyl groups as a target with Alexander Nikolaitchik**[*]

Gomberg, in an attempt to prepare hexaphenylethane by the Wurtz-Fittig reaction from triphenylchloromethyl and Ag or Hg,[27] found that yellow colour developed in solution. In the absence of oxygen this solution, which he claimed to be a solution of hexaphenylethane in benzene, remained yellow for an unlimited period of time. The yellow colour intensified upon heating, and rapidly disappeared upon addition of oxygen or halogens. From these observations Gomberg deduced that a reversible reaction was taking place. Since hexaphenylethane is too sterically crowded to be stable it must be partially dissociated to the free radical in solution. After elucidation of the structure of the

---

[*] The work in this section is taken largely from the Ph.D. Dissertation of Alexander Nikolaitchik; Center for Photochemical Sciences, 1995.

products particularly in the presence of oxygen, Gomberg proposed an existence of a new trivalent form of carbon. Chemical and spectroscopic characterization, undertaken later, proved that, though his postulate of hexaphenylethane was wrong, the other suggestion was not. What Gomberg had synthesized was the triphenylmethyl radical, $\Phi_3C\bullet$.

In 1967 Nauta and Lankamp[28] proved, using NMR, that Gomberg's putative dimer (hexaphenylethane) was really a para position coupled product, the quinoid dimer, and that this too was dissociative;

Reaction of triphenylmethyl radicals

Eventually ESR and other spectroscopic techniques probed the character of the equilibrium processes, involving other stable free radicals. Neumann et al.[29] studied a series of mono and disubstituted triphenylmethyl radicals by ESR, to determine their relative resonance stabilization. The corresponding quinoid dimers of the radicals were synthesized and their degree of dissociation was calculated by measuring the spin concentration in the samples.[30] The degree of dissociation, α, the equilibrium constants for the reverse dissociation of the quinoid dimer, and the thermodynamic parameters for formation of several stable triphenylmethyl radicals are reported in Table 1.

Table 1
Properties of p-Substituted Triphenylmethyl Radicals

R_1	R_2	α,%	$K \times 10^3$, mol/l	ΔG, kcal/mol	$ΔH_d$, kcal/mol
H	H	12	0.33	4.75	10.7
H	t-Bu	18	0.79	4.23	10.2
t-Bu	t-Bu	36	4.05	3.26	8.2
H	CN	28	2.18	3.63	10.0
CN	CN	49	9.42	2.76	
H	COPh	33	3.25	3.39	10.2
H	OMe	24	1.52	3.85	
OMe	OMe	29	2.37	3.58	7.1
OMe	CN	45	7.36	2.91	
t-Bu	CN	47	8.34	2.84	7.9

In early papers, Gomberg also reported that triphenylmethyl radical and many of its analogs bleach with sunlight. The photochemistry of the parent triphenylmethyl radical was later investigated systematically by Letsinger et al.,[31] who found 9-phenylfluorene-derived photoproducts, as well as triphenylmethane among the products. Meisel et al. carried out flash experiments and reported the formation of a radical product with absorption at 490nm, which was accompanied by bleaching of the triphenylmethyl radical.[32] This absorbance was assigned to the cyclized phenylfluorenyl radical.

Formulation of phenylfluorene

That excitation of triarylmethyl radicals in polar medium at shorter wavelengths, such as 248nm or 308nm, resulted in the photoionization and was reported by Steenken et al.,[33] who observed the corresponding trityl cation, as well as the solvated electron. Thus a mechanism suggested for the formation of 9-phenylfluorene and triphenylmethane, is an oxidation/reduction process, towards the formation of a cation/anion pair occurring on the pathway to the reported products. Excited states of all stable free radicals appear to show enhanced electron-donor and -acceptor properties compared to the ground state radicals.

iii. p-Benzoylphenyldiphenylmethyl radical

The p-benzoylphenyldiphenylmethyl radical (RAD) was first studied by Wittig et al.[34] because it was postulated to be a compound with large resonance stabilization of the unpaired electron. This radical is of interest because it contains the target benzophenone chromophore. By using it to absorb the radiation, one anticipates specific photochemistry.

The resonance structures of p-benzoylphenyldiphenylmethyl radical may be divided into three groups; those with the unpaired electron located on carbon; those with the

Resonance Structures for RAD

electron localized on the oxygen atom; and those consisting of charge-separated structures. The 'g-value' of RAD from its ESR spectrum (g = 0.0029) indicates that it is a carbon centered radical which predominates in the hybrid.

RAD was synthesized from the chloride in a like manner to the synthesis of triphenylmethyl.[35] The absorption spectrum of RAD (Figure 8), indicates that solvent polarity has little effect on the position of the maximum λ_{max} = 582-584nm. Thus there is only a small change in radical dipole moment associated with the absorption process in either CH_3CN or PhH.

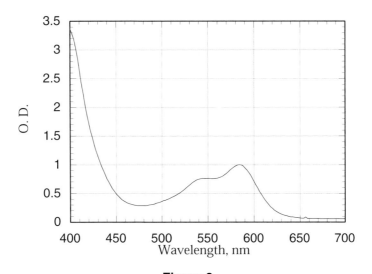

Figure 8

Absorption Spectrum of RAD in PhH ([R-Cl] = 1.12x10^{-3} M)

Emission spectroscopy indicates a very different situation with the first excited state of the radical, RAD, where a red shift is observed in the fluorescence spectra in different solvents. This indicates a significant dipole moment of the D_1 (doublet 1) excited state, as well as the possible existence of a different energy state in polar solvents. A change of spectral shape occurring along with the red shift in the emission spectrum also supports this proposal.

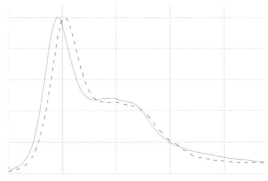

Figure 9

Fluorescence Spectra of RAD: 1 - Hexane, 2 - Mineral Oil, 3 - Benzene, 4 - CH$_3$CN (Scaled to Unity), Excitation 540nm

As is the case for other triarylmethyl radicals, RAD undergoes a reversible dimerization process RAD ---> R-R. Compound R1, formed via 1-5 H-shift after an addition of an acid to the dimer of RAD, provides further evidence for the structure of the RAD dimer.[36]

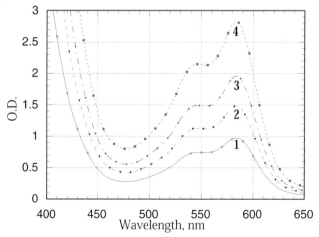

RAD Dimerization Process

The presence of the dimer can be observed indirectly when measuring the fluorescence of RAD in solution because the fluorescence spectrum is concentration dependent. As the concentration of RAD in the solution is increased, the fluorescence decreases. The absorption and fluorescence spectra as a function of concentration are in Figures 10 and 11 respectively.

Figure 10
Absorption Spectra of RAD: 1 - 7.65x10^{-4} M, 2 - 1.45x10^{-3} M, 3 - 1.92x10^{-3} M, 4 - 2.60x10^{-3} M

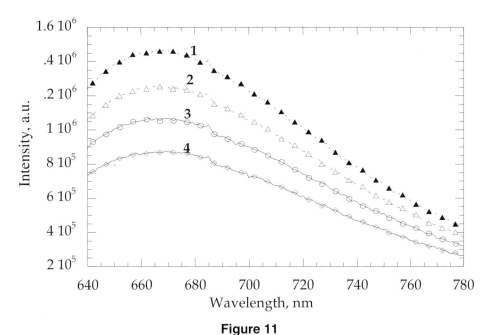

Figure 11
Fluorescence spectra of RAD in CH_3CN: 1 - 7.65×10^{-4} M , 2 - 1.45×10^{-3} M, 3 - 1.92×10^{-3} M, 4 - 2.60×10^{-3} M

The concentration of RAD can be calculated from the known values of the concentration of the chloride precursor and the dimerization equilibrium constant.[37] As the concentration of R-Cl is increased (sample #1 to sample #4), the fluorescence intensity (λ_{ex} = 400nm) decreases. This can be explained by the presence of the quinoid dimer. As the concentration of R-Cl increases so does that of the dimer. Since both RAD and dimer absorb light at 400nm, at higher [R-Cl] the ratio [RAD]/[R-R] is smaller than at lower [R-Cl]. The concentration of excited RAD formed with 400nm excitation therefore decreases and so does the fluorescence intensity. The lower fluorescence intensity at higher [R-Cl] concentration can also be explained by the quenching of the excited RAD with the dimer or self quenching of RAD.

The ground state of RAD, is a doublet, identified by D_0, as is the first excited state D_1.[38] Though intersystem crossing from the D_1 might populate the state with quartet multiplicity, (Q_1 is analogous to the decay of an S_1 to a T_1 state in a molecule having a singlet ground state) experimental evidence indicates this does not happen.

RAD dimerization is an endothermic process, thus the lower the temperature, the greater the concentration of dimer and the smaller [RAD]. When two different methods were used to generate RAD in methyltetrahydrofuran (MeTHF) glass at 77K, two very different sets of spectra resulted. The first method, treating the corresponding chloride in a solution of MeTHF at 25°C with Ag, and then rapidly cooling the solution by immersing it in liquid N_2 (spectra 1a, 1b, Figure 12). A fraction of RAD molecules are trapped in the solid by this method. In an alternative procedure, the chloride precursor of RAD was dissolved in MeTHF at 25°C, the solution was deoxygenated with Ar, cooled in liquid N_2 and irradiated with 366nm light (spectra 2a, 2b, Figure 12). The irradiation caused the chloride to dissociate forming RAD.

Samples prepared by both procedures were studied for any emission from 580nm excitation in the 600-800nm range with different delay times. Absence of such an emission at longer time delays suggests that Q_1 is not populated.

RAD synthesized in MeTHF glass at 77K via irradiation is blue (#2), while RAD synthesized at 25°C in solution is red (#1).

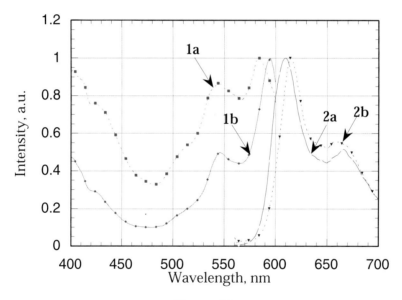

Figure 12

Fluorescence Excitation Spectra of RAD: 1a, 1b and Fluorescence Emission Spectra of RAD: 2a, 2b in MeTHF glass at 77K; a - Synthesized in Solution; b- Synthesized in Glass

A red shift of 15nm in the fluorescence excitation spectrum for RAD in sample #2 explains the difference in colour between the samples. This spectral shift originates, we suggest, from conformational differences in the RADs in the two samples. In sample #2 chloride precursor is trapped in the matrix, and it retains its conformation in RAD formed after irradiation. In sample #1 RAD was trapped in its normal room temperature conformation in solution as the sample was being frozen.

iv.   **Photochemistry of RAD in Solution**

The products of photolysis of RAD **1** with visible light and UV light were identified as: p-benzoylphenyldiphenylmethane **2** (major), benzophenone **3**, tetraphenylethylene **4**, and 1,4-dibenzoylbenzene **5**.[35] In deuterated benzene or CD_3CN, no deuterium was incorporated in either the hydrocarbon triphenylmethane or in the product benzophenone.

Photolysis of RAD in PhH with visible light

The efficiency of this reaction is low when RAD is excited with both UV and visible light. Because the presence of the dimer interferes with absorption by the radical, the quantum yield for this reaction could not be measured.

On initial examination, it appeared that the products benzophenone, tetraphenylethylene, and 1,4-dibenzoylbenzene were formed as a result of a photodissociation of RAD. Tetraphenylethylene can be formed via coupling reaction of diphenylcarbene, while 1,4-dibenzoylbenzene comes from a recombination reaction of benzophenone radical and benzoyl radical. These reactions are clearly precedent setting in that they represent the cleavage of a photoexcited free radical to another radical and a carbene.

Formulation of a benzophenone radical and a diphenylcarbene

The photoreactions of RAD in PhH and CH_3CN in the presence of CH_3OD and D_2O were also studied. Deuterated p-benzoylphenyldiphenylmethane and p-benzoyl-phenyldiphenylmethyl methyl ether in 1:1 ratio were among the products of the photoreaction. Deuterated p-benzoylphenyldiphenylmethane and p-benzoylphenyl-diphenylmethyl carbinol in 1:1 ratio were the products of the photoreaction in CH_3CN in the presence of 1 M D_2O.

Photoreaction of RAD in the presence of D$_2$O and CH$_3$OD

Though benzene solutions of RAD in the presence of 1.0 M CH$_3$OD were stable in the dark, when deuterated methanol was added to the solution of RAD in CH$_3$CN, a dark reaction took place with a formation of p-benzoylphenyldiphenylmethyl methyl ether (major) and p-benzoylphenyldiphenylmethane (minor).

GC/MS analysis of the photolysis products formed in the presence of CH$_3$OD and D$_2$O show significant presence of deuterium in the hydrocarbon p-benzoylphenyldiphenyl-methane suggesting, if not ionization, that there a significant polarity of RAD in its excited state.

v. Photochemistry and Photophysics in the Absence of Dimer

The equilibrium between RAD and its dimer is affected by the solvent and the presence of dimer likely influences RAD photochemistry. Preventing the formation of dimer is, therefore, essential in order that the photochemistry of the nascent free radical can be studied. Bulky groups, when appropriately substituted on the para positions of the radical, prevent radical dimerization.

Dissociation of dimer is an endothermic process with ΔH = 10.3 kcal/mole at 25°C in benzene. Thus, at 70°C and 10^{-5}M concentration of radical, one can assume the concentration of dimer to be negligible. The disadvantage of carrying out studies at such temperatures, however, is that the radicals are unstable in a number of solvents, CH$_3$CN, for example. Magnetic susceptibility and absorption measurements indicate that p-tri-t-butylphenylmethyl radical is 100% dissociated in solution at room temperature.[29] This predicted that the p-benzoylphenyl-di-4,4'-t-butylphenylmethyl radical should also be 100% dissociated in solution. In this radical all three para positions are blocked with bulky substituents.

The synthesis of tert-butyl substituted triarylmethyl radical (called TBR) from the corresponding chloride was accomplished by reaction with submicron Cu powder in benzene at 80°C for 2 hours, or reaction of the chloride with silver powder at room temperature in CH_3CN.

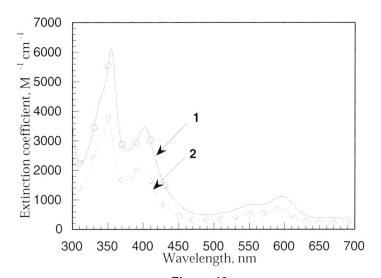

Synthesis of TBR

The absorption spectra of TBR in benzene and CH_3CN are presented below:

Figure 13
Absorption spectra of TBR in 1 - PhH; 2 - CH_3CN

A relatively small blue shift (4nm) is observed as the solvent is changed from nonpolar - benzene to polar - CH_3CN. The extinction coefficients of TBR are given in Table 2.

Table 2

The Extinction Coefficients of TBR (M^{-1} cm^{-1})

Solvent	400 nm	550 nm	594 nm
Benzene	3400	850	1120
Acetonitrile	2100	530	690

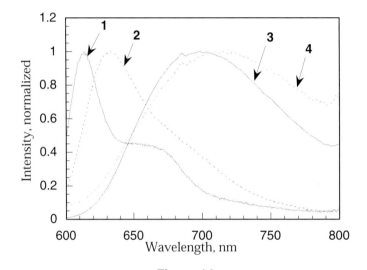

Figure 14

Fluorescence spectra of TBR in 1 - Methylcyclohexane; 2 - PhH; 3 - PhCN; 4 - CH_3CN (scaled to unity), excitation 590nm

A significant change in the shape of the fluorescence spectrum, as well as in the efficiency of fluorescence, is observed in polar solvents, Figure 14. In benzene, the fluorescence spectrum of TBR is the mirror image of the absorption spectrum. It is thus possible to conclude that the first excited doublet state (D_1) of radical TBR has a rigid structure, similar in geometry to that of the ground state D_0 in this solvent. In contrast, a broad emission is observed in CH_3CN and this is accompanied by a 50-70nm red shift of the maximum, relative to that observed in PhH. This relatively large red shift corresponds to an excited state with a significant degree of charge transfer. The quantum yields of fluorescence of TBR varied from 0.44, 2.4 x 10^{-2}, and 1x10^{-3} in PhH, PhCN, and CH_3CN, respectively. The fluorescence excitation spectrum of TBR in CH_3CN, Figure 15, shows that the same excited D_1 state is formed initially in CH_3CN.

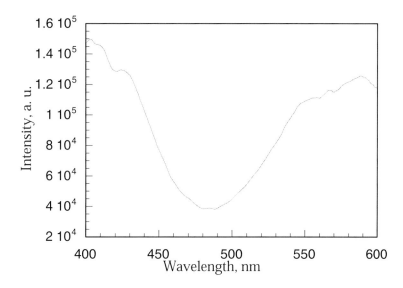

Figure 15
Fluorescence excitation spectrum of TBR in CH_3CN (observation at 714nm)

These results represent an almost classic example of the existence of a charge transfer (CT) state in the case of TBR. In polar solvents this CT state is lower in energy than the D_1 state and a rapid transition between the D_1 state and the CT state takes place. A low quantum yield of fluorescence, ϕ_{fl} and broad structureless emission, supports the notion that the emission observed in polar solvents comes mostly from this state. Molecules with flexible geometry in the excited state, exhibit low values of ϕ_{fl}, since the energy is dissipated into rotational and vibrational modes of motion.[41] A possible structural representation of the CT state is shown in Scheme 3.

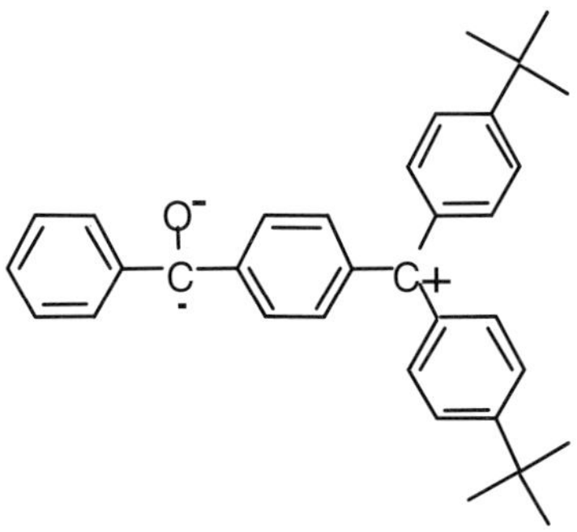

Scheme 3

Structure corresponding to the charge transfer state of TBR

vi. Fluorescence of TBR in Rigid Medium

Excitation of TBR in polar solvents, produces a charge transfer state with flexible geometry, as opposed to the rigid geometry presumed for the ground state. A number of compounds, for example those used as fluorescence probes Figure 16 (Chapter 3), exhibit a dramatic increase in quantum yield of fluorescence, with an increase of the viscosity-rigidity of the surrounding medium. Loutfy et al.[41] called such compounds 'molecular rotors', and showed that they evidence a 100 fold difference between fluorescence quantum yield in solution, and when the probe is incorporated in a rigid polymer, such as poly(methylmethacrylate).

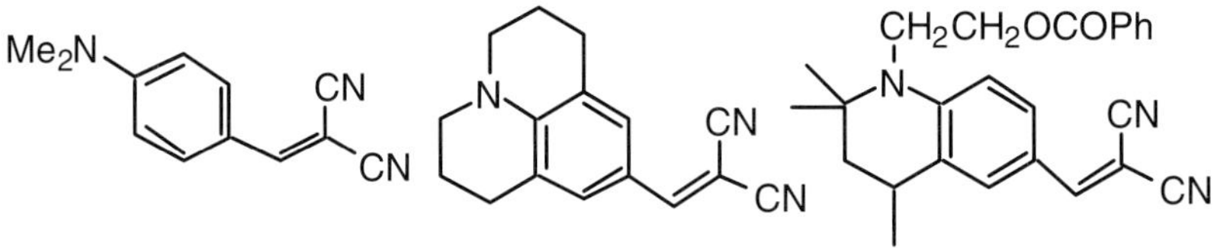

Figure 16

'Molecular Rotor' fluorescence probes

In order to determine whether TBR exhibits behaviour similar to that of molecular rotors, fluorescence spectra were compared in methylcyclohexane solution at -78°C and in a mineral oil glass, Figure 17.

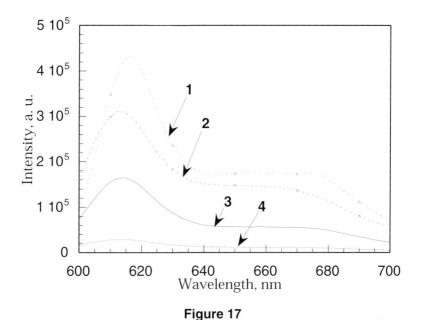

Figure 17
**Fluorescence of TBR in 1 - Methylcyclohexane, -78°C;
2 - Methylcyclohexane, +23°C; 3 - Mineral Oil Glass, -78°C;
4 - Mineral Oil Liquid, +23°C**

Fluorescence quantum yields are often larger at reduced temperatures and TBR is no exception. In methylcyclohexane, which is a mobile liquid at +23°C and at -78°C, an increase of 1.3 times in the quantum yield of fluorescence was observed at the lower temperature. In mineral oil the increase in ϕ_{fl} was 5.6 times. This increase is a manifestation of both the lower temperature and the rigid environment. Since both methylcyclohexane and mineral oil are nonpolar solvents, the maxima of TBR fluorescence is 613nm and 615nm, correspondingly, and the shape of the emission spectrum is the mirror image of the absorbance. Based on the observed behaviour of the ϕ_{fl}, it is concluded that the D_1 state of TBR has a flexible geometry.

The fluorescence spectra of TBR in 2-cyanoethyl acetate at different temperatures are shown in Figure 18.

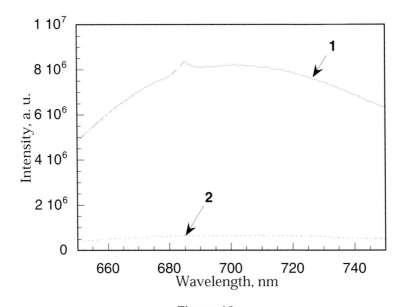

Figure 18
Fluorescence of TBR in 2-cyanoethyl acetate: 1 - at -78°C; 2 - at +23°C (excitation 590nm)

Even though, at -78°C, 2-cyanoethyl acetate is a viscous oil, an increase in the quantum yield of fluorescence of 12 times was observed when the temperature was decreased. This indicates a greater than expected sensitivity of the fluorescence quantum yield in polar solvents, toward the viscosity of the surrounding media. Thus a 'molecular rotor' model can be applied toward the emitting state of TBR in polar solvents.

No phosphorescence emission was detected in an organic glass of TBR in methylcyclohexane at 77K glass, so that there is no evidence for the population of the quartet state of TBR.

vii.   Electrochemistry of TBR

Since hydrocarbon product is also observed in radicals, employing the 'benzophenone-target principle,' the matter of excited state oxidation/reduction is also a matter worthy of consideration here. An important contributing factor in the photoreactivity of the triphenylmethyl radical is its excited state oxidation to the corresponding triphenylmethyl carbocation, the so-called trityl cation.

$$\Phi_3 C\bullet \; - e \; \text{-------}> \; \Phi_3 C^+$$

Oxidation of triphenylmethyl radical

This oxidation appears to occur through the bimolecular electron transfer. The electrochemistry of TBR was therefore studied in CH_3CN in the presence of 0.1 M tetrabutylammonium perchlorate as the supporting electrolyte. Oxidation and reduction curves were measured vs. Ag/0.01 M $AgNO_3$ reference electrode.

An oxidation potential for TBR of +0.21 V vs. SCE was calculated. This compares to an oxidation potential for 4,4',4"-trimethyltriphenylmethyl radical of +0.29 V vs. SCE in DME.[42] The reduction potential of TBR in CH_3CN (-0.87 V vs. SCE) is comparable to that of 4,4',4"-trichlorophenylmethyl radical (-0.87 V vs. Ag/AgCl in DMSO).[43]

Rehm-Weller calculations give the free energy for an electron-transfer reaction, between the first excited doublet state of radical $R(D_1)$ and the ground doublet state of radical $R(D_0)$ to form solvated ions.

$$\Delta G = E_{0x} - E_{red} - E_{0-0} + 2.6/e - 0.13$$

Using an excited state energy of TBR in benzene as E_{0-0} = 1.96 eV (λ_{max}(fluorescence) = 633 nm), in PhCN E_{0-0} = 1.77 eV (λ_{max}(fl) = 700nm), and in CH_3CN E_{0-0} = 1.74 eV (λ_{max}(fl) = 710nm), the free energy of electron transfer in benzene is $\Delta G(PhH)$ = +0.14 V, in benzonitrile $\Delta G(PhCN)$ = ~0.73 V, and in CH_3CN $\Delta G(CH_3CN)$ = ~0.72 V respectively. The proposed electron-transfer reaction is thus endergonic in PhH and exergonic in PhCN and CH_3CN. In order for such a reaction to be possible, assuming a radical concentration of 0.01M, the lifetime of D_1 state of TBR should be at least a few hundred nanoseconds. Under such conditions however, oxidation reduction can be anticipated.

viii. Transient Studies

A solution of TBR in benzene (1.3 x 10^{-3}M) was investigated using nanosecond flash-photolysis, and Figure 19 shows the decays of the observed fluorescence at 680nm and the transient absorbance at 467nm. The lifetime of the observed fluorescence at 680nm was 26 ± 2ns. The lifetime of the detected absorbance at 467nm was 25 ± 2ns, and thus equal to lifetime of observed fluorescence decay. We conclude that a transient absorbance in the 400 - 500nm range corresponds to a $D_1 \rightarrow D_2$ transition.

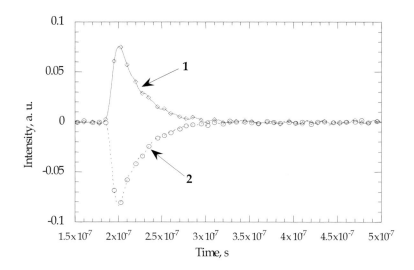

Figure 19

Transient decays of D_1 state of TBR in PhH: 1 - Absorbance (467 nm), Lifetime 26 ± 2 ns; 2 - Fluorescence (680 nm), Lifetime 25±2 ns ([TBR] = 1.3 x 10-3 M)

Fluorescence lifetimes were measured for TBR in CH_3CN (2.3 x 10^{-3}M) and benzonitrile (3 x 10^3M). The lifetime of fluorescence at 710 nm was determined to be of the order of 1-2ns.

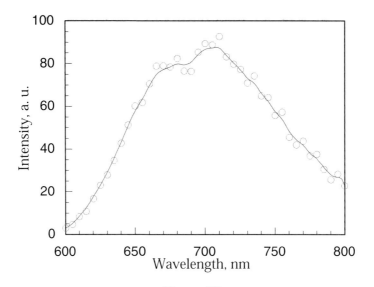

Figure 20
Fluorescence spectrum of TBR (2.3 x 10-3 M) in CH$_3$CN (delay after the laser pulse = 10ns)

The observed fluorescence lifetime in benzonitrile, yielded a value of 11 ± 2 ns for TBR.

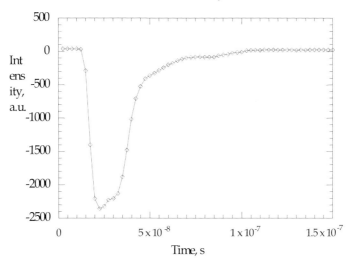

Figure 21
Transient fluorescence decay of TBR (6.8 x 10^{-3}M) in PhCN

No transient absorbance in the 400-500 nm region was detected for TBR in either CH_3CN and PhCN on the nanosecond time scale, suggesting that a state other than the D_1 state was populated, and emitting, in polar solvents.

In the picosecond time domain, an absorbance of the D_1 state was detected in CH_3CN when 532nm laser pulses with 30 ps width were employed for the excitation, Figure 22. The lifetime of the D_1 state of TBR in CH_3CN was 126 ± 5 ps. The peak at 640nm in Figure 22 corresponds to the D_1 state of TBR, since the same decay parameters were obtained at this wavelength. The lifetime of the D_1 state in PhCN was estimated at 1.60 ± 0.07ns.

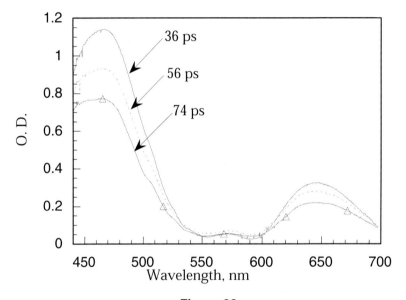

Figure 22.

Transient absorption spectra of D_1 state (TBR in CH_3CN)

Since all lifetimes for TBR were in the 125ps to 25ns range, it is possible to rule out bimolecular reactions of the D_1 state of the radical, with other molecules present in the system, except for the solvent molecules.

Upon the exposure of a TBR solution to visible or UV light, a slow solvent dependent bleaching process was observed. To determine the quantum yield of radical disappearance, the absorbances of radical solutions were measured at known times, while the solutions were irradiated with a 514nm line of an Ar-ion laser, Figure 23.

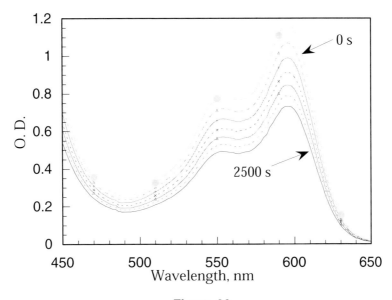

Figure 23
Photolysis of TBR in PhH (0 s to 2500 s at 360 s intervals)

From values of the extinction coefficients of TBR (Table 2), a light intensity (3.5 x 10^{-7} Einsteins/s), and the sample volumes, one calculates the quantum yields of photolysis ϕ_r. We found the quantum yields of TBR photolysis to be very low in the solvents employed, namely: 9.0 x 10^{-4}, 3.9 x 10^{-4}, and 1.0 x 10^{-4} in PhH, PhCN, and CH_3CN, correspondingly. This makes product isolation difficult. The data indicates that the higher the polarity of the media the lower the values of ϕ_r.

Because the highest value of quantum yield of radical disappearance was obtained in benzene, product isolation studies were attempted in this solvent. Tetra-4-t-butylethylene was obtained in 20% yield, but no benzophenone was observed. We propose the formation of 4,4'-di-t-butylcarbene and p-benzoylphenyl radical as a result of the photolysis. The carbene can couple to give tetra-4-t-butylethylene or react with O_2 upon exposure. Thus the photochemistry of TBR in PhH differs somewhat from that of RAD. It is proposed, that the difference arises from the presence of the quinoid dimer in the solution of RAD. The dimer acts as an H-donor, giving rise to two major photolysis products: p-benzoylphenyldiphenylmethane and benzophenone.

The photolysis products of TBR in benzene could not be formed without C-C bond cleavage in the excited radical. From an average value for a C-C bond dissociation energy (85 kcal/mole) and the energy of the light used for the excitation (50 kcal/mole) one must conclude, this requires absorption of at least two visible light photons. The low values of the quantum yield of TBR photolysis (9×10^{-4}) also support such a multiphoton photodissociation.

The systems, RAD and TBR, provide important examples of photochemical processes in which a chromophore whose photochemistry/photophysics is relatively well known is chosen. Either by using this chromophore as the target, or by identifying excited states based on it, much about the photophysics of the reacting species can be determined. Other examples of this will be pointed out in later sections.

4. Oxidation Reduction: Single Electron Transfer Involving Compounds Containing Carbonyl Groups

i. Camphore quinone

The importance of electron transfer was recognized in industry long before reaction mechanisms to account for it were proposed. Camphor quinone and tertiary amines were used to photopolymerize acrylates for dental reconstructions as long ago as the late 50's. Camphor quinone, a deeply yellow dione with a colour almost identical to that of benzoquinone, from which the name is likely derived, is produced by the selenium dioxide oxidation of camphor, Equation 17.

camphor → camphor quinone (via SeO_2)

Equation 17
Preparation of camphor quinone

Since camphor quinone undergoes no photoreduction to alcohols analogous to that observed with benzophenone (Equation 18), the mechanism for the formation of radicals by its reactions with amines was obscure, until papers by Cohen[44] and his coworkers, followed by transient studies of Kevin Peters,[45] suggested the now well accepted electron transfer mechanism.

Equation 18
Oxidation/Reduction of camphor quinone

In the presence of a tertiary amine, radicals are formed via the route outlined in Equation 19.

Equation 19
Reaction between camphor quinone and a tertiary amine

Camphor quinone/tertiary amine initiator systems are used for most dental applications which employ photoreactions. A disadvantage of camphor quinone for this and other applications is its deeply intense yellow colour. In addition, coloured products are created by the reaction, hence the quinone is not bleached entirely. This colour and the lack of bleaching severely limit the depth of penetration to which the initiating process can occur. The consequence is that in most dental applications, teeth are filled with polyacrylates in stages, a few hundred microns at a time.

ii. Electron Transfer to Benzophenone Triplets - Reduction with Amines

Electron transfer to aromatic ketones like benzophenone and the xanthones, such as iospropylthioxanthone, form the basis for many UV initiator systems including Hammond's initiator mentioned earlier. The n-π* transition in aromatic ketones produces the triplet state immediately. Aromatic ketone triplets can be easily reduced, particularly by compounds with low ionization potentials (tertiary amines, borates, certain polycyclic aromatic hydrocarbons, and certain organotin deriviatives) producing radical anions that can be detected easily by transient spectroscopy. Most electron transfer reactions are fast on a relative time scale. For example the triplet state of benzophenone is reduced to the radical anion by triethylamine, over 2000 times faster than a single hydrogen from isopropyl alcohol transferred to the same triplet. Electron transfer from triethylamine to benzophenone has a bimolecular rate constant of approximately 2×10^9 l mol^{-1}s^{-1}. Electron transfer from amines to benzophenone triplets follows the ease of their oxidation: tertiary amine > secondary amine > primary amine.

Many aromatic ketones, similar to benzophenone, fail to react in a way similar to that of the well known hydrogen atom transfer reduction of benzophenone. Fluorenone and p-aminobenzophenone undergo no photoreduction in isopropyl alcohol, for example, but are reduced rapidly to a pinacol product by electron transfer from tertiary amines. The lack of hydrogen atom reduction of p-dimethylaminobenzophenone triplets is

fluorenone   p-dimethylaminobenzophenone

best understood in that their lowest lying excited states have substantial electron transfer contributions,

Resonance structures of p-dimethylaminobenzophenone

This decreases the excited state radicaloid electron distribution on the oxygen atom, and the latter is responsible for hydrogen abstraction. In contrast, charge transfer contributions to excited state character enhance electron transfer reduction by indemnifying the already partially developed negative charge on oxygen into a fully developed negative charge in the anion (also sometimes called the ketyl) radical,

Structure of the ketyl radical

after the electron is transferred. Note that the result of a mechanism that involves electron transfer to form the radical anion and the amine radical cation, followed by deprotonation of the radical cation, forming the α-amino radical, is ultimately the same as if it were a hydrogen abstracted by the ketone triplet from the amine. Pinacol is generally a product, though the route to its formation is entirely different from that of the classical photoreduction process.

Photoreduction of p-dimethylaminobenzophenone

Other functionalities may also serve to alter the reactivity of excited state carbonyl systems by distributing the excited state energies to conjugated chromophores. 2-Acetonaphthone,

2 Acetonaphthone

undergoes no photoreduction under normal conditions because its lowest lying triplet is π-π* rather than n-π* in character. This means that the excited state electronic redistribution is mainly delocalized over the naphthalene ring, and the carbonyl group is nearly devoid of typical radicaloid reactivity.

5. Intramolecular Electron Transfer: Carbonyl Compound Chemistry Leading to Intramolecular Oxidational Reduction with Shengkui Hu

Intramolecular electron transfer reactions of "targeted" carbonyl compounds offer the opportunity to study both substituent effects and conformational effects on the reactivity of excited states. Since the regio- and stereochemistry of products formed by intramolecular cycloaddition reactions may provide significant insight into the more elusive bimolecular transition state structures, reactions between functions tethered by alkyl chains has become an important testing method.

As pointed out previously, the alpha-keto ester chromophore of phenylglyoxylates is easy to work with since its reactions can be followed both by changes in the infrared spectrum in the carbonyl region of the starting ester where the two carbonyl groups are easily distinguishable, and also in the UV spectrum where a typical n-π* absorption is found at around 325nm. For example, in the infrared spectrum, the ketone chromophore, being a conjugated aromatic moiety, absorbs at $\approx$ 6.00μ. The ester, on the other hand, absorbs in the expected area that is around 5.72μ. We have already seen an example of changes in the UV spectrum of photopolymers made from this chromophore in Section 2.iii.

As suggested above, bimolecular quenching of excited triplet states, in a reaction which results in electron transfer, can be a much more rapid process than hydrogen abstraction and it gives some unusual products in intramolecular systems. The phenylglyoxylate triplet illustrates the importance of this possibility, in a competition that occurs between an intramolecular, as opposed to an intermolecular, process.

Thiolesters of phenylglyoxylic acid, Scheme 4, gave some unusual results. The occurence, or lack of, of Norrish Type II products, depends on the position of the sulphur atom, Scheme 5.

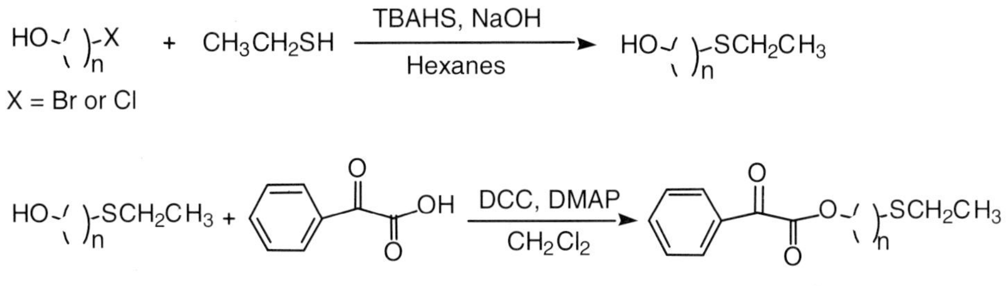

HO~()ₙ-X + CH₃CH₂SH $\xrightarrow[\text{Hexanes}]{\text{TBAHS, NaOH}}$ HO~()ₙ-SCH₂CH₃

X = Br or Cl

HO~()ₙ-SCH₂CH₃ + Ph-C(O)-C(O)-OH $\xrightarrow[\text{CH}_2\text{Cl}_2]{\text{DCC, DMAP}}$ Ph-C(O)-C(O)-O~()ₙ-SCH₂CH₃

6

DCC = dicyclohexylcarbodimide

TBAHS = Tetrabutylammonium hydrogen sulphate

a, n = 2
b, n = 3
c, n = 4
d, n = 5
e, n = 6

f, n = 7
g, n = 8
h, n = 9
i, n = 10
j, n = 11

Scheme 4
Preparation of various thiolesters of phenyl glyoxylic acid

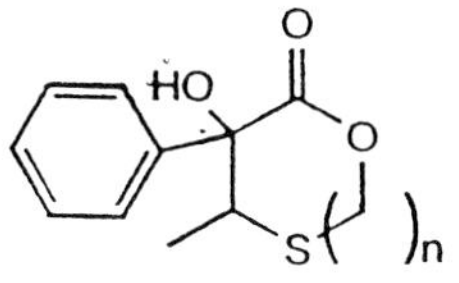

7

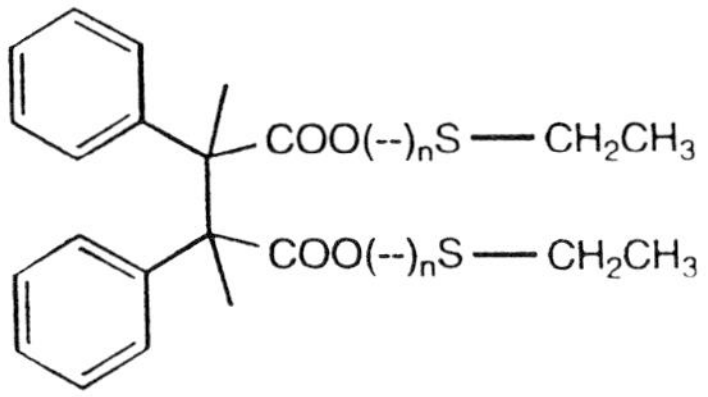

8
Ph-C(-COO(--)ₙS—CH₂CH₃)(-COO(--)ₙS—CH₂CH₃)(Ph)

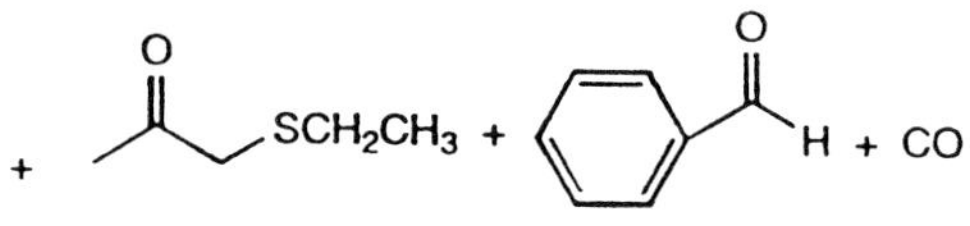

9 CH₃-C(O)-CH₂-SCH₂CH₃ + **10** Ph-C(O)-H + CO

Scheme 5
Products from the photoreaction of the thiolesters of phenylglyoxylic acid

Flash photolysis of **6d** produced two intermediates of comparable lifetimes. One was identified as the triplet state which absorbed with a maximum of 470nm. A second intermediate, whose lifetime was measured at 550nm, Figure 24, was also observed.

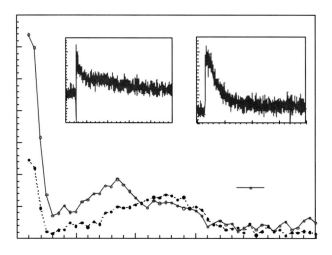

Figure 24.
Transient absorption spectra after laser flash photolysis of benzene solution (0.019M) of 6d. Insert 1 is the decay trace monitored at 470 nm. Insert 2 is the decay trace monitored at 550 nm.

The products of sulphur substituted glyoxylate esters as shown in Scheme 5 include dimer, **8**, Norrish Type II products **9 & 10**, and cyclized product **7**. The relative amounts of each depend on n, the number of methylenes separating the sulphur from the ester function, Table 3.

Table 3
Photoreaction products and yields.

Substrate 6	n	Products and Yields(%)[a]			
		7 (isomer ratio)	8	9	10
6a	2	100 (1.2 : 1)	b	b	b
6b	3	96 (3.6 : 1)	b	b	b
6c	4	89 (2.9 : 1)	trace	b	b
6d	5	71 (3.5 : 1)	19	trace	b
6e	6	53 (2.5 : 1)	30	trace	b
6f	7	43 (single isomer)	25	5	b
6g	8	32 (single isomer)	14	10	b
6h	9	30 (single isomer)[c]	15	15	b
6i	10	25 (single isomer)[c]	35	15	trace
6j	11	b	41	16	5

[a] Isolated unless notice otherwise.
[b] Not observed.
[c] From GC analysis.

The lifetime of similar glyoxylate triplet states depends on the distance of separation of the carbonyl group and the sulphur atom, Table 4.

Table 4
Kinetic data obtained by laser flash photolysis.

Substrate	Atom numbers of cyclic transition state required for Electron Transfer	Absorption maximum of triplet (nm)	Triplet lifetime (μs)	k_{et} (s^{-1})
6a	7	440	0.60	1.5×10^6
6b	8	450	0.54	1.8×10^6
6c	9	470	0.52	1.8×10^6
6d	10	470	0.25	3.9×10^6
6i	16	440	0.03	3.2×10^7

Similar cyclic products were not observed when the analogous oxygen compounds were synthesized and irradiated (see below) suggesting that the longer wavelength reaction intermediate observed in the sulphur case (550 nm) is the ketyl biradical, Scheme 6, but that the initial step leading to its formation is oxidation of the sulphur atom by the keto ester triplet. This is consistent with the relative ease of oxidation of the sulphur in a sulphide, relative to the oxygen atom of an ether.

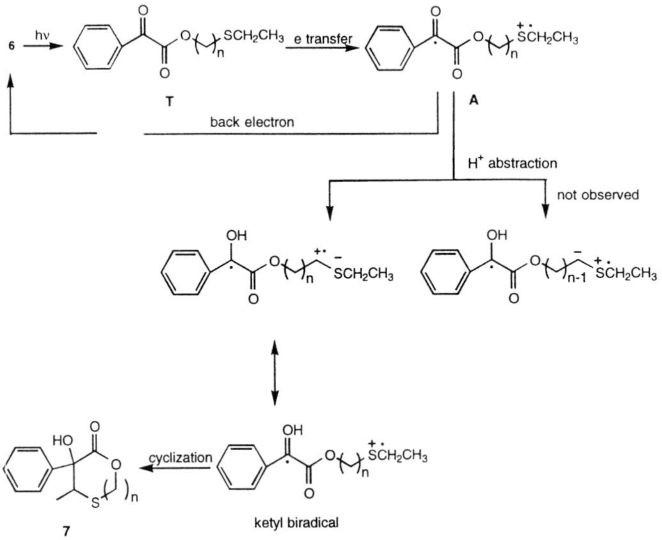

Scheme 6
Mechanism by which cyclisation of the thiolesters of phenyl glyoxylic occurs

A transition state for electron transfer (below) is suggested for the thiol ester series. The relative contributions of a cyclic transition state whose formation is driven by the interactions of the d-orbital electrons on sulphur with the oxidizing carbonyl triplet oxygen atom will depend on the number of methylene groups (n) separating the sulphur atom from the carbonyl group.

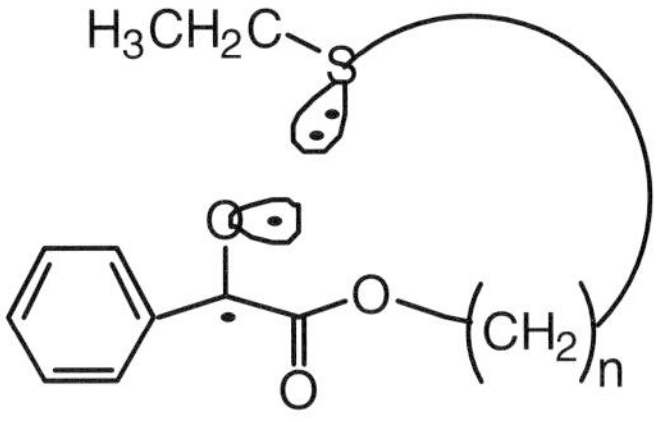

Cyclic transition state of phenyl glyoxylic acid thiol esters

As was pointed out above, one reason for the acceptance of the concept that electron transfer oxidation of the sulphur and reduction of the excited carbonyl group precede cyclization, in the case of the sulphides, is the wholly different behaviour of the corresponding oxygen compounds (Scheme 7).

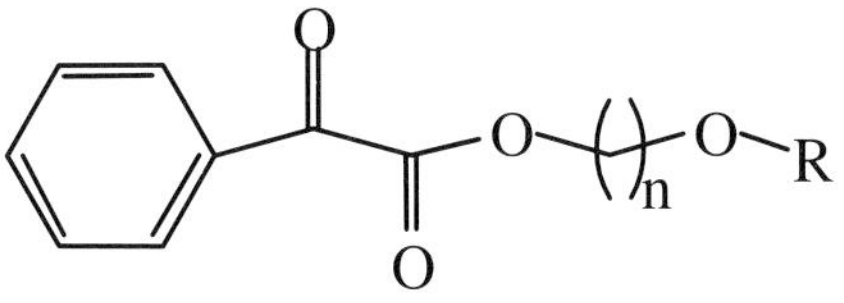

11

a, n=2, R=CH_3
b, n=2, R=CH_2CH_3
c, n=3, R=CH_2CH_3
d, n=4, R=CH_3
e, n=5, R=CH_3
f, n=6, R=CH_3

g, n=3, R=Bn
h, n=5, R=Bn

Scheme 7
Structure of phenylglyoxylates used

In the oxygen series one sees only Norrish type II and dimeric products, Scheme 8. Except in very special circumstances, where remote hydrogen abstraction can occur, albeit with low quantum yield, are cyclol products formed. The distance of separation between the oxygen atom and the carbonyl group does not effect the triplet lifetime, Table 5.

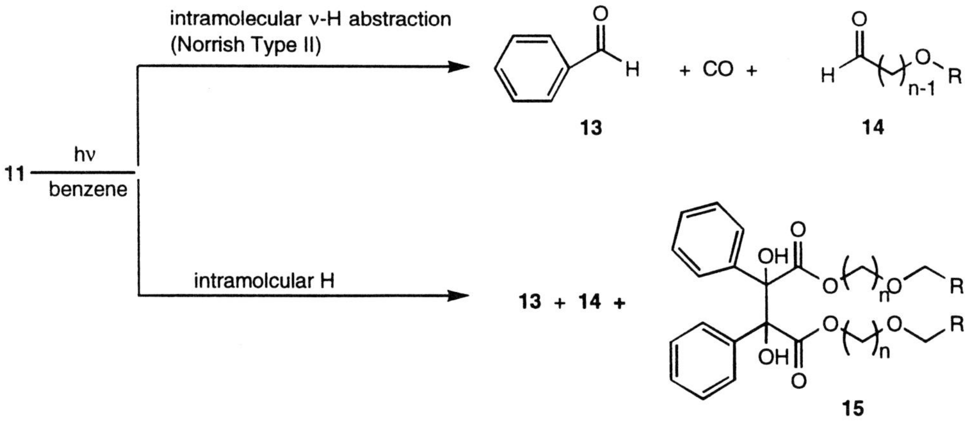

Scheme 8
Products from the photoreaction of phenylglyoxylates

Table 5

Triplet lifetimes and maximum absorption wavelengths.					
Compds	11a	11b	11e	11g	11h
τ (ns)	415	294	493	385	505
λ_{max} (nm)	440	440	450	430	450

Returning now to the case of the olefins and the regiochemistry thought to require the intervention of an exciplex, we consider the situation with olefinic esters and the consequential effect on the Paterno-Büchi reaction in the phenylglyoxylate series.

Several different olefin esters (Scheme 9) were studied. As the distance of separation of the olefin from the carbonyl triplet increases, the lifetime decreases rapidly, indicating a more favourable conformation for quenching exists at longer distance.

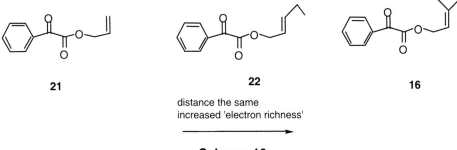

n = 1; **16**
n = 2; **18**

19

20

distance increases
excited state lifetime decreases rapidly
⎯⎯⎯⎯⎯⎯⎯⎯→

Scheme 9
Effects of lengthening the alkyl chain on the lifetimes of the phenylglyoxylate triplet states

In a second comparison, distance from the carbonyl triplet is kept the same, but additional alkyl substituents make the olefin more electron rich, Scheme 10. The lifetimes obtained for each member of the series in Schemes 9 and 10 are shown in Table 6. As is clear from the data, the structure of the alkyl chain has a sustained effect on the triplet lifetime of the components.

21   **22**   **16**

distance the same
increased 'electron richness'
⎯⎯⎯⎯⎯⎯⎯⎯→

Scheme 10
Structures used to evaluate effects of changing electron richness

The cis/trans pair isomers **23**

and

Cis /Trans isomers of pentenyl phenylglyoxylates

were also compared. The results for each are shown in Table 6.

Table 6
Triplet lifetimes detected.

Compound	16 cis	-23	trans 23	18	19	20	24	25
τ (ns)	260	230	470	170	30	<7	1,380	12,700
k_{et} (10^6 s^{-1})	3.1	3.6	1.4	5.2	32.6	>142	/	/

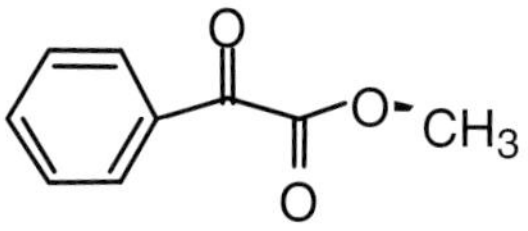

24

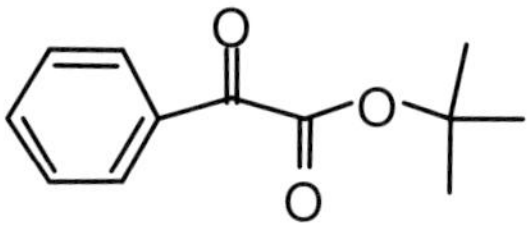

25

More phenylglyoxylates

Scheme 11 completes the mechanism for cycloaddition from the glyoxylate in the intramolecular system. In Scheme 11, one sees firstly the intervention of an exciplex and subsequent electron transfer. The latter is followed by proton loss to form the radical and ring closure.

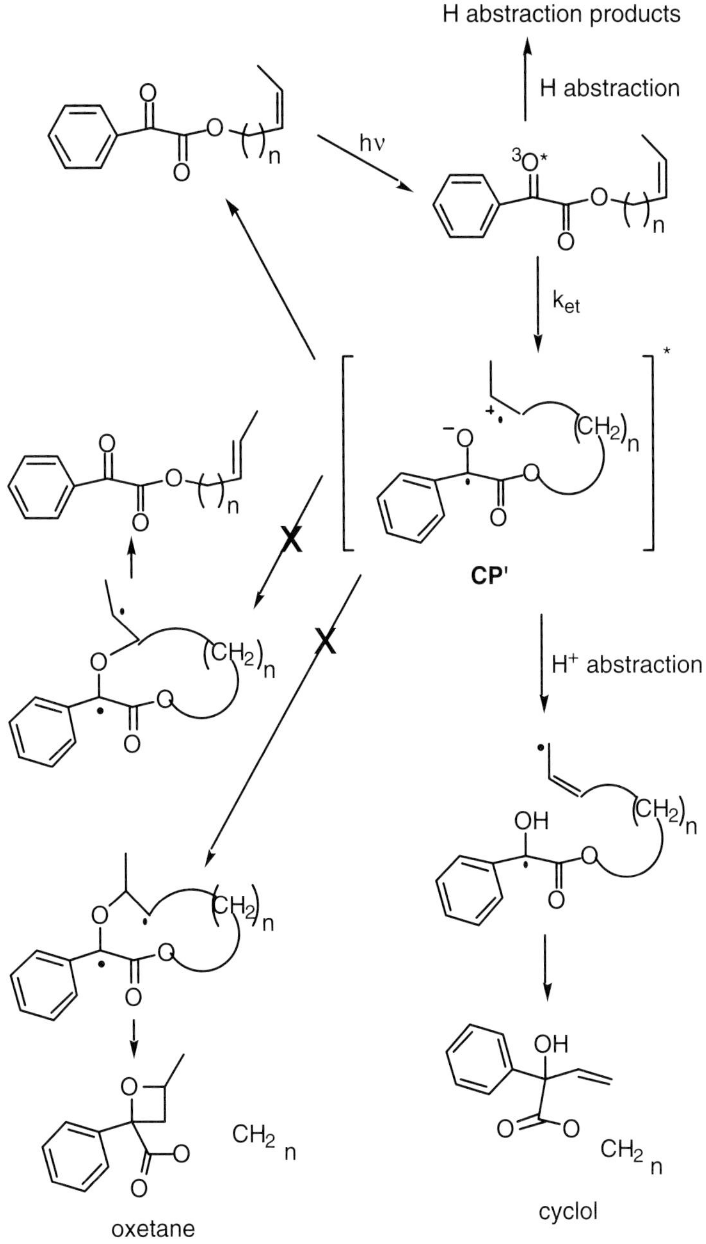

Scheme 11
Photolysis of phenylglyoxylates

6. Charge Transfer Complex Photoinitiators

Ground state charge transfer complexes result when an electron donor/electron acceptor pair, in which the charge is at least partially transferred, form a stable complex. Such complexes are usually reserved for partnerships between very strong acceptors such as tetracyanoethylene and relatively good donors like tetramethoxybenzene In the complex, the acceptor becomes negatively charged and the donor positively charged, Scheme 12.

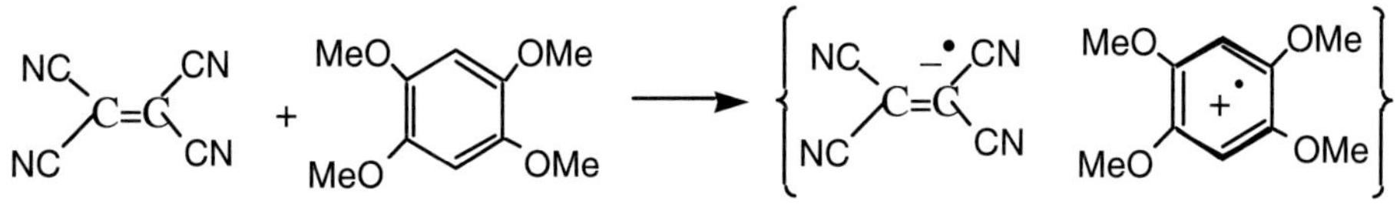

Scheme 12
Formulation of a charge transfer complex

The experimental observation, indicating that a charge transfer complex has been formed, is that the absorption spectrum of an equimolar mixture of the two compounds, presumed to make up the complex is substantially different from the additive spectra of equal concentrations of the two compounds taken individually in the same solvent. Generally the formation of a complex is indicated by a substantial shift of the absorption maximum to longer wavelengths and a significantly increased extinction coefficient for the complexed pair when compared to the individual compounds alone.

One example of a charge transfer complex reported to initiate polymerization was that reported by Jönsson, Hoyle and their coworkers.[*] They report that N-phenylmaleimide forms charge transfer complexes with acrylates, Scheme 13. Another recently reported example is the iodonium salt/borate complex.[**] Iodonium salts form stable complexes with borate anions, Scheme 14, and in addition can be used directly to initiate the polymerization of acrylates. We'll discuss the Jönsson/Hoyle initiator system (Scheme 13) first.

---

[*] These workers have reported a number of examples of this phenomenon. These are reported in the accompanying references.

[**] This is the work of Kesheng Feng.

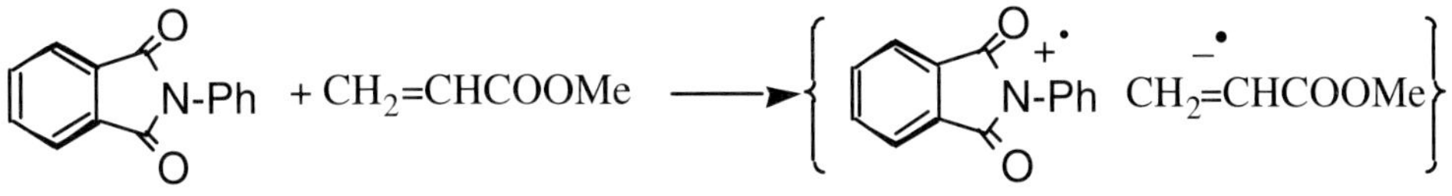

[Scheme 13 structure: N-phenylmaleimide + CH₂=CHCOOMe → charge transfer complex]

Scheme 13

Formulation of n-phenylmaleimide acrylate complex

$$Ar_2I^+, Ar_3B\text{-}R^-$$

Scheme 14

Iodium borate salt

i. **Principles of Charge Transfer Photoinitiators:**

Donor/acceptor complex photochemistry has been of continuing interest almost since the concept of the charge transfer complex was first delineated.[46] Donor/acceptor type monomers were first reported by Vogl[47], and the concept subsequently expanded by others.[48,49,50,51,52] The possibility of photopolymerization initiated by the photochemistry of such a monomer/donor complex, suggests systems that might be 'initiator free' that could avoid some of the problems common to UV initiator systems. A principle problem with most Irgacure initiators is the formation of benzaldehyde. Being of low molecular weight, benzaldehyde gives an odour to many formulations in which it is formed, mainly because UV initiators have to be used in such high concentrations to get tack free surface cure and acceptable photospeed. Since it also 'blooms' to the surface and is extractable from cured formulations, it raises health and safety issues particularly in *in vivo* or food applications of UV cured adhesives and coatings.

Jönsson and Hoyle[53] report practical UV polymerization systems based on the direct absorption of light by the donor/acceptor complex formed from both vinyl ether or styryloxy monomers and N-phenylmaleimide. The mechanism of the initiation step is not entirely obvious, however it is clear that N-substituted maleimides form ground state complexes with these monomers and that the complexes absorb at longer wavelengths than either of the individual compounds and are therefore legitimately referred to as charge transfer complexes. When irradiated at the wavelength of absorption of the complex, polymerization of the monomer occurs.

Since N-substituted maleimides serve as electron acceptors for the monomers above, the complex can be written as in equation 20, where D is any donor function.

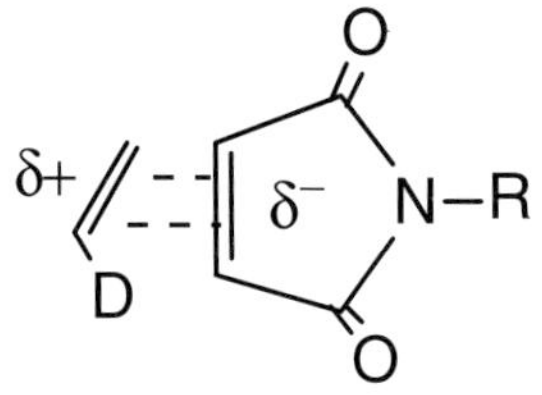

Equation 20

Vinyl/maleamide charge transfer complex

A vinyl ether charge system is shown in Figure 25, where the donor is a difunctional vinyl ether below and the acceptor a difunctional N-substituted maleimide.

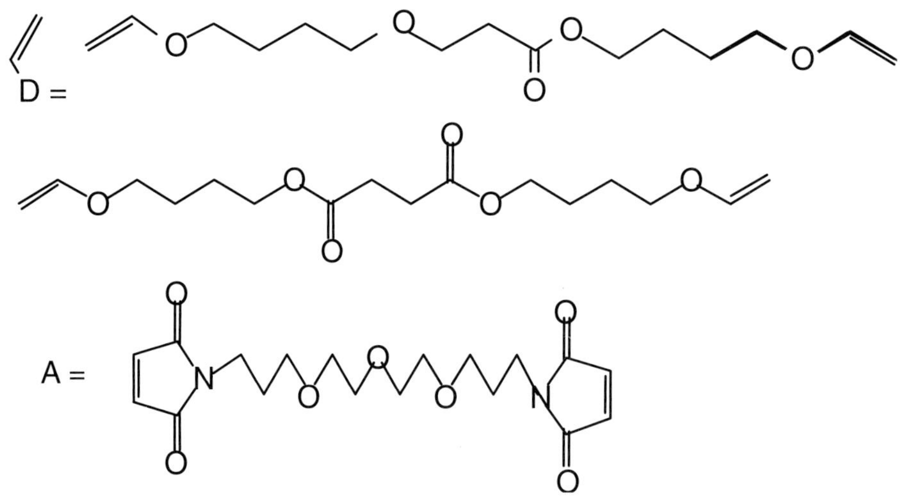

Figure 25

A vinyl etherin-phenyl N-substituted maleimide system

The system bleaches as it initiates and a decrease in UV absorption occurs throughout the course of a photopolymerization.

Direct photolysis of the maleimide/vinyl ether pairs shown in Table 7, as followed by photo-DSC, leads to conversions at various levels as shown in Table 8. Hoyle and Jönsson have looked at other donor acceptor pairs as well.[54] Some of these systems are faster than comparable acrylates, for example, no mean data in that highly functional acrylates are very reactive.

Table 7
ACCEPTORS

Increasing Acceptor Strength →

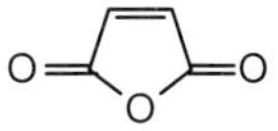

Maleicanhydride — MAN

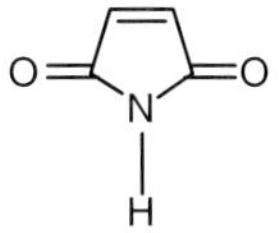

Maleimide — MI

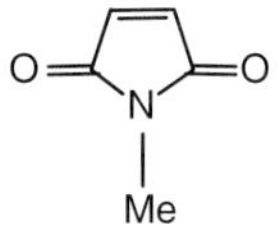

N-Methylmaleimide — MMI

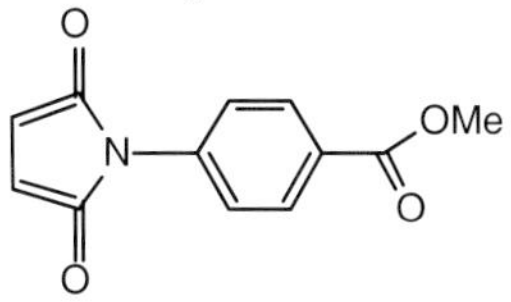

P-Acethoxyphenylmaleimide — p-APMI

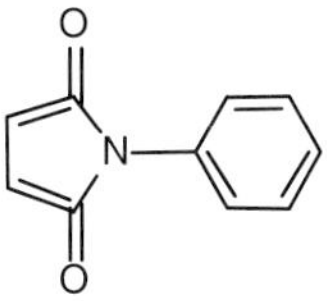

N-Phenylmaleimide — PMI

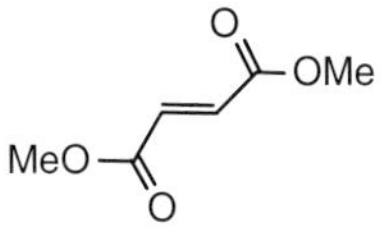

Dimethylfumarate — DMF

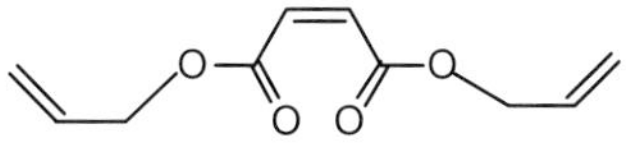

Diallylmaleinate — DAM

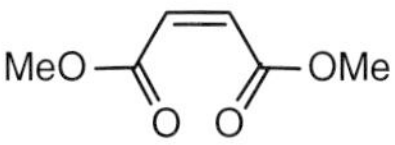

Dimethylmaleinate — DMM

Table 7 continued

DONORS   IP(eV) ↑

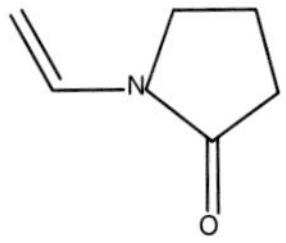

N - Vinylpyrrolidone NVP

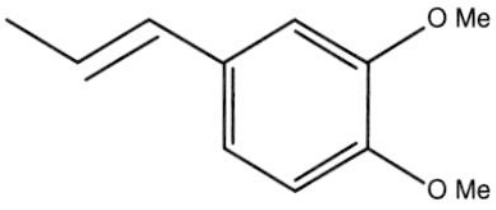

IsoEugenol iEU

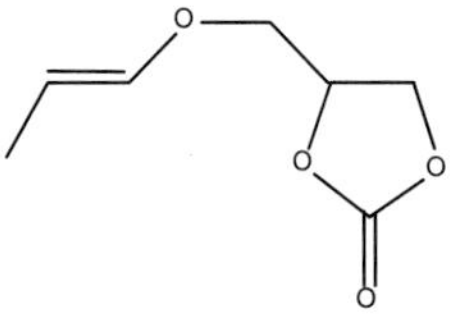

4-Propenyloxymethyl-1,3-2-dioxolanone   PEPC

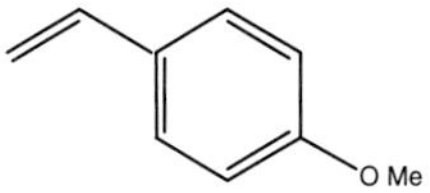

Paramethyoxystyrene   p-MOS

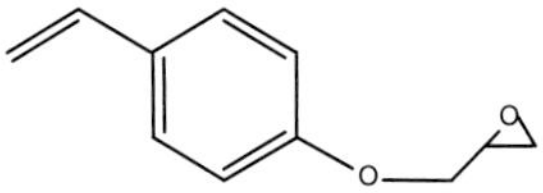

Paraglycidyloxystyrene   p-GOS

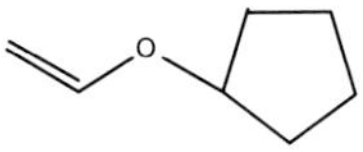

Furfurylvinylether   "diet"PEPC

Increasing Donor Strength →

Table 8
Results of Photo-DSC measurements.

Time to exotherm peak, T_p, in sec.. The maximum heat flow, Hf max, in J/g,s.
The total exotherm area, in J/g. The degree of double bond conversion, Molar %.
Light intensity was 13.6 mW/cm^2, using the pyrex filter. The temperature was 40 °C.

Donor	Acceptor	No	T_p (s)	Hfmax (J/g,s)	AREA (J/g)	CONV.^{c} %	COMMENT
TEGDVE*	EMI	1	175	2.14	358	45	60 °C
TEGDVE	EMI	2	74/70	1.47/1.75	438/382	55/48	40 °C
TEGDVE	EMI	3	---	---			Methyldiethanol-amine CT complex
TEGDVE	EMI	4	67	1.62	337	42	0.1% Water added
TEGDVE	EMI	5	---	---			In Air, No exoth, after 90s
TEGDVE	MAN	6	65	1.0	229	25	
TEGDVE	MAN	7	432	2.61			In Air
TEGDVE	HMI	8	89	2.54	527	83	
IPDBDVE	HMI	9	43/48	4.22/4.84	471/490	95/99	
IPDBDVE	HMI	10	36	4.84			In Pyrex
IPDBDVE	HMI	11	180	3.64	485	98	In Air
CHVE	HMI	12	50/38	4.90/5.05	503/456	75/68	[VE]/[MI] - 1.4
CHVE	HMI	13	65	3.29	453	67	[VE]/[MI] = 0.28

* Triethylene glycolvinyl ether (TEGDVE)

Table 8 continued

CHVE	HMI	14	90	2.46	411	61	[VE]/[MI] = 0.15
CHVE	HMI	15	---	---			Triplet quencher[a]
CHVE	HMI	16	---	---			In Air
CHVE	HMI	17	12/13	20.5/19.4	431/464	64/69	1wt % Irg. 651
CHVE	HMI	18	106	6.45	425*	63	1st% Irg. 651, *after 3 min. In Air
DDVE	HMI	19	144	0.67			Small Exotherm !
DDVE	HMI	20	360	0.35			NDF = 0.3; 50% hv Red.
DDVE	HMI	21	32	11.9	318*	70	1wt%Irg. 651, *after 2 min.
	HMI	22	77/86	4.46/4.49	494/502*	> 100[b]	1wt%Irg. 651, *after 5/4 min.
	HA/HDDA	23	11	12.0	376*	65	1wt% Irg. 651, *after 3 min.
	HA/HDDA	24	---	---			1wt% Irg. 651, *after 3 min.
	HA	25	46	2.26	399	80	1wt%Irg. 651

[c]ΔH for VE and MI has been assumed to be 90kJ/mol.
[a] 1,8 Octa (4,6)dienediacid, [b]Estimated value for homopol. of HMI, 80kj/mol, needs correction.

Data from Hoyle and Jönsson.[54]

Acrylates and styrenes can also be polymerized via light absorption by the donor/acceptor complex with N-substituted maleimides. Examples are shown in Table 9.[55]

Table 9
Monomer Structures

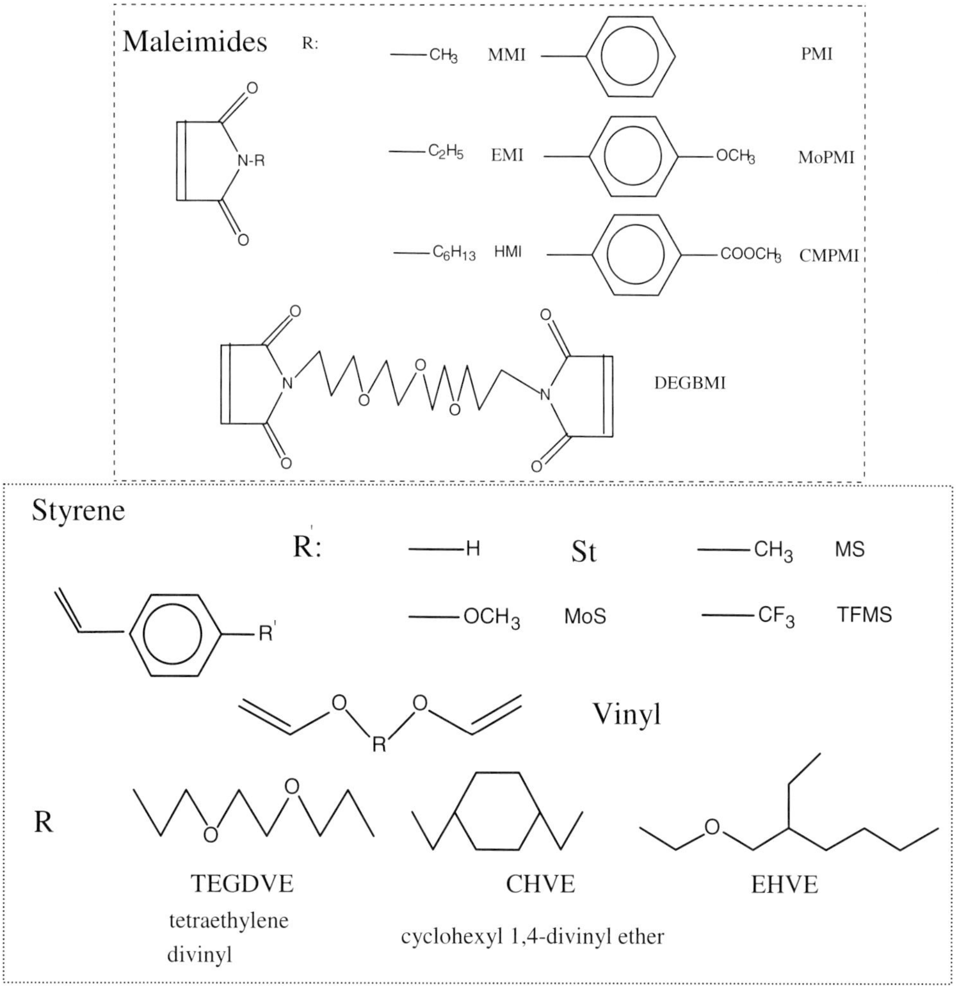

Maleimides
- R: —CH$_3$ MMI
- —C$_2$H$_5$ EMI
- —C$_6$H$_{13}$ HMI
- (phenyl) PMI
- (4-methoxyphenyl) MoPMI
- (4-methoxycarbonylphenyl) CMPMI
- DEGBMI

Styrene
- R: —H St
- —OCH$_3$ MoS
- —CH$_3$ MS
- —CF$_3$ TFMS

Vinyl
- TEGDVE — tetraethylene divinyl
- CHVE — cyclohexyl 1,4-divinyl ether
- EHVE

```
                        Acrylates

    O
    ‖
    ⁄⁄\O/C₆H₁₃        HA

    hexyl acrylate
                                                              TEGDA
                                  O
                                  ‖
    O                     ⁄⁄\/\(CH₂CH₂O)₃\/\
    ‖                                     ‖
    ⁄⁄\/(CH₂)₆\O/\⁄⁄      HDDA            O

                              tetraethylene glycol
                                  diacrylate
    1,6-hexanediol diacrylate
```

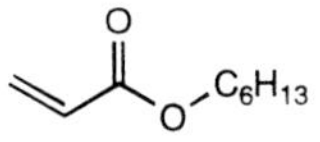

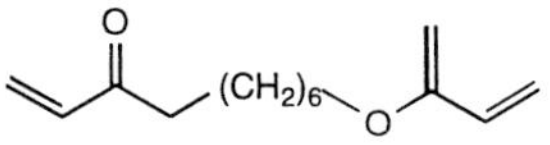

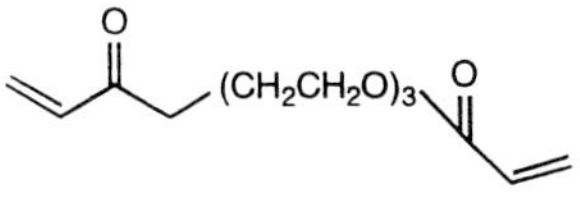

The relative reactivities of various monomer/maleimide pairs are shown in Table 10.

Table 10

1. The reactivity of maleimide/p-methoxystyrene systems

Maleimide	Monomer Consumption %		Polymer Yield (%)
	Maleimide	MeOS	
EMI	46.9	46.1	38.8
MoPMI	6.1	8.7	2.1
PMI	8.8	8.4	6.6
CMPMI	26.7	27.0	23.4

Initial concentration (Maleimide) + 0.25M (MeOS)
Irradiation time: 1 hour

2. The reactivity of N-ethylmaleimide/styrene systems

Styrene	Monomer Consumption (%)		Polymer Yield (%)
	EMI	Styrene	
MS	28.9	27.6	23.7
St	19.6	18.2	7.9
TFMS	9.7	9.3	0

Initial concentration: 0.25 M (EMI) + 0.25 M (Styrene)
Irradiation time: 1 hour

3. The reactivity of donor/acceptor monomer mixtures

Acceptor	Exotherm(Mcal/s)	Peak Maxima
	No initiator	1% Photoinitiator
N-Hexylmaleimide (HMI)	1.68	5.06
Vinylene carbonate (VC)	0	0.20
Dimethyl maleate (DMM)	0	1.01
Diethyl fumarate (DEF)	0	0.92

Light intensity: 7.8 mW/cm^2, Sample amount: 3 µL, Filter: Pyrex
Donor: CHVE, Temperature: 313K, Atmosphere: N_2

4. The reactivity of maleimide/vinyl ether systems

Monomers	Exotherm(Mcal/s)	
1:1 molar ratio	No initiator	1% Photoinitiator
HMI/CHVE	3.12	9.88
DEGBMI/CHVVE	6.23	10.38
HDDA/HA	0	10.94

Light intensity: 13.5 mW/cm^2, Sample amount: 3 µL, Filter: Pyrex
Temperature: 313K, Atmosphere: N_2

5. The reactivity of maleimide/acrylate Systems

Maleimide	Acrylate	Exotherm(Mcal/s)
PMI	HDDA	0
MMI	HDDA	0.27
MMI	TEGDA	3.70

Maleimide:MMI (10%), Light intensity: 15.5 mW/cm^2, Filter: Pyrex,
Sample amount: 3 µL, Temperature: 313K, Atmosphere: N_2

Another major driving force for the development of CT initiator systems has been to replace acrylates as monomers. In spite of their superior performance characteristics in most photopolymerization applications, acrylates have several disadvantages physical properties being one. They are also smelly and, in some cases, quite toxic. They become targets of regulatory agencies.

Some example systems studied by real time infrared spectroscopy have also been reported by Hoyle and Jönsson. Figure 26 shows the results of the polymerization of maleimide (MI)/vinyl ether (VE) in the absence of initiator in the presence and absence of air.[56] The polymerization of an Irgacure 651 initiated acrylate polymerization is compared to that of maleimide/vinyl ether in Figure 27. Photopolymerizations of copolymers Figure 28, (N-vinyl pyrrolidinone) Figure 29 TMPTA and HDDA Figure 30 also exhibit the advantages of initiator complexes.

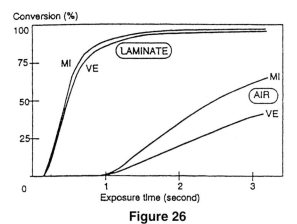

Figure 26

Polymerization profiles recorded by RTIR spectroscopy upon UV irradiation of a maleimide (MI)-vinylether (VE) combination in the absence of photoinitiator (HBVE+HPMI). Light intensity: 150 mW cm^{-2}

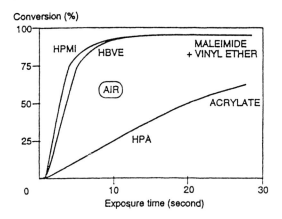

Figure 27

Comparison of the polymerisation kinetics of a monoacrylate (HPA +Irgacure 651) and a monovinyl ether-monomaleimide (HBVE+HPMI), upon UV exposure in the presence of air. Film thickness: 24µm.

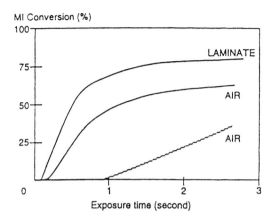

Figure 28

Photoinduced copolymerisation of difunctional maleimide-vinylether monomers (BDBMI+iPDB-DVE). Film thickness: 24μm. —: HBVE+HPMI in air.

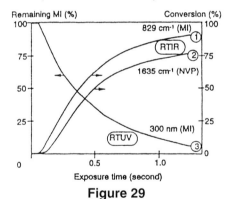

Figure 29

Photoinduced copolymerisation of a maleimide-vinylpyrrolidone mixture (HPMI+NVP), monitored by RTIR and RTUV spectroscopy. I = 150 mWcm^{-2}. laminate.

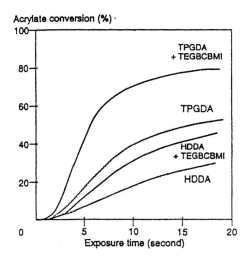

Figure 30
Influence of a bismaleimide on the photoinduced polymerisation of tripropyleneglycol diacrylate and hexandedioldiacrylate. Laminate I = 150m x Wcm^{-2}. [TEGBCBMI] = 1wt% ---: neat monomer.

Other reports[57,58,59,60,61] expand the range of monomers which can be used, they also point out some of the side reactions which occur, and deliniate the scope of donor/acceptor complex photoinitiations.

ii. Mechanism

Hoyle suggests[46] that the donor/acceptor complex itself forms the excited state (exciplex) and that the exciplex decays to biradicals, which initiate the chain processes.

$$D + A \longrightarrow [D\text{--}A] \xrightarrow{h\nu} [D\text{--}A]^* \longrightarrow \text{radicals} \longrightarrow \text{chain polymerization.}$$

Invoking exciplexes as reactive intermediates is tricky since they are generally observed as transients by short time spectroscopy. The formation of the excited complex is shown by two independent routes. Hoyle and Jönsson have written the mechanism shown below for the initiation step and this process, in which the 1,4 biradical initiates polymerization via hydrogen abstraction (probably from a molecule of monomer), is as reasonable as any other.

Equation 21a

Detailed proposal for the formulation of radicals via a charge transfer complex

There is no evidence for the exciplex in this case[62], and the mechanism could as well be that in which the biradical pair forms, directly after excitation.

As has been pointed out in this chapter, excited states have many alternatives and the excited states responsible for donor/acceptor photoreactions are no exceptions. Cycloaddition reactions of the Paterno/Büchi type but involving two olefinic double bonds could consume excited states. In the case of maleimide/donor reactions these are non-productive as far as polymerization is concerned, equation 21.

Equation 21b

Mechanism of a Paterno/Buchi reaction between a vinyl ether and a maleimide

Jönsson works for Fusion Systems and they have continually supported this work. One anticipates, therefore, a commercial announcement regarding monomer complex initiator packages at sometime in the future.

iii.   **Electron Transfer from Organoborates with Alexander Polykarpov[*],**
       Ananda Sarker, Yuki Kaneko and Kesheng Feng

Organoborates mentioned in the previous chapter provide a number of advantages as electron transfer donors. Among these is the fact that borates impart no color, either to solutions or to coatings in which they are used as coinitiators. One disadvantage is that, as salts, they can often be quite insoluble in monomers. This can be alleviated by proper functional group substitutions.

Photoinduced electron transfer from borates to benzophenone triplet states, produce alkyl radicals which yield 1:1 adducts with the appropriate electron poor olefins.[63] In a typical experiment benzophenone was irradiated in the presence of tetramethylammonium triphenylbutylborate and n-butyl acrylate in a solution of benzene/acetonitrile (1:1). This produced a 190% yield (based on the borate) of n-butyl pentanoate, after 2-4 hours at 25°C. The mechanism suggested is described below.

Ph-CO-Ph + (Et)$_3$BPh,N(Me)$_4^+$ $\xrightarrow{h\nu}$ Ph-C(O$^-$)-Ph• + Et• + Et$_2$BPh

Et• + BuOC(O)CH=CH$_2$ $\longrightarrow$ Et-CH$_2$-ĊHCOOBu

Et-CH$_2$-ĊHCOOBu $\xrightarrow{\text{Sol-H}}$ Et-CH$_2$-CH$_2$COOBu

Photoinduced electron transfer to benzophenone triplet states

Since many boranes are accessible by the addition of olefins to (BH$_3$)$_2$, the process represents a general route to the photoformation of 1:1 radical adducts to electron poor olefins. Thus

diborane + olefin $\longrightarrow$ substituted borane $\xrightarrow{\text{alkyl lithium}}$ borate salt

---

[*] From the Ph.D. Dissertation of Alexander Polykarpov, Center for Photochemical Sciences, 1997

$$(R)_xB(Ar)_y,N(Me)_4^+ + CH_2=CHEWG \xrightarrow[h\nu]{ketone} R\text{-}CH_2\text{-}CH_2EWG$$

$$x + y = 4$$

Photoinduced addition of an alkyl group from a borate to an olefin

Borate complexes have been mentioned in the context of imaging systems in Chapter 1. Photoinduced electron transfer from a borate to a chromophore, the latter of which undergoes a bond cleavage reaction, has been the subject of a number of publications and patent applications from the Neckers group.[64,65] Applications including imaging systems[66] and as both small molecule and macromolecular photoinitiators.[67]

iv.   **Intramolecular Electron Transfer**[*]

Electron transfer from borate anions complexed to ammonium ion chromophores may be followed rapidly by homolytic C-N bond scission and concomitant formation of the tertiary amine. The pathway was established by direct observation of p-benzoylbenzyl and p-acetylbenzyl radicals, after laser flash photolysis of N,N,N-tributyl-N(4-benzoyl-1-methylphenyl) tetraphenylborate and N,N,N-trimethyl-N(4-acetyl-1-methylphenyl) tetraphenylborate, respectively.

The β-naphthylmethyl and 7-methoxycoumarin 4-methyl radical were obtained from the singlet excited states of N,N,N-tributyl-N(β-naphthylmethyl) tetraphenylborate and N,N,N-tributyl-N(4-methyl-7-methoxycoumarin tetraphenylborate, respectively. These radicals were also suggested by analysis of the products. Comparison of the rate constants for quenching of the triplet states from excited N,N,N-tributyl-N(4-benzoyl-1-methylphenyl) tetraphenylborate (**a**) and N,N,N-trimethyl-N(4-acetyl-1-methylphenyl) tetraphenylborate (**b**) and those of the singlet excited states of N,N,N-tributyl-N(β-naphthylmethyl) tetraphenylborate (**c**) and N,N,N-tributyl-N(4-methyl-7-methoxycoumarin tetraphenylborate (**d**) reveals the efficiency of C-N bond cleavage, and differences are accounted for in terms of variations in the electron accepting excited states and structural characteristics of the leaving groups.

Synthesis of borates with the reactive reducing borate anion, tetraphenylborate, $R\text{-}CH_2\text{-}NR'_3{}^+,BPh_4$ and with non-reducing anion, tetrafluoroborate, $R\text{-}CH_2\text{-}NR'_3{}^+,BF_4$ is by the procedures shown below:

---

[*] Work of Ananda Sarker and Yuji Kaneko.

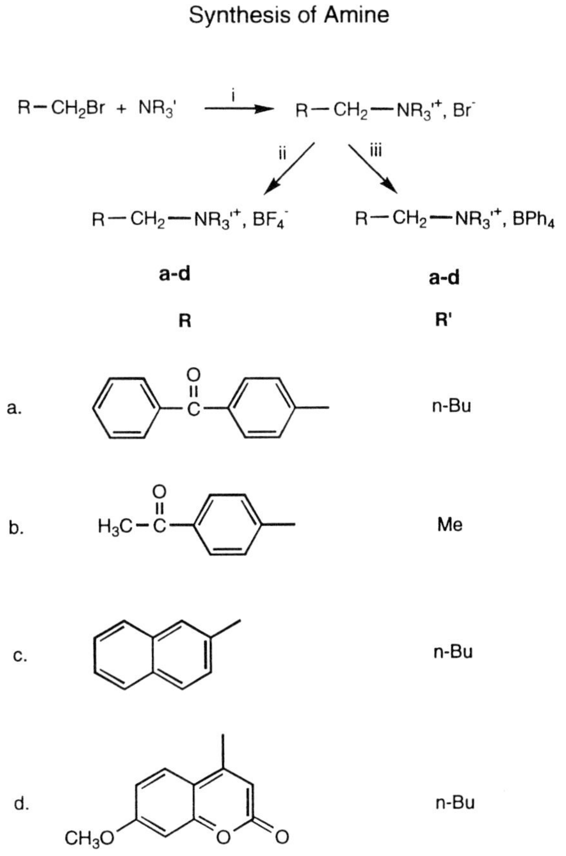

Synthesis of Amine

R—CH$_2$Br + NR$_3$' $\xrightarrow{i}$ R—CH$_2$—NR$_3$'$^+$, Br$^-$

$\swarrow$ ii   iii $\searrow$

R—CH$_2$—NR$_3$'$^+$, BF$_4^-$ R—CH$_2$—NR$_3$'$^+$, BPh$_4^-$

a-d   a-d

R   R'

a. (phenyl-C(=O)-C$_6$H$_4$—)   n-Bu

b. (H$_3$C—C(=O)—C$_6$H$_4$—)   Me

c. (2-naphthyl)   n-Bu

d. (4-methyl-7-methoxycoumarin-yl, CH$_3$O—)   n-Bu

Scheme 15
Synthesis of various borates

The photolysis products include major products that can be attributed to radicals that would be formed from the chromophore. The mechanism of their formation, with compound **b** above used as the example (tetraphenyl borate anion), is given in the scheme below:

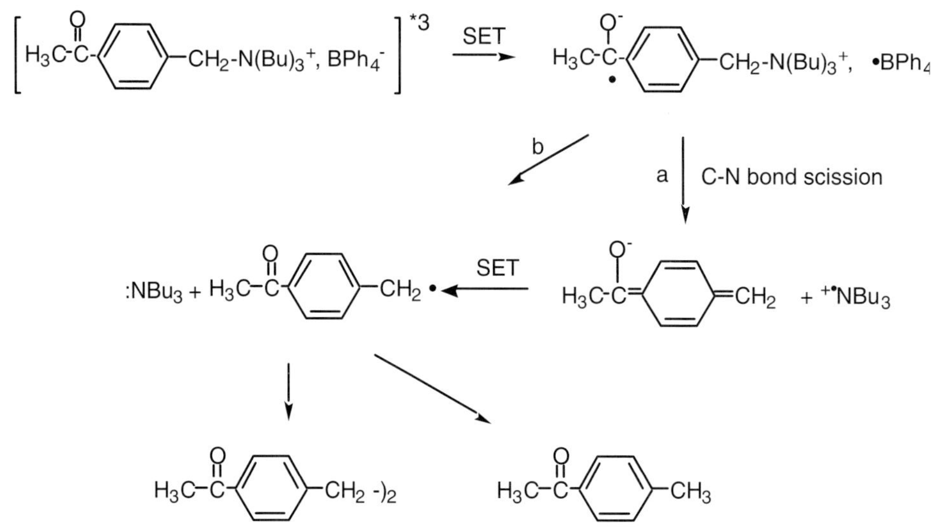

SET = Single Electron Transfer

Scheme 16
Photolysis of a tertiary amine tetraphenyl borate

The critical steps involve conversion of the chromophore to a reactive excited state, reduction of that excited state by the borate, clevage of the C-N bond in the radical cation thus formed, and single electron transfer forming, in this case, p-acetylbenzyl radical. These radicals either couple forming symmetrical p,p'-(1,2-acetylphenyl)ethane or p-acetyltoluene. The latter forms by reaction of with solvent. In the case in point, conversion is 70% and the two products are formed in 40% and 15% respectively. Similar product yields are observed in all cases though the conversions vary from being quite a bit lower in the case of β-naphthyl to substantially larger in the case of 7-methoxy-4-methylcoumarin.

In a sense its surprising that products are formed at all from these reduction reactions. The single electron oxidized products from tetraphenylborate do not form by elimination of phenyl radical and back electron transfer rates with the radical are very fast ≈ 45 psecs. Though biphenyl is observed it does not come from the reaction below:

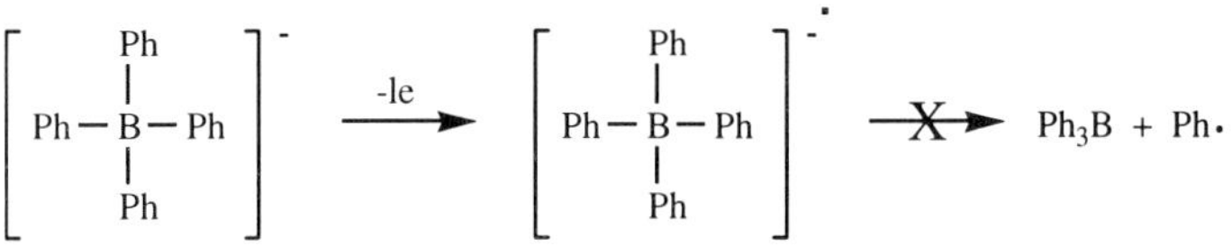

$$\left[\begin{array}{c}\text{Ph}\\|\\\text{Ph}-\text{B}-\text{Ph}\\|\\\text{Ph}\end{array}\right]^{-} \xrightarrow{-1e} \left[\begin{array}{c}\text{Ph}\\|\\\text{Ph}-\text{B}-\text{Ph}\\|\\\text{Ph}\end{array}\right]^{\bullet} \xrightarrow{\;\;\text{X}\;\;} \text{Ph}_3\text{B} + \text{Ph}\cdot$$

Biphenyl instead, results from an intramolecular process, which must also produce the corresponding borene, Scheme 17.

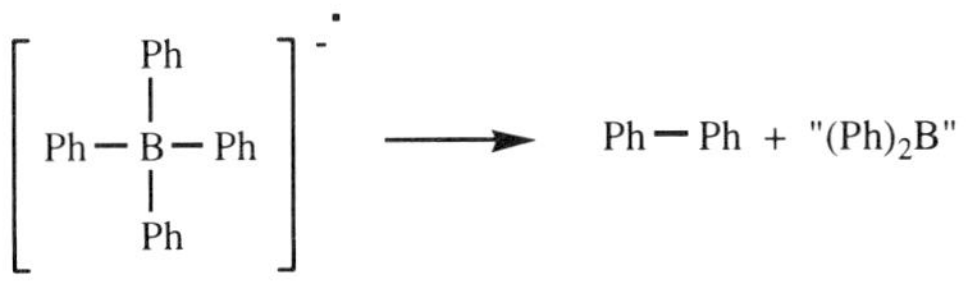

$$\left[\begin{array}{c}\text{Ph}\\|\\\text{Ph}-\text{B}-\text{Ph}\\|\\\text{Ph}\end{array}\right]^{\bullet} \longrightarrow \text{Ph}-\text{Ph} + \text{"(Ph)}_2\text{B"}$$

Scheme 17

The steady state absorption spectra of **a-b** and **d** (Scheme 15, p.171) in acetonitrile have n–π* transitions centered at 335-340nm though (R) the molar absorption coefficients differ. The spectrum of **c** only has π–π* transitions with a substantial tail at longer wavelengths. The spectra of **a-b** and **d** in nonpolar solvents are similar in shape to those in acetonitrile, however significantly red shifted, suggesting the lower excited singlet state is an n–π* state. With **c**, the lower excited state was assigned to the π–π* in acetonitrile. The molar extinction coefficient for **d** is 13,000M^{-1} cm^{-1} (334nm) in acetonitrile indicating that the π–π* transition overlaps with the n–π* transition in this region.[68] All spectra are independent of anion structure and essentially identical to those of the chromophores in the absence of complexing agent in the same solvent. Thus, there is no indication of charge transfer complex formation in **a-d** from the steady state UV absorption spectra. **Spectra** – note that in the sections that follow we refer extensively to schemes 15 and 16, pages 171, 172.

The primary processes in the reactions above can be studied using transient spectroscopy. Using a bimolecular transfer model and based on the oxidation potentials of tetraphenylborate, the reduction potentials of **a-d** and the energy of their excited states, the free energy change (ΔG°)[69] associated with electron transfer has been estimated, Table 11. The electrochemical characteristics were studied by cyclic voltammetry in acetonitrile with tetrabutylammonium perchlorate (TBAP) as the supporting electrolyte. In all cases, cyclic voltammetry measurements yield wave characteristics of irreversible one electron reduction with peaks shown in Table 11. This is taken as evidence for rapid bond cleavage reaction (on the time scale of the electrochemical experiments) of all reduced chromophore tertiary ammonium cations. Since, by transient spectroscopy, one observes the intermediate actually undergoing reaction, we have observed that the triplets of **a** and **b** above are quenched by borate in the bimolecular experiments with quenching constants of 5.6 and 5.7 x 10^9 respectively. In the case of **c** and **d**, it is the singlet states of the chromophore which are quenched by the borate. In the case of naphthyl (**c**) the reaction is about as fast as are those of benzophenone and acetophenone (**a** and **b**). In the case of 7-methoxycoumarin, however, the reaction is exceedingly fast (k_q = 5.94 x 10^{12}). The quenching constants, reduction potentials, ΔG's and rate constants for bimolecular quenching are assembled in the Table 11 below.

Table 11
Reduction potentials ($E_{red)}$), free energy changes (ΔG) for ET and quenching rate constants (k_q) for the excited states of a-d with tetraphenylborate in acetonitrile.

Compounds	E_{red} (V) vs. SCE	$E_{s,t}$ (eV)	ΔG° (eV)	k_q ($M^{-1}S^{-1}$)
a	-1.49	3.01(t)[a]	-0.62	5.65 x 10^9
b	-1.62	3.16(t)[b]	-0.64	5.76 x 10^9
c	-2.10	3.88(s)[c]	-0.88	6.25 x 10^9 [e]
d	-1.30	3.63(s)[d]	-1.43	5.94 x 10^{12} [f]

[a] ref.[70] t = triplet   t = triplet   s = singlet
[b] ref.[71] t = triplet   t = triplet
[c] ref.[72] s = singlet   s = singlet
[d] ref.[73] s = singlet
[e] $I_0/I = 1+k_q\tau_s$ [Ph_4B^-]; τ_s = 8.9 ns for 1-cyanonaphthalene.[73]
[f] $I_0/I = 1+k_q\tau_s$ [Ph_4B^-]; τ_s = 5.1 ps (Φ_f = 0.022 for 2d); τ_f = 0.23 ns for 4-methyl-7-methoxycoumarin in methanol.[74]

Quantum yields for the photodecomposition of tetraphenylborates $R-CH_2-NR'_3{}^+,BPh_4{}^-$ **a-d** (Scheme 15) in CH_3CN (5 x 10^{-3} M) were determined at 366 nm. To ensure complete absorption of the incident light, a quartz cell having a pathlength of 10 cm was used. The quantum yields of photodecomposition Φ_d of **a, b, c** and **d** (R^1 – Scheme 15)in CH_3CN are 0.49, 0.42, 0.20, and 0.07, respectively. In CH_3CN, the ammonium borates are in ionic equilibrium, so that diffusion of the ions is required for electron transfer reaction to take place. Participation of the longer lived triplet states from **a** and **b** accounts for higher values of Φ_d. A shorter lifetime of the singlet state of both **b** and **c** decreases the value of Φ_d. Photodecomposition of borate **d** in PhH (3 x 10^{-4} M) was found to proceed with Φ_d = 0.20. In nonpolar solvents, such as PhH, borate **d** exists as a tight ion pair, thus eliminating the diffusion controlled step and increasing the Φ_d.

Laser flash photolysis of tetraphenylborate **a** in degassed acetonitrile (1 x 10^{-3} M) at 355 nm leads to a transient with absorption maxima at 335 and 530nm (Figure 31). Transient absorption decay at 530nm consists of two components. The lifetime of the faster component is 200ns based on a first order kinetic law. The decay of the slower component is consistent with a second order rate law (k/ε = 1.17 x $10^6 s^{-1}$). Both absorption bands (335nm and 530nm) are quenched by oxygen, however, the time profiles are not the same (Figure 32) indicating that the absorptions occur from two different species. Under the same conditions, a solution of tetrafluoroborate **a** displays a triplet-triplet absorption with maxima at 320 and 550 nm (τ = 13.2μs)[75]. The decay of both suggest a one component system, and the behaviour toward oxygen is also similar. An analogous observation was made with 4-methylbenzophenone, whose triplet absorption spectra is very similar to that of tetrafluoroborate **a**. The triplet state of tetrafluoroborate **a** is found to be quenched by Ph_4B^- Bu_4N^+ (1x10^{-3}M, acetonitrile, k_q = 5.65 x 10^9 $M^{-1}s^{-1}$) with concomitant observation of the absorptions at 335 and 530nm. The 530 nm band shows similar behaviour to that of tetraphenylborate **a** indicating the presence of similar species.

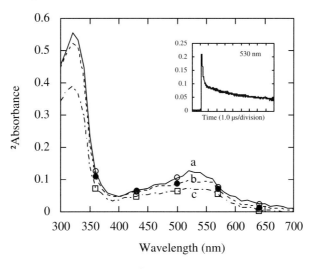

Figure 31

Transient absorption spectra obtained by laser photolysis (355 nm) of tetraphenylborate a in acetonitrile (1 x 10^{-3} M). (a) 300, (b) 590 and (c) 3200ns. Insert: time profile at 530nm

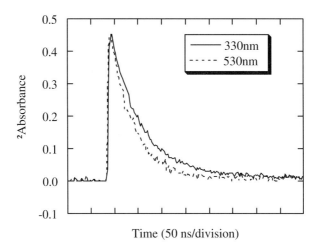

Figure 32

Time profile of transient absorption (normalized) of tetraphenylborate a (Scheme 15, R^1) in acetonitrile in presence of oxygen (20 mins bubbling).

4-(Bromomethyl)benzophenone was studied as a model compound using similar flash photolysis conditions. The only possible radical product which can result following laser flash photolysis of 4-(bromomethyl)benzophenone is p-benzoylbenzyl radical. A transient spectrum was observed with bands at 335 and 530nm. The 530nm band decays with second order kinetics (k/ε = 2.49 x 10^8 s^{-1}) which differ from those observed for the 335nm absorption This assignment is in complete agreement with the results of photolysis product studies of tetraphenylborate **a** wherein a coupled product from the p-benzoylbenzyl radical was isolated. We thus assign the reactive intermediate observed at 530nm upon laser flash photolysis of tetraphenylborate **a** and the mixture (tetrafluoroborate **a** and Ph$_4$B$^-$ Bu$_4$N$^+$) to be the p-benzoylbenzyl radical. The absorption at 335 nm is from residual absorption due to a stable coupling product of the radicals.

The time resolved transient spectrum for tetraphenylborate **b** (1 x 10^{-2}M, acetonitrile) shows (Figure 33) two absorption bands at 380nm and 520nm, respectively. Decay of the 520nm absorption is fitted by second order kinetics (k/ε = 1.75 x 10^6 s^{-1}). Laser flash photolysis of tetrafluoroborate **b** in acetonitrile gave only one absorption at 360nm, which we assigned to the triplet state.[76] This triplet state is quenched by Ph$_4$B$^-$ Bu$_4$N$^+$ with a rate constant k$_q$ = 5.76 x 10^9 M^{-1}s^{-1}. Quenching is accompanied by the appearance of a new absorption at 520nm. The decay behaviour is practically identical to that of tetraphenylborate **b** (520nm band), therefore, we conclude that we are observing the p-acetylbenzyl radical. This assignment is consistent with the results of product studies which show that the products are derived from p-acetylbenzyl.

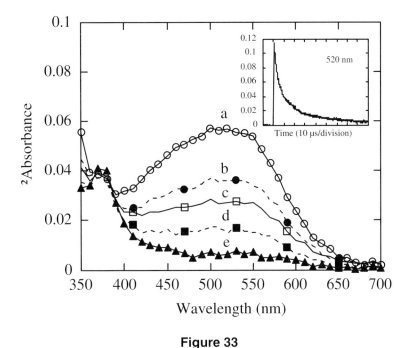

Figure 33

Transient absorption spectra recorded by 355 nm laser flash photolysis of tetraphenylborate b (Scheme 15, R^1) in acetonitrile (1 x 10^{-2} M). (a) 6, (b) 12, (c) 18, (d) 30 and (e) 75μs. Insert: time profile at 520nm.

Figure 34 shows the transient absorption of tetraphenylborate **c** (1 x 10^{-2} M, acetonitrile) with a maximum 380nm, decaying in a bimolecular reaction (k/ε = 1.07 x 10^8 s^{-1}). Triplet absorption was observed at 420nm for 2-methylnaphthalene under identical conditions. Therefore, the band at 380nm is assigned to the 2-naphthylmethyl radical.[77] This assignment was confirmed from an independently generated spectrum of the radical obtained from the photolysis of 2-(bromomethyl)naphthalene.

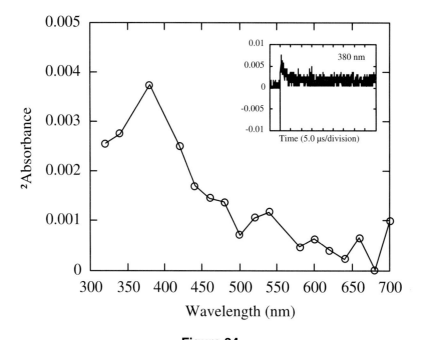

Figure 34

Transient absorption spectra of tetraphenylborate c (Scheme 15, R^1) in acetonitrile (1 x 10^{-2} M) at 2.5μs delay times after the laser pulse. Insert: time profile at 380nm.

Laser flash photolysis (355nm, 7ns, 70mJ) of tetraphenylborate **d** (Scheme 15, R^1) in acetonitrile (5 x 10^{-4}M) gives the transient absorption spectra shown in Figure 35. Following the excitation laser pulse, transient absorption bands at 480 and 600-700nm were observed. The 480nm band had an exponential decay with a lifetime of 1.5μs. The broad absorption band 600-700nm decayed with second order kinetics (k/ε = 1.3 x 10^7 s^{-1}). Both transient species are quenched by addition of oxygen. The short-lived species absorbing at 480nm is assigned to the triplet state of tetraphenylborate **d**. Further confirmation of this assignment was achieved from independent generation of transient spectra by photolysis of 4-methyl-7-methoxycoumarin in benzene (τ = 1.2μs, max = 470nm). Similar results are also obtained with tetrafluoroborate **d** (Scheme 15, R) (1 x 10^{-4}M) under identical experimental conditions, where no oxidizable group is present. The absorption at 480 nm was unaffected by the addition to tetrafluoroborate **d** of Ph_4B^- Bu_4N^+ (acetonitrile, 2 x 10^{-3}M). However, a new broad absorption at 600-700 nm appeared with decay behaviour similar to that observed for tetraphenylborate **d** around 640nm. This broad absorption at 600-700nm is assigned to the radical 7-methoxy-4-methylcoumarin by comparison to spectra generated from 4-bromomethyl-7-methoxycoumarin under similar experimental conditions.

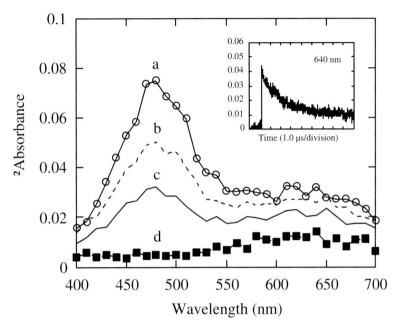

Figure 35

Transient absorption spectra of tetraphenylborate d (Scheme 15, R) in acetonitrile (1 x 10^{-4} M). (a) 0.59, (b) 1.2, (c) 1.8 and (d) 5.6µs. Insert: time profile at 640 nm.

A picosecond transient absorption study of tetraphenylborate **d** in benzene (3 x 10^{-4}M) showed the presence of the 600-700nm absorption forming along with 30ps laser excitation pulse. The presence of radical 4-bromomethyl-7-methoxycoumarin in the reaction system on this time scale indicates a fast rate of C-N bond cleavage in the reduced 7-methoxycoumarin tertiary ammonium cation.

The rate constants for bimolecular quenching of the excited states of tetrafluoroborates **a-d** (Scheme 15, R) by ⁻Ph$_4$B⁻Bu$_4$N⁺ are given in Table 11 along with reduction potentials and excited states energies. The rate constants (k_q) for the quenching of all excited states are high, and approach the diffusion controlled limits. The reduction potential of **a** is lower than that of **b**. However, the triplet state energy of **b** is higher than that of **a** compensating for a reduced reactivity toward bimolecular quenching and providing a nearly constant driving force for forward electron transfer. In spite of its high reduction potential, the reactivity of the singlet-excited state of tetrafluoroborate **c** is higher than that for tetrafluoroborates **a-b**. This difference can be accounted for in terms of the higher excited state energy of the singlet state of tetrafluoroborate **c** compared to the triplet energy of tetrafluoroborates **a-b**. The quenching of tetrafluoroborate **d** exceeds diffusion limited rates due to its low reduction potential as well as its high excited state energy. In each case **a-d**, ΔG, the driving force of electron transfer, is negative indicating a highly exothermic process. Despite the similar reactivities of the singlet and triplet excited states of each of the compounds, it seems noteworthy that the efficiency of amine formation (conversion %) of tetraphenylborate **c** is significantly lower as a result of competitive deactivation of the singlet state.[78] However, the conversion (%) and yield of products in the case of tetraphenylborate **d** are much higher, indicating that structural characteristics are also important.

Spectroscopic evidence suggests that electron transfer takes place from donor (Ph$_4$B$^-$ Bu$_4$N$^+$) to the excited triplet or singlet states depending on the acceptor. As a result of the electron transfer reaction, C-N bond cleavage occurs homolytically in the corresponding reduced tertiary ammonium cations and the radicals p-benzoylbenzyl, p-acetylbenzyl, β-naphthylmethyl, and 7-methoxycoumarin 4-methyl are formed. This bond cleavage pattern was confirmed by direct observation of the radicals by laser flash photolysis. Each radical p-benzoylbenzyl, p-acetylbenzyl, β-naphthylmethyl, and 7-methoxycoumarin 4-methyl was identified by comparison with radicals formed by independent means. All transient absorption spectra have the common decay characteristics of bimolecular reactions (see Figure 33).

Additional support for the formation of p-benzoylbenzyl, p-acetylbenzyl, β-naphthylmethyl, and 7-methoxycoumarin 4-methyl radicals comes from isolation of the photolysis products. The photolysis of tetraphenylborates **b** in acetonitrile at 350 nm in the presence of oxygen results in formation of p,p'-(1,2-acetylphenyl)ethane or p-acetyltoluene as the major and minor products, respectively. This product pattern is the same for each of the other compounds. The major product is the dimer of the arylmethyl radical and the minor the corresponding toluene. In the absence of oxygen, a similar distribution of products was observed, though yields were higher. Formation of these products indicate the corresponding radicals are present in the reaction system. The minor products, in all cases, are formed via hydrogen abstraction from CH$_3$CN, a known, though inefficient, hydrogen donor.[79] This was confirmed by using deuterated solvent in the photolysis, and thereby observing a molecular ion peak, incorporating deuterium in the mass spectra for the toluenes formed from photolysis of each of the tetraphenylborates **a-d**. That fact that radical coupling products are major components of the product mix indicates relatively long lifetimes for the radical intermediates from which they are formed.

The heterolytic C-N bond scission was completely rulled out because of the absence of products that could be attrituted to ionic intermediates.[80] The acetamide would be the expected product in acetonitrile, for example. The source of biphenyl in the product mixture is not clear. Trapping experiments employing TEMPO (2,2,6,6-tetramethyl-1-piperidinyloxy) failed to detect phenyl radical though in the absence of oxygen the yield of biphenyl was significantly lower. It seems reasonable to conclude that biphenyl[81] is formed as a result of an intramolecular coupling of two phenyl groups from tetraphenylborate, a result consistent with other experiments.[82]

Referring to Scheme 16, (page 172), on the basis of the above, a radical pair mechanism is proposed for the photocleavage of tetraphenylborate **b** to free amine. The same mechanism pertains for tetraphenylborates **a** and **c-d**. Radical p-acetylbenzyl is formed from the triplet radical pair generated after the electron transfer via pathways a and b, though we were unable to detect transient absorbance corresponding to the triplet radical pair formed on the picosecond timescale.

$$H_3C-\overset{O^-}{\underset{\bullet}{C}}-\langle\rangle-CH_2-N(Bu)_3^+, \quad \bullet BPh_4$$

Triplet radical pair

Neither could the transient be detected as suggested in pathway **a** using picosecond flash

$$H_3C-\overset{O^-}{C}=\langle\rangle=CH_2$$

photolysis. An absorption growth for radical is observed during the picosecond laser of

[structure: 4-methyl-CH_2• substituted chromen-2-one with CH_3O group]

tetraphenylborate **d** pulse, indicating a unimolecular formation process. The observation favours the direct formation of the radical after C-N bond homolysis, though it does not disprove the presence or absence of other intermediates. The bond breaking process, leading to the formation of radicals in each intramolecular electron transfer case, is coupled to the primary electron transfer process and occurs on the subpicosecond time scale. This important observation precludes back electron transfer to the tetraphenylboranyl radical which would probably be impossible because of C-N bond cleavage even though its lifetime is 45ps.[83] Back electron transfer to regenerate starting material is spin forbidden in triplet radical pairs, and in order for back electron transfer to be a probable process, the triplet radical pair must undergo intersystem crossing to the singlet state which would subsequently collapse to the ground state. Indeed, product yields reveal that back electron transfer is a minor factor for the triplet reactions of tetraphenylborates (**a-b**) where the efficiency of amine formation is only a bit lower than it is for dissociable boronyl radicals.

Back electron transfer may be significant for the photoreaction of tetraphenyl borate **c**, in which the singlet radical pair is formed.[84] This assumption is consistent with the quantum yield for disappearance tetraphenylborate **c** which is smaller than the others. The overall explanation for the lower efficiency of amine generation is the low molar extinction coefficient at 350nm for tetraphenyl borate **c** in that it shows only a tail absorption beyond 325nm.

Structural features in tetraphenylborate **d** and its efficient photocleavage invite additional comments. Although electron transfer occurs in the excited singlet state in tetraphenylborate **d**, the yield of products and rate of conversion to amine suggest a relatively minor role for back electron transfer and the importance of nuclear vibrational motion in electron transfer process. Tetraphenylborate **d** having an electron donating methoxy group in the 7-position and an electron withdrawing substituent (admittedly a marginal one - methyl) in the 4-position, provides positional push-pull substituents, and is the main reason for ultrafast electron transfer. The drastic change in reaction rate as a function of the push-pull substituents on the coumarin has been previously observed[85] wherein amino and trifluoromethyl substituents were at 7 and 4 position respectively. Enhanced reactivity was rationalized in terms of the static and dynamic roles of the rotational motion of the amino group. A similar effect of the amino group on the rate of electron transfer is often discussed in the studies of twisted intramolecular charge transfer (TICT), such as in the case of N,N-dimethylaminobenzonitrile.[86]

In the present system, rotational motion of the 7-methoxy group plays an important role. Change in the vibrational character of the methoxy group by substitution may cause a change in the electron transfer rate. A change in electron transfer dynamics effected by a change in the vibrational parameters has been explained in terms of the extended Sumi-Marcus model.[87] However, we consider that the presence of nonreversible electron transfer is due both to electronic effects as well as vibrational motion. The presence of an electron withdrawing substituent in the 4 position and an electron donating substituent in the 7 position of coumarin allows for a large value of the electron transfer rate constant. This ultrafast reaction takes place from a nonreversible state successfully competing with back electron transfer.

v. Borates as Electron Donors

Subsequent to the demise of Mead Imaging, Ciba-Geigy took a license on a number of Imaging's patents from the parent Mead Corporation. The specific interest was in the development of acid stable borates. This has resulted in patent applications in Great Britain, Europe, and presumably the US dated November, 1995. The compounds Ciba claims of interest are borates of the general structures shown below:

$$\begin{bmatrix} & R_1 & \\ & | & \\ R_4 - & B - R_2 \\ & | & \\ & R_3 & \end{bmatrix}^{-} G^{+} \qquad \begin{matrix} & R_2 & \\ & | & \\ R_4 - & B - R_{1a} - E^{+} \\ & | & \\ & R_3 & \end{matrix}$$

General structures of commercially interesting borates

Ciba's patent applications disclose a large number of potential applications. As in the case of the Irgacures, the systems appear to make good use of the sterically crowded mesityl group in achieving Ciba's objectives. One such example from which a number of borate salts could be synthesized was trimesityl borane.

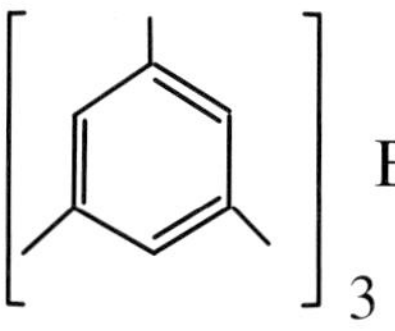

Trimethyl borane

Borates are made from this and other similar boranes by the standard route; addition of the appropriate lithium reagent or Grignard reagent followed by exchange of the metal gegen ion with the desired ammonium ion or other cation.

The borates so synthesized could be converted to cyanine salts in a manner identical to the procedures reported by Gottschalk and others at Mead[88] to produce photoreactive cyanine borates that were subsequently tested in the photopolymerization of various acrylates.[89]

Several other gegen ions in addition to cyanine were incorporated with various borates. These include Safranin O and triphenylsulphonium.[90] Other groups of compounds that appear to be important are various fluorine substituted aromatics, rather similar to the previously reported Rhone Poulenc tetraperfluorophenyl borate systems. Clearly a substantially larger number of fluorine substituents on the aromatic nucleus would decrease the electron density of boron in the borate and rended the compound less susceptible to acid. Increasing the steric compression around the borate would have the same effect. In fact another patent in the series[91] concentrates on the m-fluorophenyl as the aromatic group in the borate and describes a number of compositions in which it can be used. The gegen ions of interest in this application include cyanine, iodonium, pyrylium, methylene blue, safranin O, sulphonium, cresyl violet, brilliant green, pyronine GY, rhodamine B, eosin, camphor quinone, ITX, various coumarins and several hexaarylbisimidazoles. Ciba patents also disclose polyborates such as those that might be synthesized from polymesityldiphenylborane and 1,1-bis(dimesitylboryl)ferrocene.[92] They report the use of these systems, alone and in combination with the dyes above and others, as photoinitiators.

vi. Iodonium/borates with Kesheng Feng[*]

Iodonium salts are used commercially to initiate cationic polymerization (Chapter 1). Additionally they improve polymerization rates in radical polymerizations when used as coinitiators with dye/amine or dye/borate initiator systems.[93] In the dye/borate/iodonium system, either oxidation of the borate anion via electron transfer to the excited state of the dye or reduction of the iodonium cation by electron transfer from the excited state of the dye to the onium salt, leads to the decomposition of both anion and cation, generating butyl and phenyl radicals respectively. Both serve to initiate polymerization. Since systems prepared in the absence of absorbing dye remained excellent radical initiators, this prompted a search for stable electron acceptor/donor complexes of iodonium/borates (see Scheme 14, page 155). The calculated $\Delta G°$ of electron transfer for the pair is sufficiently negative to anticipate a ground state reaction, however such complexes can be stabilized as solids, isolated and characterized. When dissolved in formulations, such formulations remain stable as long as inhibitors are present. When the inhibitors are removed (as they would be by initial photoreactions of the iodonium borates), iodonium borate salts become reactive UV photoinitiators in their own right.

A particular question of mechanism becomes, "What role does the iodonium cation play?" "Does electron transfer directly from the alkyltriphenylborate anion to the iodonium cation lead to the simultaneous formation of alkyl and aryl radicals?" "Do both radicals initiate the polymerization?"

To study these questions, symmetrical iodonium salts, such as di-(p-methylphenyl)iodonium bromide, were prepared by reaction of toluene with potassium iodate and sulphuric acid, and then metathesized with a solution of sodium bromide.[94]

---

[*] This elegant work identifies the initiating radicals in iodonium borate photolysis. It is the work of Kesheng Feng.

Asymmetric iodonium bromides were prepared by direct coupling of an aromatic compound with iodobenzene biacetate in the presence of sulfuric acid, and then metathesized with a solution of sodium bromide. Iodonium bromide so produced is reacted with tetramethylammonium butyltriphenylborate or sodium tetraphenylborate, to form iodonium butyltriphenylborate salts or iodonium tetraphenylborate salts in high yield (Scheme 18). The individual borate/iodonium complexes are indicated by acronyms: diphenyl iodonium triphenyl butyl borate (**DPIBB**), p,p'dimethylphenyl iodonium triphenyl butyl borate (**DMPIBB**), p,p'dimethylphenyl iodonium tetraphenyl borate (**DMPITB**), p-butyloxyphenyl phenyl iodonium triphenyl butyl borate (**BPPIBB**), p-octyloxyphenyl phenyl iodonium triphenyl butyl borate (**OPPIBB**) and p-octyloxyphenyl phenyl iodonium tetraphenyl borate (**OPPITB**). It is important for the reader to note that iodonium salts differing only in the borate were synthesized and studied. Important mechanistic information resulted from studies of identical iodonium cations with triphenylbutylborate and tetraphenylborate anions respectively, Scheme 18.

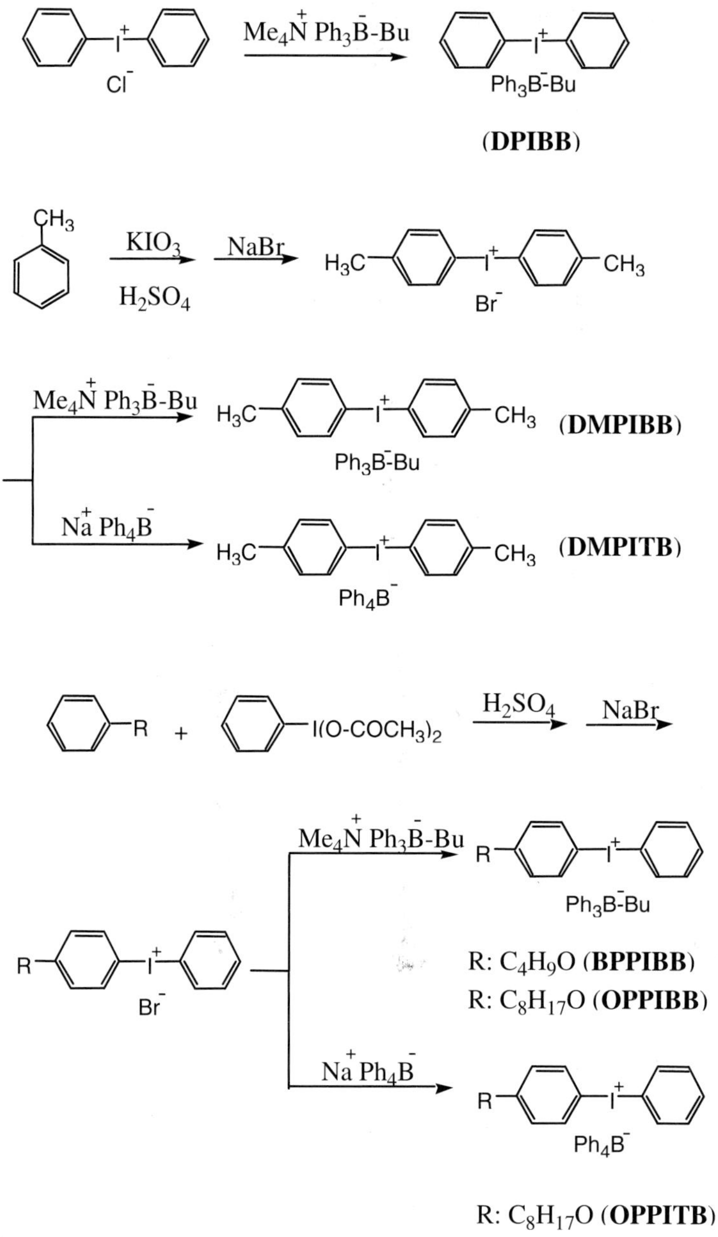

Scheme 18
Synthesis of iodonium borate salts

vii. Photoinitiating activities - Iodonium/borates

A key issue in photoinitiator performance is the solubility of the initiator in prepolymer solution. Usually, compounds which are asymmetric are more soluble than are the related symmetrical analogs. The introduction of long chain alkyl groups into a molecule is often used to increase the solubility of an initiator system into a monomer mixture. The photopolymerization of a standard acrylic resin containing different iodonium borate salts was carried out. The solubility of initiators in resin was as follows: p-octoxyphenylphenyliodonium butyltriphenylborate (**OPPIBB**) > p-octoxyphenyl-phenyliodonium tetraphenylborate (**OPPITB**) > p-butoxyphenylphenyliodonium butyltriphenylborate (**BPPIBB**) > diphenyliodonium butyltriphenylborate (**DPIBB**) and di(p-methylphenyl)iodonium butyltriphenylborate (**DMPIBB**) > di(p-methylphenyl)-iodonium tetraphenylborate (**DMPITB**). Relative initiator activity tests were carried out at the concentration of 5×10^{-3} M. Even at this low concentration, iodonium butyltriphenylborate salts have a high efficiency in initiating polymerization under UV light, and this was higher than equimolar mixtures made up from borate/iodonium salt pairs.

A mixture of p-octoxyphenylphenyliodonim hexafluroantimonate (**OPPI**), butyrylcholine butyltriphenylborate (**BB**), and the mixture of **OPPI** with **BB** were tested in control experiments. All curing experiments were performed using same monomer, irradiation source and concentration of initiator. Conversion of the acrylic double bonds was measured by FT-IR.[95]

Iodonium butyltriphenylborate salts proved to be efficient photoinitiators in resin under UV light, and double bond conversion was nearly complete within 50 seconds. Asymmetric iodonium butyltriphenylborate salts, such as **BPPIBB** and **OPPIBB**, were more active than the symmetric iodonium borate salts **DPIBB** and **DMPIBB**, and double bond conversion was more than 80% after the formulated resin was irradiated 30 seconds. Curing was also fast with a mixture of **OPPI** and **BB**, but slower than that with the isolated complexes. The results are shown in Figure 36.

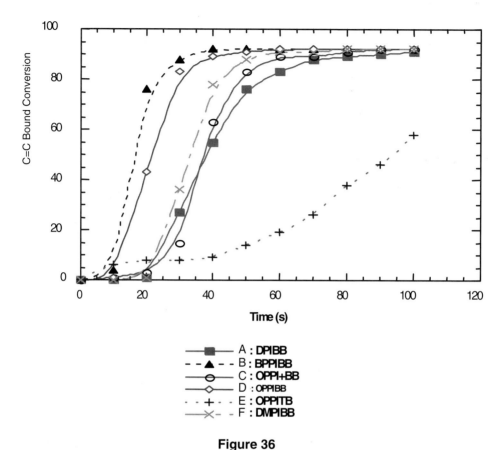

Figure 36

Photopolymerisation of SGL-1 resin irradiated by UV light (iodonium borate as photoinitiators, 0.005M), SGL1 is a standard acrylate mixture

At concentrations of 5×10^{-4}M in acetonitrile the UV absorption spectra of the iodonium borate salts showed a tail absorption above 400nm, even beyond 430nm for **OPPIBB**. In ethyl acetate where the pairing of the ions is likely to be larger, longer wavelength absorption was even more pronounced. A mixture of **OPPI** and **BB** also showed a tail absorption, but the tail absorption of **OPPIBB** was greater and shifted further to the red than the mixture. Thus both **OPPIBB** and **BPPIBB** were tested using visible radiation. At 0.01 M, double bond conversion was almost 90% for both the **OPPIBB** and **BPPIBB** systems after 60 seconds using a 700mw/sq.cm dental lamp. When 5,7-diiodo-3-butoxy-6-fluorone was included in the mixture, double bond conversion reached 91% after 5 seconds. Therefore, 5,7-diiodo-3-butoxy-6-fluorone rendered iodonium butyltriphenylborate a much more efficient initiator for visible light photopolymerization.

Products from the photolysis reaction in the absence of monomer provide clues as to how iodonium/borate complexes initiate polymerization. The photolysis of butyltriphenylborate salts in the presence of electron-acceptors is known to generate biphenyl and butyl radicals.

$$Ar_3B\text{-}Bu^- + \text{electron acceptor} \xrightarrow{h\nu} Bu\bullet + Ar\text{-}Ar$$

Though the photolysis of tetraphenylborate in the presence of electron acceptors also produces biphenyl as a major product,

$$Ph_4B^- + \text{electron acceptor} \xrightarrow{h\nu} Ph\text{-}Ph$$

this is the result of an intramolecular rearrangement rather than from the combination of phenyl radicals.[96] The evidence is that the photolysis of sodium or lithium tetra-p-tolylborate forms only p-bitolyl. The carbon atoms originally bonded to boron in the sodium or lithium tetraaryl borate salt are linked in the product, biaryl.[97]

When an aqueous solution of equal parts sodium tetraphenylborate and sodium tetraphenylborate-d20 was irradiated, mass spectrometric analysis showed the biphenyl fraction consisted only of $C_{12}H_{10}$, $C_{12}D_{10}$ and $C_{12}D_9H$. There was no evidence of $C_{12}D_5H_5$ indicating that the formation of biphenyl was essentially intramolecular.[98]

The photolysis products of selected iodonium borate salts in a model monomer, methyl acrylate, demonstrate that the decomposition processes are complex. For example, the decomposition of **OPPIBB** is very fast, and nearly 50% disappears after 10 min. irradiation according to the integration of the peaks at 4.1ppm in the nmr spectrum. Similar results were obtained for the photolysis of **OPPITB**, but the decomposition of **OPPITB** was slower than that of **OPPIBB** as shown in Figure 37. The decomposition of a mixture of **OPPI** and **BB** is not as fast as is that of **OPPIBB**. In control experiments, the decomposition of **OPPI** under similar conditions is very slow while no decomposition occurred at all with **BB** after 60 min. irradiation. These results compare well to the rate of photopolymerization when the compounds were used as photoinitiators.

If the iodonium borate salts form tight-ion pairs in solvent, electron transfer should occur directly from anion to cation within the solvent cage. With a mixture of **OPPI** and **BB**, since there are two anions and two cations in the system, one would expect the common ion effect to impact the decomposition rate. Thus both the decomposition and the polymerization the initiator induced in acrylate monomers was faster with the isolated complex **OPPIBB** than was that with the mixed 2 salt initiator system. In the isolated complex, the only ion pair is that from which a productive electron transfer can be expected. In the mixture of borate/iodonium salt, each also carries a non-reactive gegen ion.

From the electrochemistry, tetraphenylborate anion is not as easy to be oxidize as the butyltriphenylborate anion. This accounts for the observation that the decomposition of **OPPITB** is slower than is that of **OPPIBB**. A study of the reactions of iodonium borate with methyl methacrylate (see below) indicates that the mechanism of photolysis of iodonium butyltriphenylborate salt and iodonium tetraphenylborate also differs.

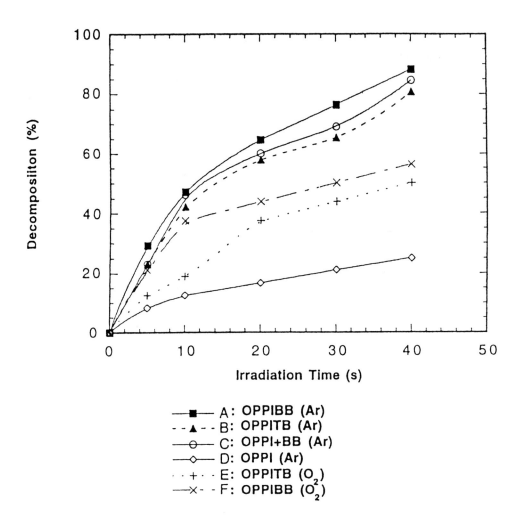

Figure 37
Photolysis of iodonium borate salts

When photolysis was carried out after purging with oxygen, decompositions were slower than those of identical systems which had been purged with argon. The effect of oxygen is unclear. Though the result might suggest the intervention of a triplet state, it is more likely to be the result of oxygen quenching of radical intermediates which slows further chain decompositions. GC-MS indicated that aryl iodide and biphenyl were the major products from the photolysis of all the iodonium borate salts.

viii. Photochemical Reactions of Iodonium Borate Salts with Methyl Methacrylate

There is a significant difference in the photopolymerization rates induced by iodonium butyltriphenylborate and the iodonium tetraphenylborate systems. In order to study the initiating steps in these polymerizations, the products from the photochemical reactions of each of the iodonium borates with methyl methacrylate were isolated and identified, Scheme 19.

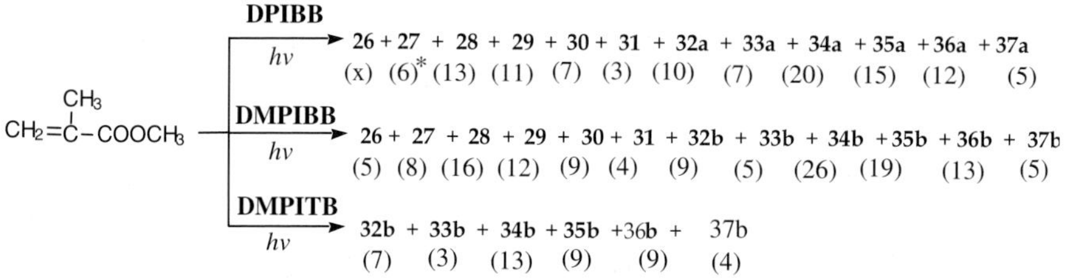

$$CH_2=\underset{CH_3}{\underset{|}{C}}-COOCH_3 \quad \begin{array}{l} \xrightarrow[hv]{DPIBB} \begin{array}{l} 26 + 27 + 28 + 29 + 30 + 31 + 32a + 33a + 34a + 35a + 36a + 37a \\ (x)\ (6)^*\ (13)\ (11)\ (7)\ (3)\ (10)\quad (7)\quad (20)\quad (15)\quad (12)\quad (5) \end{array} \\ \xrightarrow[hv]{DMPIBB} \begin{array}{l} 26 + 27 + 28 + 29 + 30 + 31 + 32b + 33b + 34b + 35b + 36b + 37b \\ (5)\ (8)\ (16)\ (12)\ (9)\ (4)\quad (9)\quad (5)\quad (26)\quad (19)\quad (13)\quad (5) \end{array} \\ \xrightarrow[hv]{DMPITB} \begin{array}{l} 32b + 33b + 34b + 35b + 36b + \quad 37b \\ (7)\quad (3)\quad (13)\quad (9)\quad (9)\quad\quad (4) \end{array} \end{array}$$

* Relative ratios (%) to biphenyl (100%).

The relative ratio of iodobenzene was 140% in the case of **DPIBB**. The relative ratios of iodotoluene were 250% in the case of **DMPIBB** and 69% in the case of **DMPITB**, respectively.

x Covered by the peak of butylbenzene.

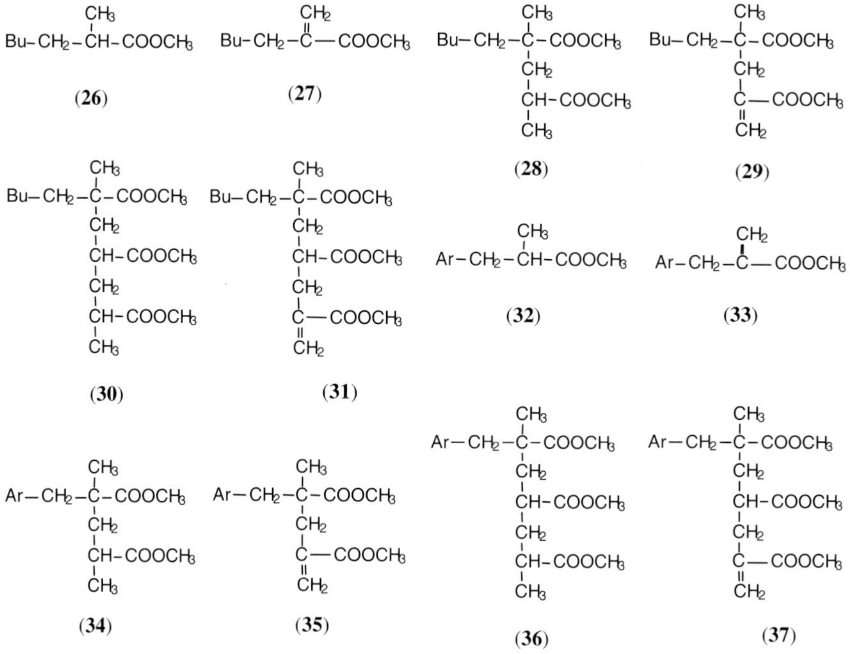

(26) Bu–CH$_2$–CH(CH$_3$)–COOCH$_3$

(27) Bu–CH$_2$–C(=CH$_2$)–COOCH$_3$

(28) Bu–CH$_2$–C(CH$_3$)(COOCH$_3$)–CH$_2$–CH(COOCH$_3$)–CH$_3$

(29) Bu–CH$_2$–C(CH$_3$)(COOCH$_3$)–CH$_2$–C(=CH$_2$)–COOCH$_3$

(30) Bu–CH$_2$–C(CH$_3$)(COOCH$_3$)–CH$_2$–CH(COOCH$_3$)–CH$_2$–CH(COOCH$_3$)–CH$_3$

(31) Bu–CH$_2$–C(CH$_3$)(COOCH$_3$)–CH$_2$–CH(COOCH$_3$)–CH$_2$–C(=CH$_2$)–COOCH$_3$

(32) Ar–CH$_2$–CH(CH$_3$)–COOCH$_3$

(33) Ar–CH$_2$–C(=CH$_2$)–COOCH$_3$

(34) Ar–CH$_2$–C(CH$_3$)(COOCH$_3$)–CH$_2$–CH(COOCH$_3$)–CH$_3$

(35) Ar–CH$_2$–C(CH$_3$)(COOCH$_3$)–CH$_2$–C(=CH$_2$)–COOCH$_3$

(36) Ar–CH$_2$–C(CH$_3$)(COOCH$_3$)–CH$_2$–CH(COOCH$_3$)–CH$_2$–CH(COOCH$_3$)–CH$_3$

(37) Ar–CH$_2$–C(CH$_3$)(COOCH$_3$)–CH$_2$–CH(COOCH$_3$)–CH$_2$–C(=CH$_2$)–COOCH$_3$

32a -37a : Ar = Ph;   32b -37b : Ar = CH$_3$-Ph.

Scheme 19
Addition products from photoreactions

The products which form, result from the addition of butyl radicals and phenyl radicals, respectively, to acrylate. Radical addition, when followed by chain propagation and chain termination, produces the observed products.

Consider the case of **DPIBB**. When irradiated the butyl or phenyl radicals formed from the partners are observed as addition products to the double bond of methyl methacrylate to form radical **38** (Scheme 20) which then abstracts a hydrogen to form product **26** or **32a**. Radical **38** could continue to add to the double bond of a second methyl methacrylate to form radical **40** which could also abstract a hydrogen to form product **28** or **34a**. Radical **40** repeated the reaction observed for radical **38** forming product **30** or **36a**. If methyl methacrylate loses hydrogen from the α methyl group, stable radical **39** would form. This radical can couple with a butyl or phenyl radical to form product **27** or **33a**, couple with radical **38** to form **29** or **35a**, or couple with radical **40** to form product **31** or **37a**. The relative ratios formed by butyl radical addition were not significantly different from those formed by phenyl radical addition (Scheme 20), indicating that both butyl and phenyl radicals each efficiently initiate the polymerization of acrylates.[99]

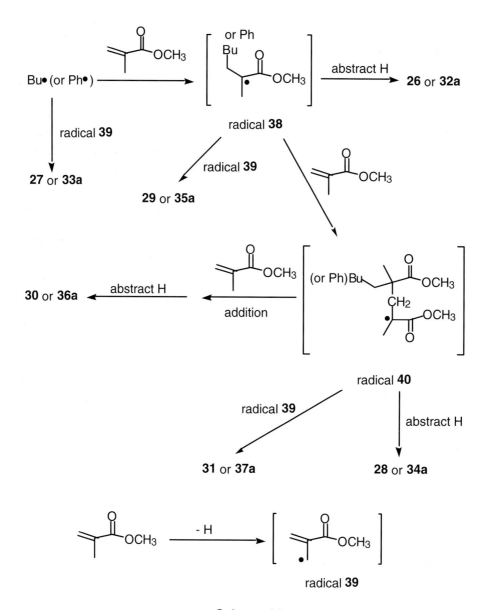

Scheme 20
The mechanism of product formation

When **DMPIBB** was used in the same photoreaction, iodotoluene and biphenyl were the major products. A series of butyl radical addition products **26** to **31** similar to the case of **DPIBB** were also found. No detectable phenyl radical addition products were found. However, a series of products formed which were obtained by the addition of a p-methylphenyl radical to the double bond of methyl methacrylate. These products are identified **32b** to **37b** to distinguish them from the phenyl case. The quantities of methylphenyl radical addition products were also not significantly different from those obtained by butyl radical addition (Scheme 19). *The experiment reveals that the butyl radical comes from the borate anion, while the phenyl or methylphenyl radical derives exclusively from the iodonium cation.*

When a solution of **DMPITB** and methyl methacrylate was irradiated, GC-MS indicated that the only radical addition products obtained were those of the p-methylphenyl radical, **32b** to **37b**. The amount of phenyl radical product detected was small in comparison. The other major products were iodotoluene and biphenyl. The result indicates the only aryl radical released is by the iodonium cation. This result is the same for both the iodonium butyltriphenylborate and the iodonium tetraphenylborate. In accord with Schuster's results the tetraphenylborate anion releases few, if any, phenyl radicals.

These results indicate that the mechanism of photolysis of the iodonium butyltriphenylborate salts differs from that of the iodonium tetraphenylborate salts, and that this is the principle reason for the difference in polymerization rates between the two systems. Iodonium butyltriphenylborate salts, when irradiated, undergo electron transfer from borate anion to iodonium cation causing the decomposition of both partners to produce butyl radicals and aryl radicals respectively. Any biphenyl formed *from the borate* is the result of non-radical processes. In the case of the iodonium tetraphenylborate salts, back electron transfer is faster than decomposition of the boranyl radical. In this case, the decomposition of the iodonium cation occurs at a rate similar to that of other iodonium salts forming aryl radicals and aryl iodide. The tetraphenylborate anion produces biphenyl by the known intramolecular process and the rate of radical production from this partner is limited.

7. Oxidation/Reduction Reactions with Visible Initiator Systems: The Case of Organic Dyes

Among important oxidation/reduction reactions leading to radicals are those of organic dyes with partner oxidants or reductants. Such reactions are, in fact, the major route for forming polymers in photoinitiated reactions wherein the light source causing reaction emits mainly visible radiation.

The history of visible initiators for vinyl polymerization has been reviewed elsewhere[100] and needs little further discussion here. The first report of dyes that initiated polymerization was Oster's[101] who described three dye series, the acridines, the xanthenes and the thiazenes, which were photoreducible. Coinitiators were required in Oster's formulations and these included tertiary amines such as those we have mentioned previously or ascorbic acid (Vitamin C).[102] The dyes mentioned in Oster's work, for example thiazenes such as methylene blue, were well known to organic chemists

Methylene Blue

of Oster's day and used both in commerce and research. A major problem with them was that they were 'fugitive,' i.e., they bleached. However, bleaching in a reaction which releases radicals and initiates polymerization is a major advantage for deep cures, which represent a major use of visible light initiator systems in photopolymerization.

Among the xanthenes of Oster's reports were Eosin, Rose Bengal and Erythrosin.

X = Br, Y = H; Eosin
X = I, Y = H; Erythrosin
X = I; Y = Cl; Rose Bengal

Various xanthenes

These dyes bleach in the presence of reducing agents, though not as rapidly as methylene blue. The bleached products from Rose Bengal have been carefully delineated, both in the presence of reducing agents[103] and upon irradiation in polar solvents, in the absence of reducing agent with 'white light'.[104]

A common feature of most organic dyes of these various older series is the presence of a quinomethine or azoquinomethine unit. These functions are somewhat more easily reduced than the parent quinone.[105]

Thus Phillips and Read isolated such products from the reduction of Eosin with tribenzylamine.[106] Zakrzewski isolated products from Rose Bengal which were those that resulted from the addition of two hydrogen atoms, or a hydrogen atom and an another group, across the quinomethine unit.

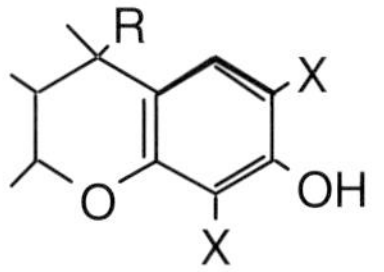

R = H; alkyl

Result of addition across quinomethine group

The quinomethine unit, which deepens the conjugation of the aromatic system, is removed in the reduction process producing a product that is less colored than the starting material. The accepted mechanism of reduction of the xanthenes involves single electron transfer[107] to the triplet from a proximate electron donor.[108]

Reoxidation of the bleached product may regenerate the quinomethine causing a colour reversion of bleached materials products. Several functionalities lessen the rate of reoxidation. For example bleached products with R = alkyl rather than hydrogen (above) and salts of phenols, other than the protonated phenol, are harder to oxidize.[109]

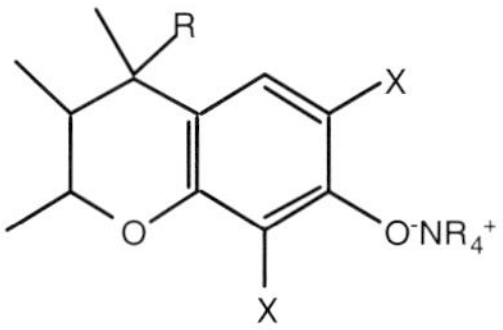

Tertiary amine salt of phenolic hydroxy group of quinomethine

The mechanism of the reduction of the Rose Bengal derivative, 6-acetoxy-21H, "RBAX"

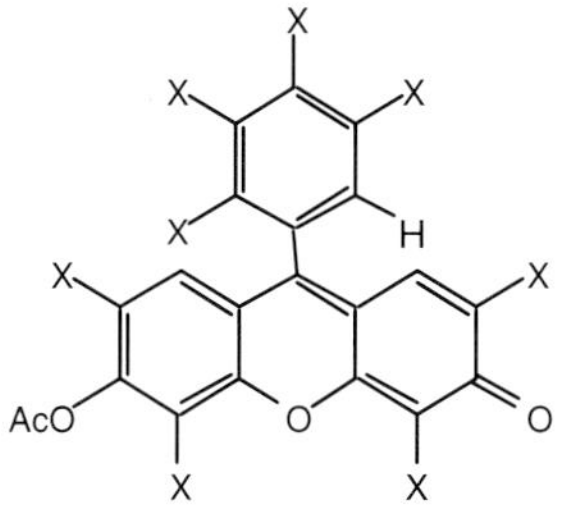

RBAX

was shown by Valdes-Aguilera. The radical anion of RBAX loses acetyl about 10% of the time generating the dye from which RBAX was originally synthesized. This makes the reaction easy to follow. Acetyl, being an initiator

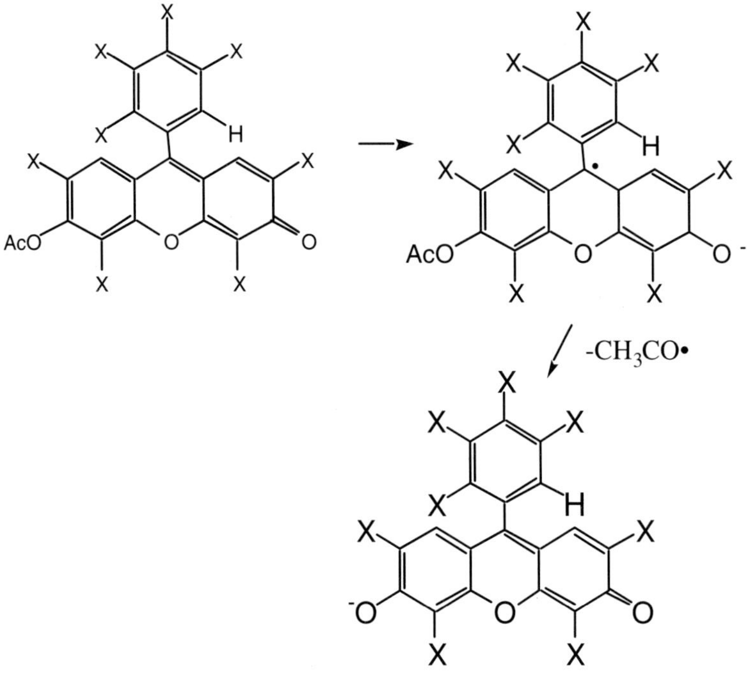

Scheme 21
Reaction of RBAX

in its own right, also facilitates polymerization when RBAX is used as an initiator.

The overall reaction mechanism of reduction of Rose Bengal and its derivatives has been assumed based on steady state kinetic arguments. With the fluorones, however, the spectroscopic mechanism has been clearly delineated.

Fluorones were originally synthesized by Shi[110] and shown to be effective catalysts for polymerization in the presence of electron donors.[111] Many fluorones have now been reported and substantial evidence of their use as polymerization initiators indicated.[112]

The fluorone series consists of both polar, ionic compounds and the less polar fluorone ethers. Functionality (**X**) about the aromatic nucleus, is like that in the xanthones. The functional groups can be halogens, hydrogen or alkyl depending on the spectroscopic properties required of the dye, the solubility characteristics desired, and subsequent chemical functionalizations that may be of interest. The group, **W**, consists of electron withdrawer groups which have a substantial effect on the absorption maximum of the dye system. Substituting CN for H (**W**) shifts the absorption maximum almost 100 nm to the red.

General formula for fluorones

X = H, halogen, Alkyl; W = H, CN

Also important are compounds substituted at C-6 with ether substituents of which 2,4-diiodo-6-butoxy-3-fluorone series is among the most important.

General formula for fluorone ethers

The absorption spectrum of DIBF is shown in Figure 33. These compounds are more soluble in monomer than their charged analogs, though they have many other properties which are the same.

Most of the known fluorones, with their spectroscopic properties, are reported in Table 12.

Spectral characteristics of these compounds in Table 12

Table 12 The Photo-properties of dyes

T	U	V	W	X	Y	Z	λmax (nm)	ε	λfl (nm)	τfl (ns)	Φfl	λph^{77k} (nm)	τph^{rt} (ms)	pKa	Eox	Ered	dye
O	H	H	H	H	H	OH	504	24700	513	3.58	0.95			5.97	1.04	-0.95	1
O	H	H	CN	H	H	OH	548	24700	608	1.56	0.38						2
O	Br	Br	H	Br	Br	OH	530	39300	539	2.33	0.52		140	3.29	1.09	-0.95	3
O	Br	Br	CN	Br	Br	OH	576	24500	638	1.45	0.20					-0.53	4
							594	51400									
O	I	I	H	I	I	OH	536	91200	548	0.65	0.13	690	100	4.08	1.34	-0.99	5
O	I	I	CN	I	I	OH	638	80000	654	0.69	0.02					-0.57	6
							586	35000									
O	I	I	H	COC7H15	I	OH	538	84800									7
O	I	I	CN	COC7H15	I	OH	636	17100									8
							590	9900									
O	I	I	H	C8H17	I	OH	534	92600									9
O	I	I	CN	C8H17	I	OH	630	40100									10
							580	23300									
O	I	I	H	H	H	OMe	470	23500	544		0.03(?)	676				-0.72	11
									568								
O	I	I	H	H	H	OEt	470	21100			0.058						12
O	I	I	H	H	H	OBu	470	30200			0.020					-0.88	13
O	I	I	H	H	H	OOc	472	31600			0.024						14
O	I	I	H	H	H	OC4H9	470	22300			0.015						15
O	I	I	H	H	H	OC5H11	470	25500			0.010						16
O	I	I	H	H	H	OC6H13	472	26600			0.015						17
O	I	I	H	H	H	OC8H17	472										18
O	t-Bu	H	H	t-Bu	H	OH	518	101000									19
O	t-Bu	H	CN	t-Bu	H	OH	614	47400									20
							564	23220									
O	t-Bu	I	H	t-Bu	I	OH	532	90800									21
O	t-Bu	I	CN	t-Bu	I	OH	636	68300									22
							582	33600									
O	H	H	H	C3H6CO2H	H	OH	598	74000			0.44						23
S	H	H	H	H	H	OH	524		562		0.33	642					24
O	I	I	CN	H	H	OBu	524										25

Table 12: continued

Photo-properties of dyes

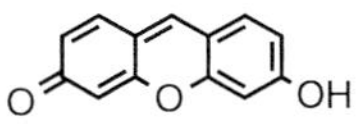

1 a)

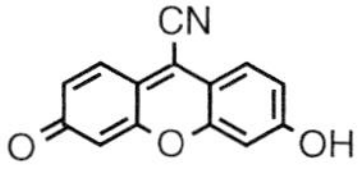

2 a)

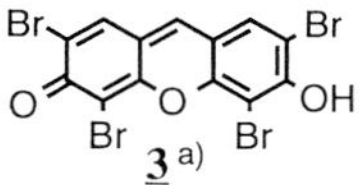

3 a)

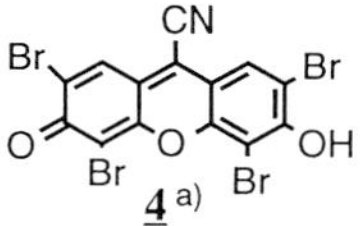

4 a)

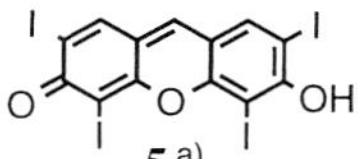

5 a)

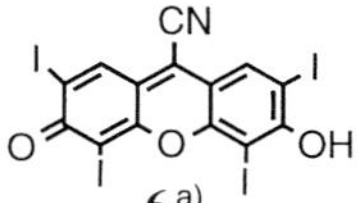

6 a)

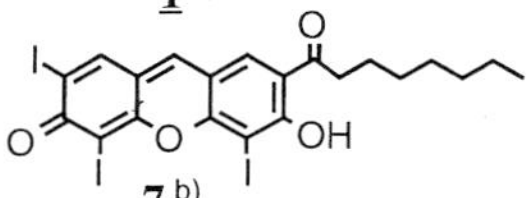

7 b)

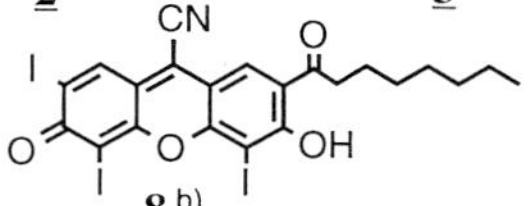

8 b)

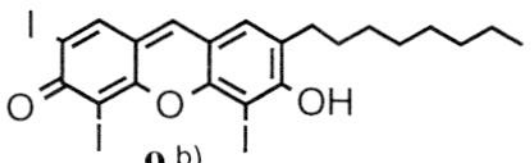

9 b)

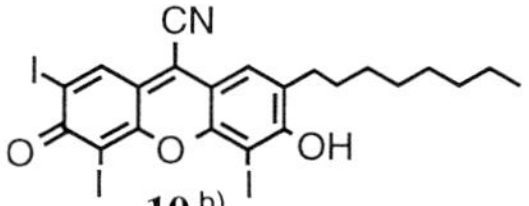

10 b)

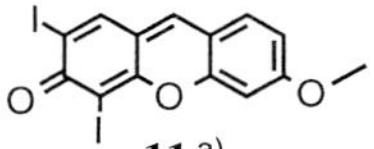

11 a)

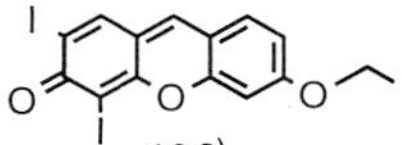

12 a)

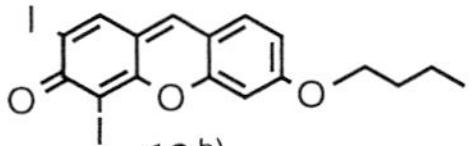

13 b)

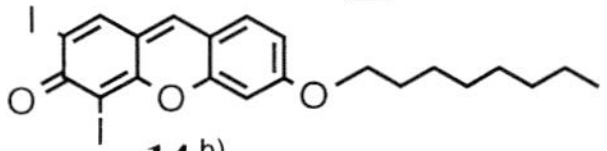

14 b)

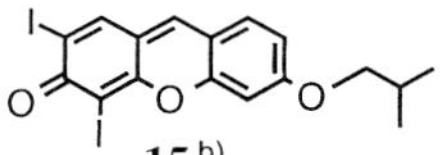

15 b)

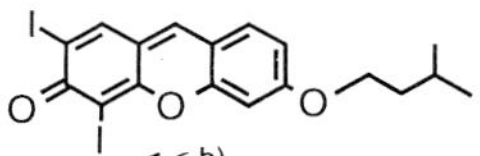

16 b)

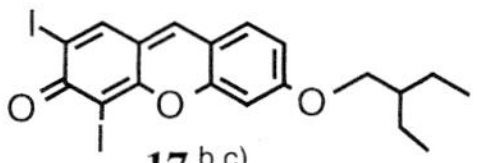

17 b,c)

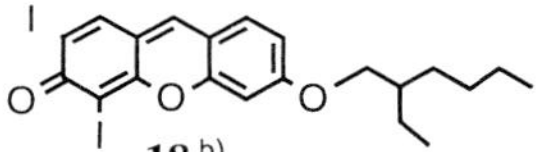

18 b)

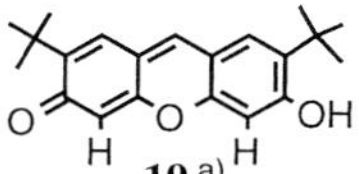

19 a)

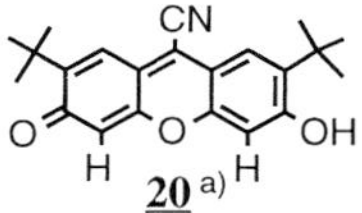

20 a)

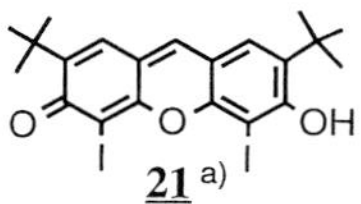

21 a)

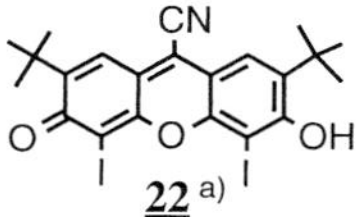

22 a)

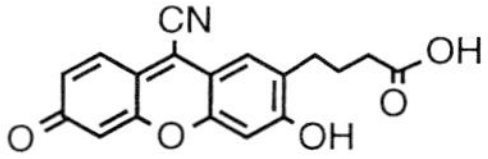

23 d)

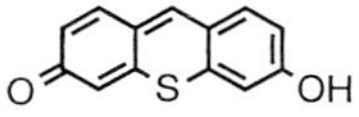

24 e)

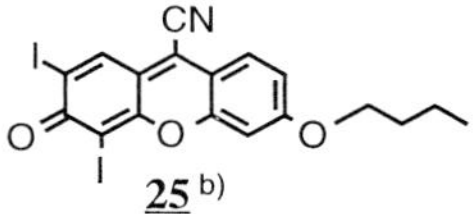

25 b)

a) J. Shi, b) T. Tanabe, c) A. Sarker, d) X. Liu, e) A. De Laaff

Table 12: continued

a) J. Shi, b) T. Tanabe, c) A. Sarker, d) X. Liu, e) A. De Laaff, f) R. Bao

Table 12: continued

T	U	V	W	X	Y	Z	λmax (nm)	ε	λ_fl (nm)	τ_fl (ns)	Φ_fl	$\lambda_{ph}^{?}_{k}$ (nm)	$\tau_{ph}^{r}_{t}$ (ms)	pK_a	E_ox	E_red	dye
O	I	I	H	H	H	~O-Si(tBu)	468	24100	544								26
O	I	I	H	H	H	~O-Si	470	18700	545								27
O	I	I	H	H	H	~O-Si(Et)₃	468	25200	545								28
O	I	I	H	H	H	~O-Si(Et)₃	468	25100	545								29
O	I	I	H	H	H	~O-Si-cyclohexyl	470	25600	546								30
O	I	I	H	H	H	~O-Si-CH₂-cyclohexenyl	470	27400	546								31
O	I	I	H	(CH₂)₃CO₂H	I	OH	478	19300	545								32
O	I	I	H	H	H	methacrylate	470	21200	549								33
O	I	I	H	H	H	methacrylate ester	470	13100	545								34
O	I	I	H	H	H	allyl ether	472	9920	544								35
O	I	I	H	H	H		470	19400	545								36
							472	23500	570								37

i. Photopolymerization with the Fluorones

Photopolymerization with the fluorones is highly efficient and competitive, in every sense, with polymerization initiated by UV light and UV initiators. Much of the mechanistic work establishing single electron transfer as the initiating step has been done on 2,4-diiodo-3-butoxy-6-fluorone (DIBF),

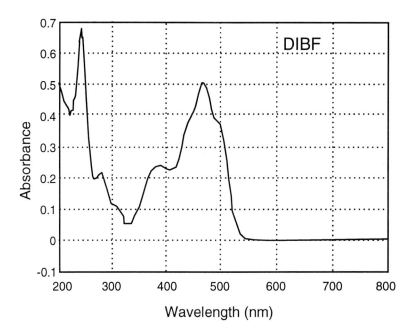

the most soluble photoinitiator in the series. Although it was first synthesized to be used with the argon ion laser, the broad absorption of this orange compound means that much of the light emitted by sources of various kinds from 300 to 500nm can be captured by the dye. The absorption spectrum of DIBF is shown in Figure 38

Figure 38
Absorption spectrum of DIBF

ii.   Mechanism of Photoreduction of DIBF[*]

The mechanistic hypothesis that tertiary amines reduced excited states by single electron transfer in producing radicals to initiate acrylate polymerization, was basically accepted from work with benzophenone, before the work of Hassoon,[113] who demonstrated unequivocally that the radical anion was formed from the triplet when DIBF was flashed in the presence of a tertiary aromatic amine. The transient absorption spectrum in acetonitrile [2 x 10^{-5}] M taken at several different delay times is shown in Figure 39. The appearance of a new, positive broad absorption with peaks at 320 nm and 520 to 800nm is characteristic of the triplet/triplet absorption seen with many other xanthenes[114] and assigned to that. The decay is accompanied by recovery of the ground state absorption. The occurrence of two isobestic points means that the triplets are the only product. At long times, a small positive absorption remains at 420 nm and this Hassoon assigned to the absorption of the radical anion. Klimtchuk[115] had shown dye triplet self-quenching to be an important reaction with other fluorone dyes.

$$Dye*^3 \;+\; Dye \;\text{-----}\!\!> \;Dye^{+\bullet} \;+\; Dye^{-\bullet}$$
$$\text{or}$$
$$Dye*^3 \;+\; Dye*^3 \;\text{------}\!\!> \;Dye^{+\bullet} \;+\; Dye^{-\bullet}$$

This formation of the radical anion in small amounts from the DIBF triplet was attributed to a similar electron transfer from the amine. The trace, Figure 39, shows the negative bleaching of the initiator and the growing in of absorptions above 500nm.

---

[*] The work herein is that of Salah Hassoon.

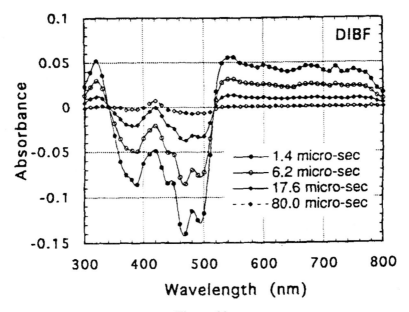

Figure 39
Transcient absorption spectra recorded at different delay times obtained from DIBF $(2 \times 10^{-5}M)$ deaerated solution in MeCN.

The traces can be separated into the absorption spectra for the triplet and for the radical anion. The decay of the triplet is first order with a lifetime of $15 \pm 1\mu sec$ in argon. As expected, oxygen quenches the DIBF triplet reducing the lifetime to 200nsec when DIBF is flashed in acetonitrile in air. Though triplet quenching by oxygen leads to singlet oxygen, the matter has not been pursued as a synthetic source of this important reaction intermediate.

Table 13
Singlet Oxygen Decay and Quenching

	$k_d(\text{sec}^{-1}) = k_1 + k_q[Q]$	k_q (mol^{-1}.sec^{-1}.lit)
1. DIBF(air)	$(1.35 \pm 0.06) \times 10^4$ [t = (73 ± 4) µsec]	
2. DIBF(air) + DIDMA*	$(2.45 \pm 0.1) \times 10^4$ [t = (41 ± 2) µsec]	$(3.96 \pm 0.2) \times 10^4$
3. DIBF(air) + BORATE	$(1.71 \pm 0.08) \times 10^4$ [t = (58 ± 3) µsec]	$(1.89 \pm 0.1) \times 10^4$
4. DIBF(air) + OPPI	$(1.39 \pm 0.06) \times 10^4$ [t = (72 ± 4) µsec]	~ 0

DIBF: 2×10^{-5} M/L, DIDMA: 2.78×10^{-3} M/L, BORATE: 1.9×10^{-3} M/L,
OPPI: 1.9×10^{-3} M/L Solvent : MeCN
*2,6-diisopropyl-N,N-dimethylaniline

Quenching with tertiary amine leads to a lifetime of 2.5 ± 0.13µsec under argon. This is shown in the transient spectrum, Figure 40, below. Although no radical anion is observed in the transient spectrum under these conditions, a new product absorbing at 360 nm and assigned to the DIBF radical is observed. Its rising time is comparable with the decay of the triplet suggesting that proton transfer after SET is extremely rapid (Scheme 22).

Though the quenching rates are a bit slower with borate than with tertiary amine (lifetime = 5 ± 0.4µ sec) the reaction in the presence of borate is also informative. Formation of the radical anion is documented when DIBF is flashed in the presence of triphenylbutyl borate by the appearance a new absorption with a maximum at 420 nm (Scheme 23). The rising time of this absorption, and the appearance of two isobestic points at 340nm and 510nm (Figure 40), confirm that it is formed exclusively from the triplet state.

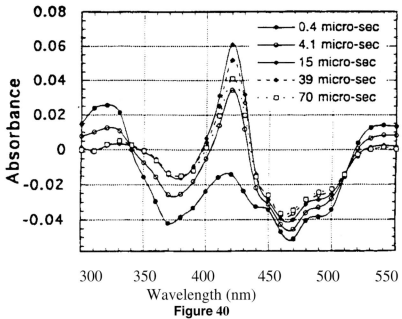

Figure 40
Transient absorption spectra recorded at different delay times of DIBF (2 x 10^{-5} M) in the presence of (a) BORATE (1.9 x 10^{-3} M), (b) DIDMA (2.78 x 10^{-3} M), and (c) OPPI (1.9 x 10^{-3} M) in MeCN.

The reactions thus are as follows:

Scheme 22
Reaction of DIBF with tertiary amine

In the presence of borate, the product observed is the radical anion.

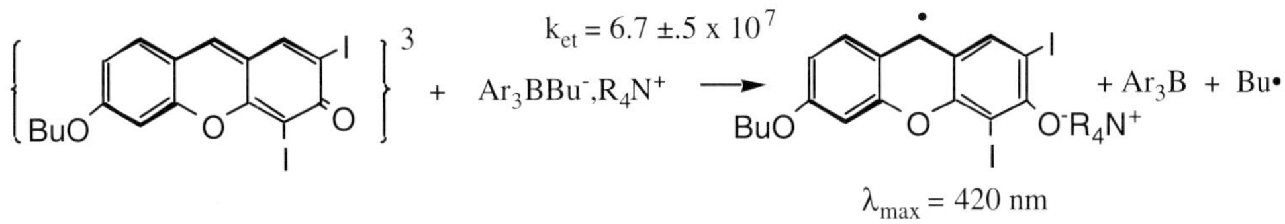

$$\{{}^3\text{[DIBF]}\} + Ar_3BBu^-, R_4N^+ \xrightarrow{k_{et} = 6.7 \pm .5 \times 10^7} \text{[reduced DIBF]} + Ar_3B + Bu^\bullet$$

$\lambda_{max} = 420$ nm

Scheme 23
Reaction of DIBF with borate salt

The difference in gegen ion at C-3, tetraalkyl ammonium in the case of reduction by borate and H$^+$ in the case of reduction by tertiary aromatic amine, leads to different products in the two cases and this leads to different rates of reoxidation, a fact which impacts on the rate of colour reformation from the reduced dye.

Table 14
DIBF Triplets, Lifetimes and Quenching Constants*

	$k_{d(sec^{-1})} = k_1 + k_q[Q]$	$k_q(mol^{-1}.sec^{-1}.lit)$
1. DIBF (Argon)	$(6.91 \pm 0.30) \times 10^4$	
	[t = (15 ± 1) μsec]	
2. DIBF + Air-O$_2$	$(4.92 \pm 0.25) \times 10^6$	$(2.55 \pm 0.15) \times 10^9$
	[t = (200 ± 10) nsec]	
3. DIBF+DIDMA (Argon)	$(4.06 \pm 0.20) \times 10^5$	$(1.21 \pm 0.08) \times 10^8$
	[t= (2.5 ± 0.13) μsec]	
4. DIBF+BORATE (Argon)	$(1.97 \pm 0.16) \times 10^5$	$(6.71 \pm 0.50) \times 10^7$
	[t= (5 ± 0.4) μsec]	
5. DIBF+OPPI (Argon)	$(6.25 \pm 0.30) \times 10^4$	~0
	[t= (16 ± 1) μsec]	

* DIBF-2 x 10^{-5}M/L; O$_2$-1.9 x 10^{-3}M/L; DIDMA-2.78 x 10^{-3}M/L; BORATE-1.9 x 10^{-3}M/L; OPPI-1.9 x 10^{-3}M/L; Solvent - MeCN

iii.   **DIBF Radicals: Formation, Decay and Absorption Spectra**

Table 15
DIBF.- and DIBF. Radicals Formation and Decay Constants

	kformation	Lifetime	kdecay[b]	Lifetime[b]
	$10^{-5}(sec^{-1})$	t (μsec)	$10^{-4}(sec^{-1})$	t(Nsec)
1. DIBF	2.44 ± 0.20	4.11 ± 0.30	0.77 ± 0.07	130 ± 10
(BORATE)	(1.97 ± 0.16)	(5.00 ± 0.40)[c]		
2. DIBF	3.88 ± 0.20	2.50 ± 0.13	1.20 ± 0.10	83 ± 7
(DIDMA)	(4.06 ± 0.20)	(2.50 ± 0.13)[c]		
3. DIBF	4.30 ± 0.30	2.33 ± 0.15	2.10 ± 0.16	48 ± 4
(DIDMA+OPPI)	(4.06 ± 0.20)	(2.50 ± 0.13)[c]		

a)   DIBF-2 x 10^{-5}M/L; DIDMA-2.78 x 10^{-3}M/L; BORATE-1.9 x 10^{-3}M/L; OPPI-1.9 x 10^{-3}M/L; Solvent- MeCN

b)   The decay of the radicals did not exactly fit first order kinetics. However, the above decay rates and lifetimes are from first order kinetics, and are given for comparison only.

c)   DIBF triplet decay and lifetime from Table 14.

The triplet of DIBF is unaffected by the presence of iodonium salts. In the presence of (p-octyloxy)phenyliodonium hexafluoroantimonate (OPPI), for example, the lifetime of DIBF triplet is 16 ± 1 μsec in acetonitrile solution purged with argon, a lifetime identical to the lifetime of the triplet in the absence of the oxidizing agent. This rules out oxidative electron transfer under these conditions {[DIBF] = 2 x 10^{-5}M; [OPPI] = 1.9 x 10^{-3}M}.

[DIBF structure *3] + Ar_2I^+ ⟶ Ar• + ArI + [DIBF radical cation structure]

The situation with the charged dyes, TIHF, TIBF, TIHCF and TIHBF differs.

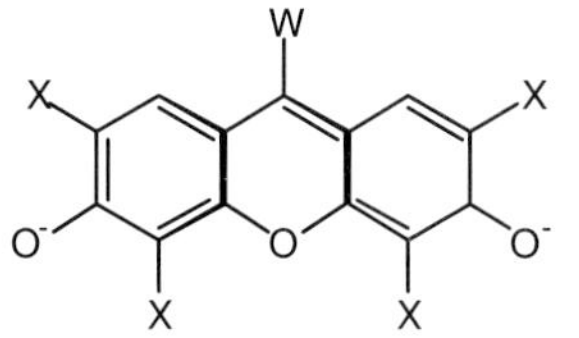

X = 1; W = H TIHF
X = 1; W = CN; TIHCF
X = Br; W = H; TBHF
X = Br; W = CN; TBHCF

Structures of various dyes

In the case of TIHF a triplet triplet absorption with a peak centered at 360nm and a broad absorption starting at 580nm is observed. Dye triplet self quenching was important in that the lifetime of the triplet was strongly dependent on the concentration of the dye used for the experiment. The corresponding radical anion is produced in the presence of tertiary amines, and this absorbs at 410nm.

Iodonium salt quenching of the dye triplet at a rate approximately equal to that of reductive quenching has been postulated before.[116] In the case of the neutral dye DIBF, however, quenching by the oxidizing agent is not observed. The implications of this result for the mechanisms of both radical polymerizations accelerated by iodonium salts, and cationic chains initiated by onium salt, DIBF mixtures will be explained shortly.

Irradiation of a three component mixture, DIBF, onium salt, amine or DIBF, onium salt, borate under similar conditions gives the transient spectra shown in Figure 41 a, b. The addition of OPPI, though it has no effect on the lifetime of the triplet, quenches the DIBF radical anion completely. OPPI has a smaller effect on the lifetime of the DIBF radical, the reactive intermediate formed in the presence of amine.

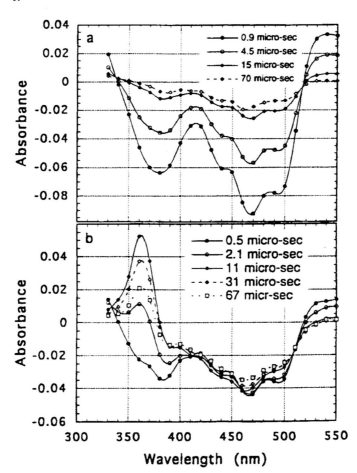

Figure 41 a & b
Transient absorption spectra recorded at different delay times of (a) DIBF-BORATE-OPPI and (b) DIBF-DIDMA-OPPI solutions in MeCN. (The concentration ratios are the same as the given in Figure 40.)

OPPI is a strong oxidizing agent and should easily accept electrons from DIBF radical anion.

Reaction of DIBF radical anion with diaryl iodonium ion

This reaction must be slow relative to the proton transfer rate however since DIBF radical forms from DIBF triplet in the presence of tertiary amine, this reaction proceeding more rapidly than the above oxidation. Regeneration of the dye is also observed thus indicating that dye bleaching is not complete and slow with respect to reactions which feed the intermediates back to the starting material.

Taken together, the results confirm that dye photoreduction in the presence of amines and borates is a one electron transfer process resulting in formation of the DIBF radical anion initially. Proton transfer in the case of the amine is fast with the end product from the single electron transfer being the neutral radical. The formation of the radical anion was observed on the nanosecond time scale in the case of the borate but could not be observed in the case of the amine and occurred at the same rate as the quenching of the dye triplet. There is no back electron transfer in the case of the borate and, as as expected from the above discussion, when it quenches the triplet the dissociation of the boranyl radical is faster than back electron transfer. In fact that latter is not observed.

The proposed mechanisms of reduction are delineated in both the amine reducing agent and borate reducing agent cases below:

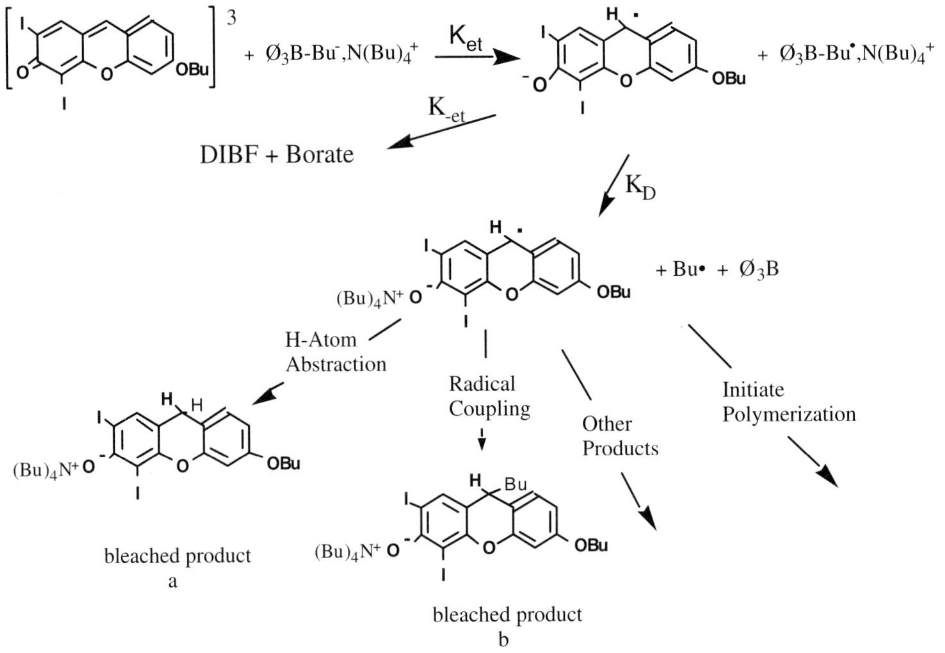

Scheme 24
Proposed mechanism in the case of DIBF/BORATE

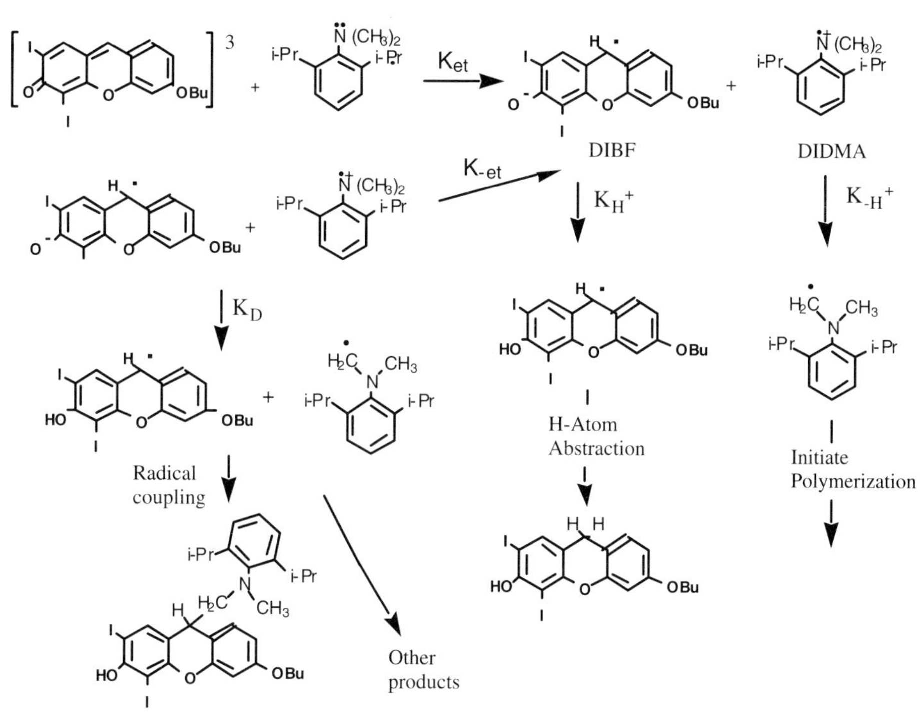

Scheme 25
Proposed mechanism in the case of DIBF and DIDMA

The free energy changes for the two systems can be estimated from the Rehm Weller equation. $E_{ox}(D/D^{+\bullet})$ is the oxidation potential of the donor borate or amine, $E_{red}(A^{+\bullet}/A)$ is the reduction potential of the acceptor dye and E^T_{0-0} = the triplet energy of the dye and $Ze^2/\varepsilon a$ the Coulombic interaction term.

$$\Delta G = E_{ox}(D/D^{+\bullet}) - E_{red}(A^{+\bullet}/A) - E^T_{0-0} - Ze^2/\varepsilon a$$

The reduction potential of DIBF measured by square wave voltammetry[117] is -0.9v (vs. standard calomel electrode - SCE). The triplet energy of the dye is determined from the phosphorescence spectrum to be 2.3 V. The oxidation potentials of the amines[118] is ≈ +0.5 v and of the borates[119] ≈ +0.7v. Assuming the Coulombic energies are small, the calculated ΔG_{ET} values are -0.7eV for the borate case and -0.9eV for the amine case respectively. This compares to a value of -0.36eV for the reduction of TIHF triplet by the Lewis base amine reported by Klimtchuk.[115]

It remains to suggest that for the mechanism occurring in the tertiary system, dye/onium salt/reducing agent. In that case, the initial step for all dye triplets is reduction either by borate or amine. In the presence of amine deprotonation of the ammonium radical cation is fast and the neutral radical forms. In the presence of the borate, subsequent reactions of the radical anion, save for oxidation, are so slow as to be essentially non-existent. Interception of the radical anion by oxidation reduces the iodonium salt and produces phenyl radicals which can contribute to subsequent radical chain initiations.

The bleached products have been isolated both in the case of reduction of DIBF by a tertiary amine, DIDMA, 2,6-diisopropyl-N,N-dimethylaniline and triphenylbutyl borate, tetramethylammonium salt. In the former case, the bleached product is the quinomethine reduced salt.

It reverts to a coloured form by oxidation.

The latter reaction is impossible in the case of the salt, and hence the species is produced upon upon reduction by the borate.

8. Oxidation/Reduction Chains Cationic Polymerization Initiated by Dye/Onium Salts

The principle commercial use of iodonium salt photochemistry is in the formation of acidic catalysts that induce subsequent cationic polymerization in epoxides, vinyl ethers and the like. Iodonium salts, however, absorb in the mid to deep UV region of the spectrum and this has hampered their use for many applications. This inconvenience can be circumvented in several ways. One way is to sensitize iodonium salt decomposition. An important contribution to the understanding of the sensitized decomposition of iodonium salts and the genesis of cationic chains therefrom derived from the work of Bi who studied ethyl erythrosin as a sensitizer for the cationic polymerization of cyclohexene oxide. Sensitized decompositions of iodonium salts had been reported both by Pappas[120] who used neutral sensitizers without reactive groups such as anthracene and xanthone and by Crivello and Lam[121] who screened nearly 75 dyes, to find that only five were effective. Ledwith first proposed what we shall term free "radical chain/cation chain polymerizations."[122] Ledwith proposed that radicals such as those obtained by the reactions of benzophenone with alcohols could, in the presence of iodonium salts, be oxidized to cations and the latter serve to initiate cationic chain processes.

Bi demonstrated this mechanism to be operative with ethyl erythrosin, a tertiary aromatic amine, and an iodonium salt by isolating and identifying all of the products from the reaction of the dye with the iodonium salt. In addition he identified the end groups in the formation of poly(cyclohexene oxide) in the 3 component initiator system above. Though, in contrast to the case with DIBF, there is competitive oxidation and reduction of ethyl erythrosin triplet by the tertiary amine and iodonium salt, the rate of the former is slightly larger than the latter. Polymer end group analysis indicates it is the tertiary amine carbocation that generates the cationic chain. The mechanism is outlined schematically in Scheme 26.

$$Dye \longrightarrow Dye^3$$
$$Dye^3 + ArN(Me)_2 \longrightarrow \longrightarrow Dye^\bullet + ArN(Me)CH_2\bullet$$
$$ArN(Me)CH_2\bullet + Ar_2I^+ \longrightarrow ArN(Me)CH_2^+ + ArI + Ar\bullet$$
$$ArN(Me)CH_2^+ + ROR \longrightarrow ArN(Me)CH_2 - O^+(R)_2$$
$$ArN(Me)CH_2^+ + ROR \longrightarrow ArN(Me)CH_2\text{-O-cyclohexyl}$$

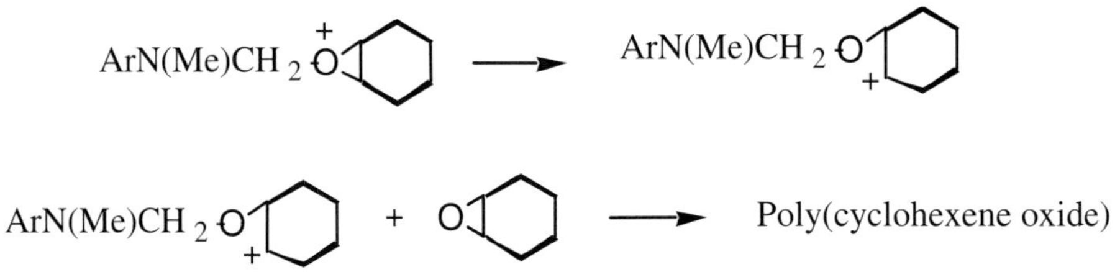

$$\text{ArN(Me)CH}_2\overset{+}{\text{O}}\diagdown\!\!\bigcirc \longrightarrow \text{ArN(Me)CH}_2\text{O}\diagdown\!\!\overset{+}{\bigcirc}$$

$$\text{ArN(Me)CH}_2\text{O}\diagdown\!\!\overset{+}{\bigcirc} + \text{O}\diagdown\!\!\bigcirc \longrightarrow \text{Poly(cyclohexene oxide)}$$

Scheme 26

Mechanism for Polymerization induced by a xanthene dye a tertiary amine and an iodium salt.

The system dye, onium salt, amine fails to polymerize *epoxy esters*, in contrast to simple epoxides. This is probably because the carbocation is trapped by the ester carbonyl group in a reaction that is favoured relative to reaction with epoxide.

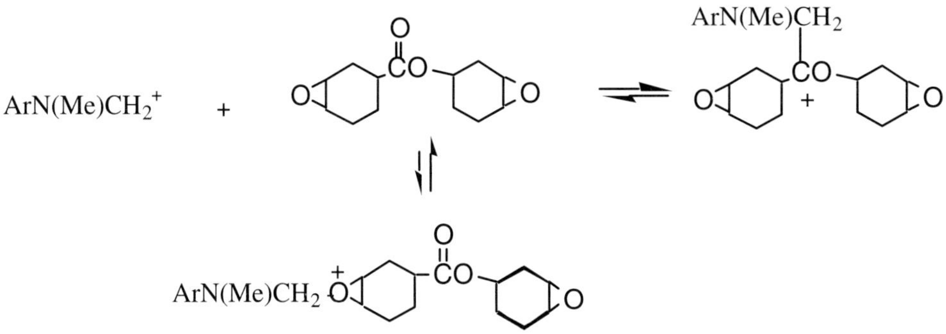

Scheme 27

Carbocation entrapment by ester carbonyl group

This latter reaction can be prevented by leaving out the amine. In this case, radicals are formed from the epoxy ester which subsequently add to the epoxide and catalyze polymerization. In this case the equilibrium concentration of the carbonyl group/cation addition product is small and the carbocation addition product to the epoxide reactive enough so that polymerization proceeds with visible light and rather rapidly (Scheme 28).

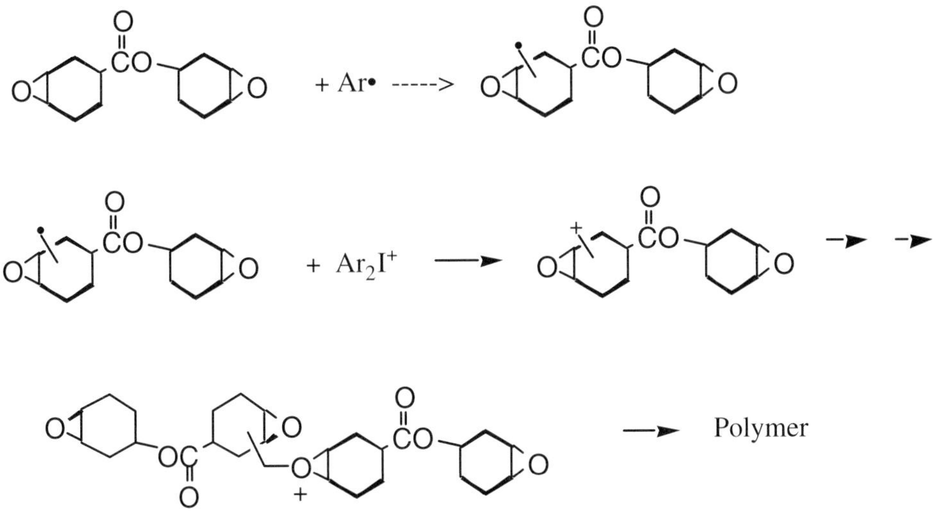

Scheme 28
Dye sensitized polymerization of cycloalphartic epoxide

The practical implication of this is that sensitizers do accelerate the polymerization of epoxy esters. Amines inhibit the reaction. However they do not do so by offering a base to the reaction mixture which traps acids putatively formed by the decomposition of the onium salt.

9. Summary

In Chapter 2 we have considered the many methods to initiate polymerization with UV and visible light. Extensive discussion of recent results describing the reactivity of aromatic ketone excited triplet states further clarify what one can expect from such a functional group when it is presented with a competitive situation. Some of the factors influencing reactivity have also been outlined.

The reactions of charge transfer species, molecules which form complexes when mixed together in a 1:1 ratio, project significant use as a new type of UV initiator. Two specific examples of such systems have been described.

Details of the reactivity of various organic dyes in the presence of reducing agents have been been outlined, and their competitive reactivity in the presence of partner oxidizing agents considered. A three part system, dye/onium salt/reducing agent has been described and the route to radicals or carbocations from it discussed.

References

[1] Paczkowski, J.; Pietrzak, M.; Kucybala, Z. *Macromolecules* **1996**, *29* 5057: Paczkowski, J.; Kucybala, Z. *Macromolecules* **1995**, *28* 269; Paczkowski, J. Rad Tech Europe, 8th Conference on Radiation Curing, **1997**, p. 77-84.

[2] Valdes-Aguilera, O; Pathak, C. P.; Shi, J.; Watson, D.; Neckers, D. C. *Macromolecules*, **1992**, *25*, 541.

[3] Linden, S. M.; Scaiano, J. C.; Wintgens, V.; Neckers, D. C. *J. Org. Chem.*, **1989**, *54* 5242.

[4] Mateo, J. L.; Bosh, P.; Lozano,A. E. *Macromolecules* **1994** *25*, 7794.

[5] Sarker, A. M., Polykarpov, A.Y., deRaaff, A. M., Marino, T. L., Neckers, D. C. *Polym. Sci., Chem. Ed,* **1996**, *34,* 2817.

[6] Details of this, and other, early results in mechanistic organic photochemistry can be found in many basic texts and references in organic photochemistry. For example, Turro, N. J. Contemporary Organic Photochemistry, University Science Books, Sausalito, CA.; Gilbert, A. C.; Baggett, J.; Organic Photochemistry; Neckers, D. C. Mechanistic Organic Photochemistry, Reinhold, NY, **1967**,

[7] Rabec, J.; Ranby, B.; Photochemistry of Polymers, Wiley, London

[8] Muzycho, T. M.; Jones, T. H.; U.S. Patent 3,888,671, July 29, 1972; via F. H. U.S. Patent 4,475,999 June 6, 1983.

[9] Hu, S.; Neckers, D. C. *J. Org. Chem.*, **1996**, *61*, 6407.

[10] Hu. S.; Neckers, D. C. *J. Org. Chem.,* **1997**, *62*, 755.

[11] Hu, S.; Neckers, D. C. *Tetrahedron,* **1997**, *58,* 2751.

[12] Hu, S.; Neckers, D. C.; *J. Org. Chem.,* **1997**, 62, 6820.

[13] Reiser, A.; Shih, H-Y.; Yeh, T-F.; Huang, J-P. *Agnew. Chem. Int. Ed. Engl.* **1996**. 35, 2428-2440.

[14] Azoplate Corp.; US-Pat 2 766 118. **1956**; US-A 3 046 122, **1962.**

[15] Reichmanis, E.; Thompson, L.F. *Chem. Rev.* **1989**, *89*, 1273-1289.

[16] Reiser, A. *Photoreactive Polymers: the Science and Technology of Resists,* Wiley, New York, 1989, chapter 2.

[17] The photolysis of monomeric 1 has also been studied. The results, mechanisms and utilities will be reported elsewhere. Hu, S.; Neckers, D. C. manuscript in preparation.

[18] **Poly-1b** produces essentially the same results as **Poly-1a. Poly-1a** was chosen for more detailed study.

[19] D_g measured as described will be slightly higher than it actually is since the first insolubilization of the negative resist occurs at the surface of the resist rather than at the substrate. The best way to measure D_g is to cast the resist on a transparent substrate and irradiate through the uniform substrate from the back side. The technical difficulties in cutting the film into smaller pieces prevented us from using a substrate other than the opaque silicon wafer.

[20] Other definitions for D_g have been reported, see Thompson, L. F.; Wilson, C. G.; Bowden, M. J. *Introduction to Microlithography*, 2nd ed.; ACS Professional Reference Book; American Chemical Society: Washington D.C. 1994.

[21] McDonald, S. A.; Wilson, C. G.; Frechet, J. M. *J. Acc. Chem. Res.* **1994**, *27,* 151-158.

[22] Thijs, L.; Gupta, S. N.; Neckers, D. C. *J. Org. Chem.* **1979** *44* 4123.

[23] Morlino, E.; Bohorques, M.; Neckers, D. C.; Rodgers, M. A. *J. Amer. Chem. Soc.* **1991**, *113* 2599.

[24] Gupta, S. N.; Gupta, I.; Neckers, D. C. *J. Poly. Sci., Polymer Chem. Ed.,* **1981**, *19*, 103.

[25] Abu-Abdoun, I. I.; Thijs, L. B.; Neckers, D. C. *J. Poly. Sci., Chem Ed.,* **1983**, *21*, 3129.

[26] D. C. Neckers, "Aromatic Ketone Peresters as Photoinitiators"; Uses Thereof", U.S. #4,416,826; D. C. Neckers, "Photopolymerizable Composition Containing Perester Photoinitiator and Photopolymerization Process", U.S. #4,498,963.

[27] Gomberg, M. *J. Am. Chem. Soc.,* **1900**, *22,* 757.

[28] Gomberg, M.; Cone, L. H. *Leibig's Annanlen,* **1909**, *370,* 142.

[29] Neumann, W. P.; Uzick, W.; Zakadis, A. K. *J. Am. Chem. Soc.,* **1986**. *108,* 3762.

[30] Neumann, W. P.; Penenory, A.; Stewen, U.; Lehning, M. *J. Am. Chem. Soc.,* **1989**, *111,* 5845.

[31] Letsinger, R. L.; Collat, R.; Magnusson, M. *J. Am. Chem. Soc.,* **1954**, *76,* 4185.

[32] Bromberg, A.; Schmidt, K. H.; Meisel, D., *J. Am. Chem. Soc.,* **1985**, *107*, 83.

[33] Faria, J. L.; Steenken, S., *J. Am. Chem. Soc.,* **1990**, *112*, 1277.

[34] Wittig, G.; Kairies, W.; Hopf, W.; *Ber.,* **1932**, *65* 767.

[35] Neckers, D. C.; Rajadurai, S.; Valdes-Aguilera, O.; Zakrzewski, A.; Linden, S. M. *Tetrahedron Lett.,* **1988**, 20(40), 5109.

[36] Takeuchi, H.; Nagai, T.; Tokura, N. *Bull. Chem. Soc. Jpn.,* **1971**, *44*, 753.

[37] Zarcadis, A. K.; Neumann, W. P.; Uzick, W. W. *Chem. Ber.,* **1985**, *118,* 1183.

[38] The spin mulitplicity is given by n + 1 where n is the number of free spins (see ref. 1).

[39] Selwood, P. W.; Dobres, R. M. *J. Am. Chem. Soc.,* **1950**, *72*, 3860.

[40] Turro, N. J. In *Modern Molecular Photochemistry,* University Science Books: Mill Valley, 1991.

[41] Loutfy, R. O., *Pure and Appl. Chem.,* **1986**, 58(9) 1239.

[42] Bank, S.; Ehrlich, C. L.; Zubieta, J.A. *Org. Chem.* **1979**, *44*, 1454.

[43] Breslow, R.; Chu, W.; *J. Am. Chem. Soc.,* **1973**, *95,* 411.

[44] Cohen, S. G.; Siddiqui, M.H. *J. Am. Chem. Soc.,* **1964**, *86*, 5047.

[45] Freilich, S.C.; Peters, K.S.; *J. Am. Chem. Soc.* **1981**, *103*, 6255.

[46] Neckers, D.C. *Mechanistic Organic Photochemistry*, Rheinhold, NYC 1967.

[47] Xi, F.; Bassett, W.; Vogl, O. *J. Polym. Sci., Polym. Chem. Ed.* **1983**, *21*, 891.

[48] Premantine, G. S.; Jones, S. A.; Tirrell, D. A. *Macromolecules* **1989**, *22*, 770.

[49] Bohnacek, R.S.; Strauss, U. P.; Jernigan, R. L. *Macromolecules* **1991**, *24*, 731.

[50] Olson, K. G.; Butler, G. B. *Macromolecules* **1984**, 17, 1884; ibid; 2480; ibid; 2486.

[51] Lapin, S. C.; Noren, G. K.; Schouten, J. J. *Conference Proceedings, Radtech Asia,* **1993.**

[52] Lee, C.; Hall, H. K. Jr. *Macromolecules* **1989**, *22* 21.

[53] Jönsson, S.; Hoyle, C. E. *Conference Proceedings, Radtech North America,* **1996**, 377.

[54] Jönsson, S.; Schaeffer, W.; Sundell, P-E.; Shimose, M; Owens, J; Hoyle, C. E. *Conference Proceedings, Radtech North America,* **1994**, 194.

[55] Shimose, M.; Hoyle, C. E.; Owens, J.; Jönsson, S.; Sundell, P-E. *Proc. SPIE, Int. Conf. Photopolymers,* **1994**, 424.

[56] Decker, C.; Morel, F.; Jönsson, S.; Clark, S. C.; Hoyle, C. E. *Polym. Mater. Sci. Eng.* **1996**, *75,* 198.

[57] Jönsson, S.; Sundell, P-E.; Shimose, M.; Owens, J.; Hoyle, C. E. *Polym. Mater. Sci. Eng.* **1995**, *72,* 470.

[58] Shimose, M.; Hoyle, C. E.; Jönsson, S.; Sundell, P-E.; Owens, J.; Vaughn, K. *Polym. Prep.* **1995**, *36* 485.

[59] Jönsson, S.; Sundell, P-E.; Shimose, M.; Owens, J.; Miller, C.; Clark, S.; Hoyle, C. E. *Polym. Mater. Sci. Eng.* **1996**, *74*, 319.

[60] Miller, C. W.; Hoyle, C. E.; Howard, C.; Jönsson, S. *Polym. Prep.* **1996**, *37*, 346.

[61] Clark. S. E.; Jönsson, S.; Hoyle, C.E. **1996**, *37,* 348.

[62] Hoyle, C. E. private communication.

[63] Polykarpov, A. Y.; Neckers, D. C. *Tet. Letters* **1995,** *36,* 5483.

[64] PCT/US/96/17561 - October 19, 1995.

[65] (a) Hassoon,D.; Sarker, A.; Rodgers, M. A. J.; Neckers, D. C. *J. Am. Chem. Soc.* **1995,** *117,* 11369. (b) Hassoon, S.; Sarker, A.; Polykarpov, A. Y.; Rodgers, M. A. J.; Neckers, D. C. *J. Phys. Chem.* **1996,** *100,* 12386.

[66] (a) Sarker, A. M.; Mejiritski, A.; Wheaton, B. R.; Neckers, D. C. *Macromolecules,* **1997,** *30,* 2268. (b) Mejiritski, A.; Polykarpov, A. Y.; Sarker, A. M.; Neckers, D. C. *Chem. Mater.* **1996,** *8,* 1360. (c) Mejiritski, A.; Sarker, A. M.; Wheaton, B. R.; Neckers, D. C. *Chem. Mater.* **1997,** *9* 1488; Sarker, A.; Lungu, A.; Mejiritski, A. and Neckers, D.C. Tetraorganyl borate Complexes as Convenient Precursors for Photogeneration of Organic Bases". Unpublished.

[67] Sarker, A. M.; Lungu, A.; and Neckers, D. C. "Synthesis and Characterization of a Novel Polymeric System Bearing a Benzophenone Borate Salt as a New Photoinitiator for UV Curing", *Macromolecules,* **1996,** *28,* 8047.

[68] (a)Traven, V. F.; Vorobjeva, L. I.; Chibisova, T. A.; Carberry, E. A.; Beyer, N. J. *Can. J. Chem.* **1997,** *75,* 365. (b) Song, P. S.; Gordon, W. H. *J. Phys. Chem.* **1970** *74,* 4234.

[69] Rehm, D.; Weller, A. *Isr. J. Chem.* **1970** *8,* 259.

[70] Leigh, W. J.; Arnold, D. R. *J. Chem. Soc. Chem. Commun.* **1980,** 406.

[71] Gallivan, J. B. *Can. J. Chem.* **1972** *50,* 360l.

[72] Selinger, B. L. *Aust. J. Chem.* **1966** *19,* 825.

[73] Mantulin, W. W.; Song, P. S. *J. Am. Chem. Soc.* **1973,** *95,* 5122.

[74] Heldt, J. R.; Heldt, J.; Ston, M.; Diehl, H. A. *Spectrachim. Acta* Part A. **1995** *51,* 1549.

[75] (a) Devadoss, C.; Fessenden, R. W. *J. Phys. Chem.* **1990***, 94,* 4540; *J. Phys. Chem.* **1991***, 95,* 7253. (b) Yamaji, M.; Kiyota, T.; Shizuka, H. *Chem. Phys. Lett.* **1994***, 226,* 199.

[76] Carmichael, I.; Helman, W. P.; Hug, G. L. *J. Phys. Chem. Ref. Data* **1987***, 16,* 239.

[77] (a) Christope, D. R.; Schuster, G. B. *Organometallics* **1989***, 8,* 2737. (b) Slocum, G. H.; Kaufman, K.; Schuster, G. B. *J. Am. Chem. Soc.* **1981***, 103,* 4625.

[78] Nau, W. M.; Cozens, F. L.; Scaion, J.C. *J. Am. Chem. Soc.* **1996***, 118,* 2275.

[79] Ratcliff, M. A.; Kochi, J. K. *J. Org. Chem.* **1971,** *36,* 3112.

[80] Inre, C.; Modro, T. A.; Rohwer, E. R.; Wagener, C. C. P. *J. Org. Chem.* **1993***, 58,* 5643.

[81] Geske, D. H. *J. Phys. Chem.* **1959***, 63,* 1062.

[82] (a) Bancroft, E. E.; Bount, H. N.; Janzen, E. G. *J. Am. Chem. Soc.* **1979***, 101,* 3692. (b) Wilkey, J. D.; Schuster, G. B. *J. Am. Chem. Soc.* **1991***, 113,* 2149.

[83] Murphy, S. T.; Zou, C.; Mierj, J. B.; Ballew, R. M.; Dlott, D. D.; Schuster, G. B. *J. Phys. Chem.* **1993***, 97,* 13152.

[84] Back electron transfer within singlet ion pairs takes place on the picosecond time scale; (a) Ashai, T.; Mataga, N. *J. Phys. Chem.* **1989***, 93,* 6575. (b) Kikuchi, K.; Taahushi, Y.; Hoshi, M.; Niwa, T.; Kataaagiri, T.; Miyashi, T. *J. Phys. Chem.* **1991***, 95,* 2378. (c) Mataga, N.; Okada, T.; Kanda, Y.; Shioyama, H. *Tetrahedron,* **1986***, 42,* 6143. (d) Asahi, T.; Mataga, N. *J. Phys. Chem.* **1991***, 95,* 1956.

[85] (a) Nagasawa, Y.; Yartsev, A. P.; Tominagu, K.; Johnson, A. E.; Yohihara, K. *J. Am. Chem. Soc.* **1993**, *115,* 7922. (b) Nagasawa, Y.; Yartsev, A. P.; Tominagu, K.; Bisht, P. B.; Johnson, A. E.; Yoshihara, K. *J. Phys. Chem.* **1995**, *99,* 653.

[86] (a) Kata, S.; Amataatsu, Y. *J. Chem. Phys.* **1990**, *92,* 7241. (b) Lippert, E.; Rettig, W.; Bonacic-Koutecky, V.; Heisel, F.; Miehe, J. A. *Adv. Chem. Phys.* **1987**, *68,* 1.

[87] Walker, G. C.; Akesson, E.; Johnson, A. E.; Levinger, N. E.; Barbara, P. F. *J. Phys. Chem.* **1992**, *96,* 3728.

[88] P. Gottschalk, G. B. Schuster, D. C. Neckers and P. C. Adair, for The Mead Corporation and Bowling Green State University, "Photosensitive Compositions and Materials", E.P. #89303687.1.
P.Gottschalk, D. C. Neckers and G. B. Schuster, "Ionic Complexes as Photoinitiators", U.S. Patent #4,772,530, E.P. #90201194.9, Canadian patent applied for, and Japanese patent applied for.
P. Gottschalk, G. Schuster, G. Benjamin, D. C. Neckers and P. C. Adair, for The Mead Corporation, "Photopolymerizable and Photocurable Compositions", E.P. #339,841, U.S. appl. #180, 915.
P. Gottschalk, D. C. Neckers and G. B. Schuster, "Photocurable Compositions Containing Photobleachable Ionic Dye Complexes", U.S. Patent #4, 895, 880 Jan. 23, 1990; P. Gottschalk, D.C. Neckers, G. B. Schuster, P. C. Adair and S. P. Pappas, "Photosensitive Materials and Compositions Containing Ionic Dye Compounds as Initiators and Thiols as Autooxidizers" U.S. Patent 4, 937, 159; June 26, 1990.
P. Adair, P. Gottschalk, D. C. Neckers, G. B. Schuster and P. Pappas, "Photosensitive Materials Containing Ionic Dye Compounds as Initiators". U.S. Patent #4, 977, 511 (December 11, 1990), Canadian patent (June 11, 1991), Japanese patent applied for, Korean patent applied for, European patent applied for.
P. Gottschalk, D. C. Neckers and G. M. Schuster, "Photosensitive Matrials Containing Ionic Dye Compounds as Photoinitiators", U.S. Patent #4, 842, 980 June 27, 1989; P. Gottschalk, D. C. Neckers, G. B. Schuster, P. C. Adair and S. P. Pappas, "Method of Dental Treatment", U.S. Patent #5, 035, 621, July 30, 1991; P. Gottschalk, D. C. Neckers and G. B. Schuster, "Cationic Dye-triarylmonoalkylborate Anion Complexes" U.S. Patent #5, 151, 520, Sept. 29, 1992.

[89] Cunningham, A. F.; Kunz, M.; Kura, H.; GB 2,307,474 A, November, 1995.

[90] Cunningham, A. F.; Kunz, M.; Kura, H.; GB 2,307,472 A, November 1995.

[91] Cunningham, A. F.; Kunz, M.; Kura, H.; GB 2,307,473 A, November 1995.

[92] Cunningham, A. F.; Kunz, M.; Kura, H.; EP 0 775, 706, A2, November 1995.

[93] Marino, T. L.; de Raaff, A. M.; Neckers, D. C. *Rad Tech '96 North America UV/EB Conference Processing*, Rad Tech International North America, Nashville, TN, April 1996, pp. 7-16.

[94] (a) Crivello, J. V.; Lam, J. H. W. *Macromolecules*, **1977**, *10,* 1307. (b) Crivello, J. V.; Lee, J. H. W. *J. Polym. Science: Part A: Polymer Cchem.* **1989**, *27,* 3951.

[95] Torres-Filho, A; Neckers, D. C. *J. Appli. Polym. Sci.,* **1994**, *51,* 931.

[96] Eisch, J. J.; Tamao, K.; Wilcsek, R. J. *J. Amer. Chem. Soc.* **1975**, *97* 895.

[97] Williams, J. L. R.; Doty,J. C.; Grisdale, P. J.; Searle, R.; Regan, T. H.; Happ, G. P.; Maier, D. P. *J. Am. Cchem. Soc.* **1967** *89* 5153.

[98] Grisdale, P. J.; Williams, J. L. R.; Glogowski, M. E.; Babb, B. F. *J. Org. Chem.* **1971**, *36,* 544.

[99] Cregge, R. J.; Herman, J. L.; Lee, C. S.; Richman, J. E.; Schlessinger, R. H. *Tet. Lett.* **1973**, 2425.

[100] Valdes-Aguilera, O.; Neckers, D. C. *Advances in Photochemistry;* Volman D. H.; Hammond, G. S.; Gollnick, K.Eds Wiley, NY, 1993, Volume 18 pp 313-394; Eaton, D. F. *Advances in Photochemistry;* Volman, D. H.; Hammond, G. S.; Gollnick, K. Eds. Wiley, NY, 1986, Volume 13, pp 427-486.

[101] Oster, G. *Nature,* **1954,** *173* 300.

[102] Oster, G. U.S. Patent 2, 850, 445 (1958); 2, 875, 047(1959); 3, 097, 096 (1960).

[103] Zarzewski, A.; Neckers, D. C. *Tet. Letters,* **1987,** *43,* 4507.

[104] Paczkowski, J.; Paczowska, B.; Neckers, D. C. *J. Photochem & Photobio. A: Chem.,* **1991**, *61, 131.*

[105] Oster, G.; Oster, G. K.; Kang, G. *J. Phys. Chem.,* **1962**, *66,* 2514.

[106] Phillips, K.; Read, R. *J. Chem. Soc., Perkin Trans.,* **1986**, *1,* 671.

[107] Valdes-Aguilera, O.; Pathak, C. P.; Shi, J.; Watson, D.; Neckers, D. C. *Macromolecules,* **1992**, *25,* 541.

[108] Hassoon, S.; Neckers, D. C. *Phys. Chem.,* **1995**, *99,* 9416.

[109] Tanabe, T.; Torres-Filho, A.: Neckers, D. C. *J. Polym. Sci., Chem Ed.* **1995**, *33* 1691.

[110] Shi, J; Zhang, X.; Neckers, D. C. *J. Org. Chem.,* **1992**, *57.* 4418; Shi, J.; Zhang, X.; Neckers, D. C.*Tet. Lett.* **1993**, *34,* 6013.

[111] Neckers, D. C.; Shi, J. U.S. Patent #5, 451, 343, 9-19-95.

[112] Marino, T. L.; Martin, D. B.; Neckers, D. C. *Radtech Proc.,* **1994**/169.

[113] Hassoon, S.; Neckers, D. C. *J. Phys. Chem.* **1995**, *99,* 9416.

[114] Wintgens, V.; Scaiano, J. C.; Linden, S. M.; Neckers, D. C. *J. Org. Chem.,* **1989**, *54* 5242; Mills, A.; Lawrence, C.; Douglas, P. *J. Chem. Soc. Faraday Trans. 2* **1986**, *82* 2291; Fisher, G. J.; Lewis, C.; Madil, D. *Photochem. Photobio.* **1976** *24,* 223; Kasche, V.; Lindqvist, L. *J. Phys. Chem.* **1967**, *71,* 817; Kasssche, V.; Lindqvist, L. *Photochem. Photobio,* **1965**, *4* 923; Zwicker, E. F.; Grossweiner, L. I. *J. Phys. Chem.* **1963**, *67,* 549; Grossweiner, L. I.; Zwicker, E. F. *J. Chem. Phys.* **1959**, *31,* 1141; Grossweiner, L. I.; Zwicker, E.F. *J. Chem. Phys.* **1961**, *34,* 1411.

[115] Klimtchuk, E.; Rodgers, M. A. J.; Neckers, D. C. *J. Phys, Chem.* **1992**, *96,* 9817.

[116] Bi, Y.; Neckers, D. C. *J. Photochem & Photobiol., A: Chem.* **1993**, *74,* 221; Bi, Y.; Neckers, D. C. *Macromolecules,* **1994**, *27,* 3683.

[117] Osteryoung, J; O'Dea, J. J. in *Electroanalytical Chemistry* Bard, A. J. Ed.; Dekker: New York 1986; Vol. 14 pp 209-308.

[118] Murov, S. L.; Carmichael, I.; Hug, G. L. *Handbook of Photochemistry* 2nd ed. Dekker; New York, 1993: 289.

[119] Chatterjee, S.; Davis, P. D.; Gottschalk, P.; Kurz, M. E.; Sauerwein, B.; Yang, X.; Schuster, G. B. *J. Amer. Chem. Soc.* **1990**, *112* 6329.

[120] Pappas, S. P.; Pappas, B. C.; Gatechair, L. R. *J. Polym. Sci. Polym. Chem. Ed.* **1984**, *22,* 77.

[121] Crivello, J. V.; Lam, J. H. W. *Polym. Sci. Polym. Chem. Ed.* **1978**, *16,* 2441.

[122] Ledwith, A. *Polymer* **1978,** *19*, 1217.

CHAPTER III

PHOTOPOLYMERIZATION PROCESSES

Chapter Three

III Analytical Techniques for Monitoring Photopolymerization Processes

1. Introduction

Accurate analytical methods for characterizing photoformed polymer films and coatings are critical in many areas of the industry, from basic research to production. In characterizing polymeric films, the conversion of monomeric units is commonly referred to as the cure of a resin, though the term cure is misleading, it is so ubiquitous that we'll use it. Many other physical properties of photocured materials generally are derived from the cure. In this chapter we have divided analytical techniques into four classes: (1) general research techniques, (2) techniques for real time monitoring, (3) techniques for off line monitoring and (4) techniques for on line monitoring (Table 1).

Table 1
Classification of analytical techniques used in this chapter

	Table 1, Classification of analytical techniques used in this chapter			
Technique:	general research (section 3.2)	real time monitoring (section 3.3)	off line monitoring	on line monitoring (section 3.4)
Environment	research development	research development	production	production
Purpose:	determine cure, structure morphology	determine kinetics polymerization mechanism	quality control (random sampling)	quality control (continuous)
Requirements	obtain data general and detailed	no interference irradiation and measuring technique	reveal specific (relevant) data cheap	fast non-destructive non-contact
Examples	Solid State NMR spectroscopy	FTIR spectroscopy Photo-DSC	stain test (iodine solution)	fluorescence spectroscopy

The physical properties of photoformed materials depend upon the exact conditions[1] for photopolymerisation and the resins and diluents employed. Analytical techniques are used to characterize photoformed products. Changing the photoinitiator or the intensity of the light which is being employed can cause significant changes. Research techniques require high sensitivity since small changes in cure, composition and morphology need to be detected. Such techniques need to provide universal data, such as the degree of cure and the exact chemical composition of a polymer. Research techniques may be destructive, i.e. destroying the shape, the morphology or even the chemical composition of the sample. They can be time consuming, as long as they provide valuable data. For example, CPMAS solid state NMR spectroscopy[2] is very useful despite the fact that obtaining spectra involves an elaborate procedure.

Recently techniques have been developed which monitor polymerization processes in real time. These techniques monitor certain properties of the photoformed material during the photopolymerization process. Instead of analyzing samples which were generated by means of a photopolymerization process afterwards, the process itself is also analyzed. Real time monitoring can be employed for determining kinetic parameters and understanding polymerization mechanisms. In combination with general research techniques, some procedures yield a detailed picture of a specific photopolymerization process. Some real time monitoring techniques can be used to investigate photopolymerization processes under conditions which closely resemble those in a production environment. Detailed information concerning the process, as it takes place in production, can be obtained. In development work, real time monitoring can be used for testing the rate of photopolymerization for new formulations and for the quality control of commercial formulations. The amount of inhibitor, initiator and the presence of contaminants can be determined from a kinetic profile. The main requirement for real time monitoring techniques is that the reaction conditions, notably the irradiation of the sample, do not interfere with the measurements. In order to follow photopolymerization processes under "realistic" conditions, these techniques need to be fast, with short response times.

In a production facility analytical techniques can be employed for controlling the quality of photoformed products. Though the relationship between the irradiation parameters (such as composition and intensity of the light and the exposure time), resin composition and the cure for any system can be obtained easily, it is important actually to measure the real cure achieved. Fluctuations in the composition of the light (intensity and wavelength), differences in formulation composition, (notably the amount of inhibitor), photoinitiator, oxygen (radical polymerizations) or water (cationic polymerizations) are commonly encountered in real systems and result in deviations between measured cure and the cure which is desired. In most practical situations, one overcomes these problems by over irradiating the formulation, since too much cure is considered less disadvantageous than too little. In general, too low a cure will result in soft, sticky and smelly polymers. Due to the inhomogeneous nature of chain polymerization processes, extractable monomers can still be present in these materials. The overcure of a resin is thus generally preferred although it sometimes results in polymers which are (too) hard and brittle. Therefore overcure generally is not a problem as far as the quality of the photoformed products is concerned. None the less, since overcuring decreases line speeds, it may be undesirable from an economic point of view.

Two types of analytical techniques for quality control can be distinguished, namely off-line and on-line monitoring. Off-line monitoring consists of random sampling of the photoformed product. In general, this procedure does not guarantee that all products meet the required quality standards. Additionally, response times, the intervals between sampling and corrective action can be extensive. Combining these factors, lack of quality guarantee and the possibility of producing large quantities of waste, in the case of a malfunction, brings us to the conclusion that off line monitoring is suited only for relatively inexpensive products. It should be noted that in systems where off line monitoring is used, irradiation times are choosen very liberal, so only large malfunctions and instrument failure can cause problems of undercured formulations.

In many cases, especially for more sophisticated applications, it is desirable to continuously monitor the quality of the material produced. This can be achieved by means of on line monitoring, where the products which leave the production line are continuously probed. Inferior (undercured) products will be detected immediately and separated from the others, while malfunctioning equipment can be fixed or adjusted instantaneously. Due to early detection, the amount of waste produced will be reduced and products which leave the production line will meet desired quality standards. Line speeds can be increased since massive overcure is no longer required.

Analytical techniques for off line monitoring are to be fast, inexpensive, reliable and must focus on those properties which are critical for the product. In general, off line techniques may be destructive, since only a small fraction of the products are sampled. Depending on the application, critical properties need not always be directly derived from the overall cure of a resin. For instance, stain resistance is critical for coatings of floor tiles. For this property, a good cure and a high cross link density at the surface is needed to make the coating impenetrable for dirt and solvents. Hence good resistance to staining by iodine solutions is a specific requirement. For other applications such as the coating of bottles and food packaging, the permeability to gases, carbon dioxide and oxygen, respectively, is relevant. In order to obtain a low gas permeability a complete through cure is critical. In this case, a lower cure at the surface can be tolerated, as long as the material is fully cured at depth. Therefore a different testing procedure is required for these materials.

On line monitoring techniques are to be fast, non-destructive, preferably non-contact and capable of application to products as they exit a production line. Since photoformed products may be of a variety of shapes and are attached to a variety of substrates, techniques which can monitor substrates of many shapes and are not interfered with by other materials are preferred. Thus only spectroscopic techniques are suitable. As an illustration of the importance of strict quality control, imagine the cost of replacing a bundle of optical fibres in a telecommunication line at the bottom of the ocean. Incomplete cure of the cladding of an optical fibre can result in malfunctioning. In principle just one inch of incomplete cure of the cladding in one of the many optical fibres can necessitate replacement of an entire section of cable.

In this chapter we will begin with a discussion of the techniques which are employed in research and development to study photoformed polymer networks. Subsequently, techniques used for real time monitoring will be discussed. This chapter will end with a discussion on techniques for on line monitoring. Techniques for real time and on line monitoring will be discussed in more detail, since these techniques are gradually gaining importance and significant developments in these areas have been reported during the last decade. Off line monitoring will not be discussed as a separate subject. The reason for this exclusion is that many techniques used for off line monitoring are too specialized, e.g. they measure physical quantities that are poorly defined and only relevant to one specific type of application. Off line monitoring techniques that are of a general interest, however, are discussed but will be classified as general research techniques.

Discussions in this chapter are limited to techniques that analyze solid polymeric networks. Most photoformed coatings and adhesives consist of insoluble polymeric networks, since multifunctional resins, or mixtures containing multifunctional resins, are employed. Techniques that analyze polymers in solution cannot be applied to these systems. Another argument to limit ourselves to the techniques mentioned above, is the fact that they are universal; all polymer films, networks as well as linear polymers, can be analyzed using them. In addition, techniques that can be employed for characterizing polymers in solution are far too numerous and are extensively reviewed elsewhere.[3]

2. General Research Techniques for Analyzing Polymeric Films

i. Introduction

The main purpose of this section is to give a brief overview of those research techniques available for the analysis of photoformed materials.[4] Many of the techniques discussed in this section will also be applicable in the field of quality control, but emphasis will be on techniques for research and development. Some of the techniques are applicable for real time monitoring of photopolymerization processes and on line monitoring, and will also be discussed in sections 3 and 4.

The general research strategy for investigating samples prepared by photoinitiated polymerization is to irradiate samples for fixed periods of time and determine their cure and other physical properties. In many cases the main objective is to find relationships between irradiation parameters, (irradiation time, intensity, spectral distribution) formulation composition, (monomers, photoinitiators, additives) and the cure and physical properties of the photoformed material. Special attention is given to the relation between the cure of a formulation and its physical properties. It should be pointed out that in many cases the aftercure or dark cure of a formulation can be considerable. If a series of measurements are taken from a sample after certain irradiation periods, these should be fast and performed in a prescribed order. Alternatively one could perform measurements after a certain period of time, allowing the sample to equilibrate and relax. Besides the aftercure which changes the conversion of a formulation after the irradiation has taken place, samples also undergo changes in physical properties due to other processes. A well known process is physical ageing in which the free volume in the polymeric matrix decreases as a function of time.

Analytical techniques can be classified in different categories, such as calorimetric, mechanical, surface analysis methods and spectroscopic techniques. Calorimetric techniques (2.ii.) and spectroscopic techniques (2.iii.) will be discussed in this chapter. Emphasis will be on spectroscopic techniques since they offer such advantages as short response times, a high sensitivity and non destructive and non-contact. As we will see later, spectroscopic techniques are particularly suited for real time and on line monitoring.

ii. Calorimetric Techniques; Differential Scanning Calorimetry

Differential Scanning Calorimetry as an analytical tool for analyzing polymeric films will be discussed.[5] Other calorimetric techniques are known but are not standardized. Results obtained are often hard to compare with each other when using these techniques. Therefore they will not be discussed here.

Differential Scanning Calorimetry, or DSC, is a calorimetric technique that measures the amount of heat flowing to or from a sample. The sample is contained in a holder, usually a small metal cup, and the heat flow to this measuring cell is compared to an empty sample holder in a reference cell. Experiments are performed under well controlled conditions, usually under a controlled atmosphere with an increasing temperature gradient. The amount of sample needed is typically in the order of a few milligrams depending on the type of information one wants to extract. Information obtained from a photoformed material strongly depends on the properties of the material under study.

One application of DSC in the study of photopolymerization processes is for determining the degree of cure of a formulation. In many photocured systems, polymerizable groups are still present after irradiation has ceased. This can be caused by a shortage of photoinitiator, limited exposure time or by vitrification of the polymer matrix at the reaction temperature. A cure well below unity is very common for systems consisting only of small multifunctional monomers (reactive diluents).[6] In many cases, the reactive groups (radicals or cations) remain active and when the temperature of a sample is raised the mobility of these groups is increased and the polymerization process can be led to completion. In many cases a spontaneous thermal polymerization occurs above a certain temperature.

In a typical experiment a known quantity of resin, to which a small quantity of a thermal initiator is added, is polymerized thermally. From the total amount of heat evolved in the thermal polymerization, one can calculate the reaction enthalpy of the reactive group. The *relative* cure of a sample of identical composition that is photocured to a degree that is not known can be calculated, using Equation 1. Thermal polymerization does not necessarily reach a 100% conversion, and the absolute cure of the unknown sample can be determined only if the cure of the thermally polymerized sample is determined by some independent technique. For many commercial formulations, consisting of functionalized oligomers and reactive diluents, it is appropriate to assume a 100% cure after thermal treatment. Alternatively one can determine the cure of a resin by using literature values for the reaction enthalpies of the reactive group in the polymer. Table 2[7,8,9,10] summarizes the reaction enthalpies of well known polymerization processes. Other endothermic processes are not allowed in the temperature domain used.

Equation 1

$$\alpha = \{1-(\Delta H_{therm} - \Delta H_{res}/\Delta H_{therm})\} \alpha_{therm}$$

where

α = cure of the photochemically cured sample

α_{therm} = cure of the thermally cured sample

ΔH_{res} = residual heat evolved after heating the photocured sample (in J.g^{-1})

ΔH_{therm} = heat evolved after thermal polymerization of the resin (in J.g^{-1})

Table 2
Heats of formation used to monitor the degree of cure of various resins.

Functional group heat of formation in kJ/mol		Reference
acrylate	77.9	7
methacrylate	54.8	8
epoxide	104 (oxirane)	9
	139 (cyclohexene oxide)	9
vinyl ether	75	10

For fully cured samples, DSC can be used to measure phase transitions such as Tg, the glass transition temperature and Tm, the melting point of the polymer. For resins that are not fully cured and contain reactive functionalities, polymerization will resume at higher temperatures. In most cases, the glass transition temperature of photoformed polymers is determined by Dynamic Mechanical Analysis (DMA). Small amounts of heat evolved by the reacting residual reactive groups do not interfere with DMA measurements. One should, however, be aware that if thermal aftercure occurs, each heating cycle in DSC or DMA measurements will increase the degree of cure *and* the Tg of the sample.

DSC is a reliable and robust technique for directly determining the cure of a formulation by monitoring the amount of residual reactive groups that remain in it. In a heating cycle, the polymerization process is forced to completion at elevated temperatures. Therefore, determining the cure at high conversions is not very accurate. In determining the cure of a resin, a major advantage is that DSC measurements are not being disturbed by the presence of additives. Absorption or emission by minor amounts of additives can seriously hamper UV and IR absorption spectroscopy and fluorescence spectroscopy. Only additives that undergo strongly exothermic decomposition reactions in the temperature domain of interest might interfere with DSC measurements.

iii.   Spectroscopic Techniques

For characterizing polymer films, spectroscopic techniques are the most useful and comprehensive. Spectroscopic techniques are non-destructive by nature, but in practice this does not always imply that sample preparation is not needed. For example, solid state NMR samples need to consist of small particles, which are usually obtained by crushing or milling samples which are cooled in liquid nitrogen. Since conventional IR spectroscopy requires thin samples on a transparent substrate, sample preparation can also be necessary to use this technique. Various reflective techniques are available which can obviate the need for a substantial amount of preparation as indeed does the photoacoustic technique. Despite the fact that the dimensions of certain objects need to be adjusted, spectroscopic techniques are non-destructive in the sense that the chemical composition and morphology of substrates do not change during their use. This implies that most spectroscopic techniques can be used prior to other (destructive) techniques for investigating the same sample.

Speed is another advantage of many spectroscopic techniques. Optical techniques such as IR and UV/Vis absorption and fluorescence spectroscopy are particularly fast with response times of the order of micro seconds or less. Those techniques that are fast and are not interfered with by the radiation needed for photoinitiated polymerization processes, are suitable for real time monitoring of such processes.

Most spectroscopic techniques are non-contact techniques. Samples interact with electromagnetic irradiation and this results in absorption, emission or diffraction of electromagnetic irradiation. Those techniques that are fast and are capable of monitoring photopolymerized materials in the shape at which they leave the production line are suited for on line monitoring. Many spectroscopic techniques can be used also for remote sensing and some can be employed simultaneously with other techniques.

Spectroscopic techniques for monitoring polymeric films can be classified in numerous ways. One criterion is the physical quantity that is measured. Many techniques measure the cure of a formulation, either directly or indirectly, by measuring a cure related property of the polymer. IR spectroscopy is a direct technique, which measures the cure of a formulation by monitoring absorptions that are specific for the monomer. Fluorescence spectroscopy on the other hand is an indirect technique which measures the increase in (micro)viscosity that occurs during a polymerization process. Techniques that measure the detailed structure and morphology of the polymer are also available. Solid state NMR spectroscopy and dielectric spectroscopy are good examples of techniques which monitor changes in the structure and morphology of polymeric films. Another classification can be made on the basis of whether a technique detects the polymer itself or whether a probe molecule needs to be added. In the first case, the polymer itself can be investigated by an intrinsic or direct technique. In the second case, a probe molecule needs to be added and the results obtained will be specific for every probe resin combination. Good examples of techniques for which probes are generally added to a formulation are UV/Vis absorption and fluorescence spectroscopy.

iii. a) Infrared Spectroscopy

One of the best known and most employed techniques to analyze photoformed polymeric networks is infrared spectroscopy which monitors the absorption of radiation in the region of the electromagnetic spectrum from 4000 to 500 cm^{-1}. Such absorptions give rise to vibrational transitions which can be designated to specific stretch or bend vibrations in the absorbing molecule. Today most spectra are taken in the Fourier Transform mode recording the whole spectrum with just a few pulses of "white" infrared light. This technique allows one to record spectra faster and more accurately than does continuous wave monitoring. FTIR spectroscopy has almost completely replaced continuous wave IR spectroscopy for most research applications.

During a photopolymerization process, the chemical structure of the resin changes and so does the IR spectrum. For virtually all formulations, specific absorptions can be found which disappear after polymerization. For example, the 810 cm^{-1} absorption which can be assigned to the alkene CH stretching in (meth)acrylates is not present in poly (meth)acrylates. By monitoring the disappearance of this specific absorption, the cure of (meth)acrylates can be determined. The IR transitions that are used for determining the cure in other resins are summarized in Table 3[12,13,114,15,16,17,18,19,20,21,22]

Table 3
IR absorptions used to monitor the degree of cure of various resins.

Functional Group	Absorption Wavenumber in cm^{-1}	Type of Vibration	Reference
acrylate	810	vinyl CH bend	12
methacrylate	810	vinyl CH bend	
epoxide	720	epoxide ring stretch	13
	780 and 860		14
	850		15
	920		16
vinyl ether	1616-1622	C=C stretch	17
	960	vinyl CH bend	18
ketene acetals	845-860	ketene CH bend	19
propenyl ethers	1620-1678	C=C stretch	17
	cis 1670	C=C stretch	20
	trans 1650	C=C stretch	21
butenyl ethers	1620-1685	C=C stretch	21
maleimide	829	vinyl CH bend	18,22
n-vinyl pyrolidine	1635	C=O stretch	22

Most IR spectra are recorded by monitoring the absorption of thin films irradiated by the IR beam at right angles in the configuration depicted in Figure 1. The absorption of the sample is calculated using equation 2. For measurements at right angles, strict geometric requirements apply. The thickness of samples needs to be in the order of 10-50 μm and depends on the type of formulations being employed.[23] This is a fairly normal thickness for coating applications. In order to relate IR absorptions to cure, films need to be of a good optical quality and uniform in thickness. Another requirement is that the substrate on which the coating is applied is transparent at the appropriate wavelengths. This means that the only a limited number of substrates, such as monocrystalline silicon, inorganic salts like NaCl and polyethylene or polypropylene films, can be used. Most other substrates absorb IR radiation, at the regions of interest and for this reason, the cure in laminates cannot be determined in most cases. Other components that are added to the formulation, such as pigments and fillers, will scatter or absorb in the IR region of the spectrum close to where the active parts absorb. However, it is often possible to obtain data from filled systems despite this complication.

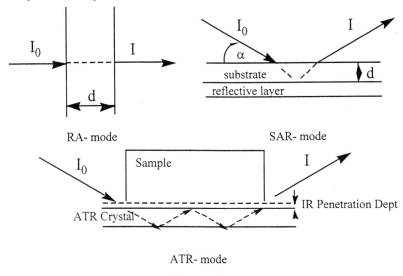

Figure 1
Different modes of IR spectroscopy that are employed for analyzing polymeric networks.

$$A = \log I_0/I = \varepsilon \cdot c \cdot d$$

A = absorption
I,I$_0$ = irradiation intensity, indicent and after absorption, respectively
ε = molar extinction coefficient (in 1000 cm^2/dm^3)
c = concentration (in mol l^{-1})
d = path length (in cm)

Equation 2

Most resins show considerable volume shrinkage during photoinitiated polymerization processes. In many cases shrinkage is limited to 10% which is a reasonable value for acrylate formulations but larger shrinkage volumes up to 30% have been reported.[24] The shrinkage influences the magnitude of all absorptions in the sample. Such shrinkage effects need to be compensated for in order to determine the cure of a resin accurately. This is done by monitoring the IR absorption that is specific to the monomer (A$_{mon}$) relative to a reference absorption (A$_{ref}$). The reference absorption is an absorption that does not change during the photopolymerization. Usually the reference absorption is ascribed to a functional group that is far away from the reactive groups and which is not altered during the polymerization process. The cure α is determined according to equation 3.

$$\alpha = 1 - \{A_{mon}(t) * A_{ref}(0) / A_{mon}(0) * A_{ref}(t)\}$$

α = cure of the resin
A$_{mon}$(t) = specific absorption of the monomer at time t
A$_{ref}$(t) = reference absorption (unchanged during polymerization) at time t

Equation 3

In Figure 2 the FTIR spectra of the donor-acceptor compound **1** before and after photopolymerization are depicted.[18] The monomer forms a polymeric network in which the main chains are composed of alternation copolymers upon irradiation, see Scheme 1.[25,26] From these spectra, it is obvious that the cure of this resin can be followed by using the 960cm^{-1} absorption from the CH bending of the vinyl ether and the 830cm^{-1} absorption originating from the CH bending in the maleimide part of the molecule. The 770cm^{-1} CH bending from the phenyl moiety was used as the reference signal. By using this method, the cure of **1** can be determined accurately at any stage of the polymerization process. An example of a profile of the cure versus the irradiation time is given in Figure 3. In this figure the double bond conversion of ethylene glycol dimethacrylate (EGDMA) is monitored by following the disappearance of the 810cm^{-1} acrylate CH bending. The aliphatic CH stretch vibrations around 2900cm^{-1} were used as a reference signal in this case.

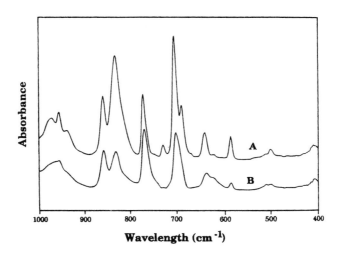

Figure 2

FTIR spectra of before (A) and after, photoinitiated polymerization (B). Taken from reference 18, with permission.

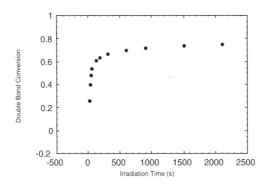

Figure 3

The double bond conversion of ethylene glycol dimethacrylate EGDMA as a function of the irradiation time. Taken from reference 70, with permission.

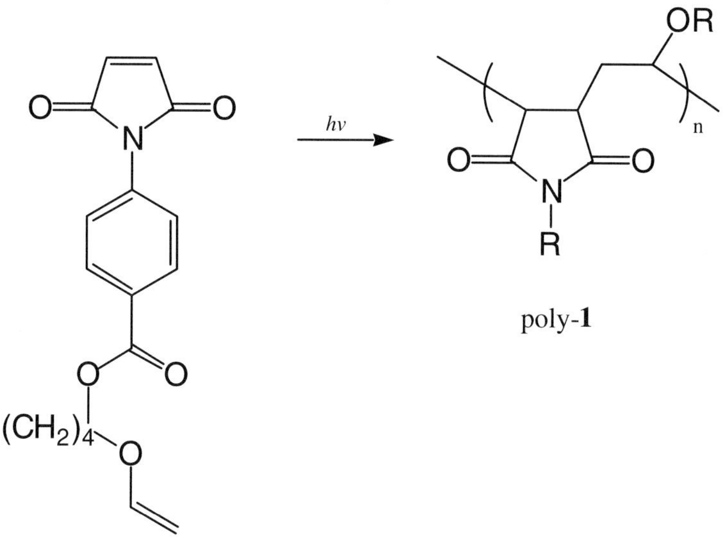

Scheme 1
Photoinitiated polymerization of the donor-acceptor compound 1.

Apart from the normal mode, which measures absorption at right angles, IR absorption spectroscopy can be employed in the reflection mode. In this technique, called Single Reflection Absorbance Fourier Transform Infrared Spectroscopy (SRA-FTIR), films are deposited on a reflecting metal surface, usually gold although other bright surfaces can also be employed. The IR beam is reflected at the metal layer but, since the outcoming beam has gone through the substrate twice, the absorption in the substrate is measured, see Figure 1. For taking spectra in the SRA mode, the requirements concerning film thickness and film quality are more stringent than for IR at right angles. Since the path length of the IR beam through the samples is increased, the samples need to be thinner, typically 5-20µm. A higher optical quality (surface smoothness, dust particles) of the films is also required.

A fundamental difference between IR in the SRA mode and IR in the normal mode at right angles is the polarization of the absorbed irradiation. For spectra taken in the reflection mode, the electric component of the absorbed irradiation is perpendicular to the surface of the film and the direction of propagation of the incident beam. For spectra taken at right angles the electric component of the absorbed light lies in the plane of the sample. By using polarizing optics, two independent polarized IR beams can be generated. It should be pointed out that the differences in polarization of the absorbed irradiation do not show up in the IR spectra of "ordinary" photoformed film. Different absorptions in SRA and normal mode IR spectra are detected only in cases in which anisotropic samples are examined. Such ordered films obtained by the photopolymerization of liquid crystalline monomers[18,27,28,29,30] and Langmuir-Blodgett films. Figures 4-6 demonstrate the effect of polarization in the IR spectrum of a LB film of octanediolacrylamide (ODA, 2).[31]

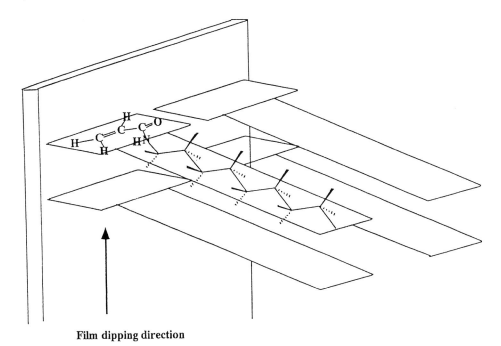

Figure 4
Schematic representation of Langmuir-Blodgett films of octanediolacrylamide ODA (2). Taken from reference 31, with permission.

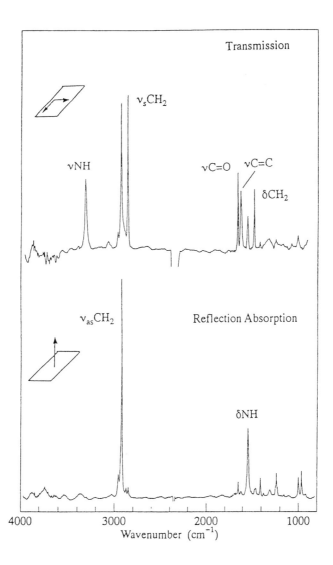

Figure 5
FTIR spectra of Langmuir-Blodgett films of octanediolacrylamide ODA (2) in the transmission and in the reflection absorption (SRA) mode. Taken from reference 31, with permission.

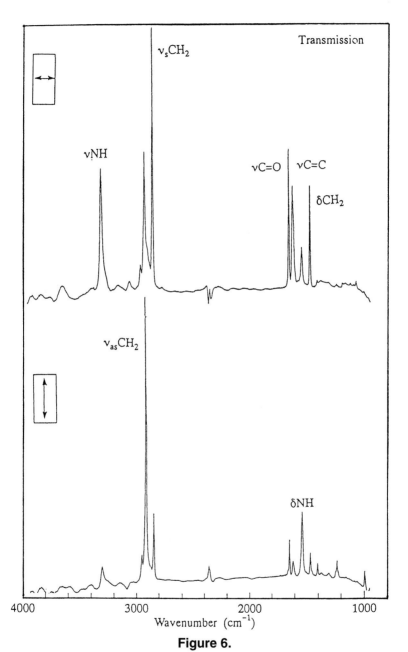

Figure 6.
Polarized FTIR spectra of Langmuir-Blodgett films of octanediolacrylamide ODA (2). Taken from reference 31, with permission.

Recently another technique for acquiring FTIR in reflection has been applied in the field of radiation cured coatings. This technique is called Attenuated Total Reflectance-Fourier Transform Infrared Spectroscopy (ATR-FTIR).[32] An IR beam is passed through an ATR crystal which is in intimate contact with the sample under investigation, see Figure **1**. The penetration depth of the IR beam in the sample (the evanescent light) is typically in the order of 0.1 to 0.6 µm depending on the exact conditions. There is no upper limit for sample thickness present due to this small penetration depth. IR absorbing substrates can be attached to the sample without interference as long as the thickness of polymer film exceeds the penetration depth of the IR beam. Using the ATR technique, samples are irradiated from above while IR spectra are taken from the bottom layer. By systematic variation of the sample thickness, in depth cure profiles of a resin can be obtained. As was the case with SRA-FTIR spectroscopy, the beam is polarized parallel to the plane of the sample and perpendicular to the direction of propagation of the IR beam.

In addition to the different modes by which IR spectroscopy can be applied to an entire sample, IR spectroscopy can also be applied to small sections of a sample. This technique is called IR microscopy[33] and is generally applied to sections of 50µm in diameter. Apart from the normal mode, IR microscopy is also reported in the ATR mode. In both cases, the IR bundle or the ATR crystal is focussed on a section of sample that has been selected by visual inspection using light microscopy. The main purpose of IR microscopy is to identify the composition of small sections of films, preferably those of a different appearance.

In general, IR spectroscopy is an excellent research tool for monitoring (photo)polymerization processes in that it detects the degree of cure of a resin directly from the disappearance of a specific IR absorption. The technique is accurate and fast, especially when using the Fourier Transform (FT) mode. The limited sample thickness (except for ATR-FTIR) and the requirement that the material needs to be uniform and flat limits its use to thin coatings on flat substrates. For measurements carried out in the normal mode at right angles, films must be coated on IR transparent surfaces. Apart from a few selected materials such as monocrystalline silicon wafers, polyalkenes and inorganic salts such as sodium chloride, most substrates absorb IR radiation. IR measurements in reflection can be performed only when thin films are deposited on a reflecting surface. For FTIR measurements in the ATR mode, thicker samples can be used but they need to be in close contact to an ATR crystal. Therefore perfectly planar samples, preferably photopolymerized on the ATR crystal itself, are required. Finally, in all cases, measurements are seriously disturbed if components in the formulation absorb at the same wavenumber as the absorption that is to be employed for monitoring the photopolymerization process. For example, some pigments have absorptions around 810 cm^{-1} which hampers monitoring the cure of pigmented acrylates. However other absorptions associated with the acrylate moiety such as the 1410cm^{-1} absorption could be employed in these instances.

Due to the speed of acquiring spectra and the insensitivity for UV irradiation, (FT)IR spectroscopy is particularly suitable for monitoring photopolymerization processes in real time, see Section 3. iii. Due to the requirements concerning the thickness of films and substrates to which they can be attached, IR and FTIR spectroscopy is not an appropriate technique for on line monitoring.

iii. b) UV/Vis Spectroscopy

UV/Vis spectroscopy is a well established technique for characterizing polymeric materials, but this technique is not widely employed for characterizing photoformed polymeric films. UV/Vis spectroscopy monitors the absorption of radiation in the ultraviolet and the visible part of the electromagnetic spectrum between 200 and 800nm. These absorptions give rise to electronic transitions in molecules, i.e. those transitions in which electrons are promoted to higher molecular orbitals. Many compounds are comprised of single bonds only and absorb in the deep UV below 250nm. UV/Vis transitions that absorb above 250nm are observed only if molecules contain chromophores that contain double bonds often in combination with heteroatoms.

In general, it is not possible to determine the cure of commonly used formulations by direct UV/Vis measurements. UV spectra of pure monomers that contain vinyl moieties and the polymers derived from them can be clearly distinguished, as is shown in Figure 7.[36] Most monomers and polymers, have absorptions that are exclusively at short wavelengths below 300nm. Therefore absorptions originating from the monomer and the polymer in common formulations, are completely overshadowed by absorptions from inhibitors, photoinitiators, other additives and impurities. Figure 7 illustrates this point for tetraethylene glycol diacrylate (TEGDA).[34] As a result, UV monitoring of the cure of most resins is possible only in the case of polymerization of extremely pure samples. This can be achieved by employing techniques that do not require photoinitiators. Electron beam (EB) curing and γ induced polymerizations are examples of such techniques.

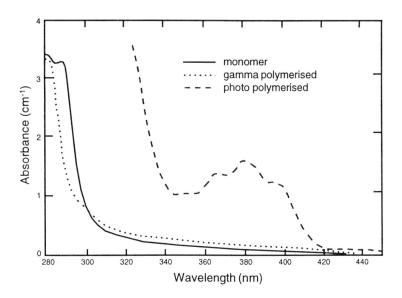

Figure 7.
UV spectrum for pure TEGDA, pure poly-TEGDA and for TEGDA containing a photoinitiator. Taken from reference 34, with permission.

The aforementioned limitations, which prevent direct monitoring of photoinitiated polymerization processes by UV spectroscopy do not apply to recently developed formulations based on donor and acceptor alkenes.[26,35,36] These formulations can be photopolymerized without adding photoinitiators since the complex formed between the electron rich alkene and the electron poor alkene is capable of initiating a polymerization reaction after excitation by light. In a formulation in which N-vinylpyrrolidone (NVP) is the electron donor and 5-hydroxyphenylmaleimide is the electron acceptor, the cure was directly monitored by the disappearance of the maleimide absorption at 300 nm, Scheme 2.[22] Though donor-acceptor based formulations are gaining importance and are expected to find commercial application soon, the use of formulations without inhibitors and photoinitiators is an exceptional case. Also, it is not expected that commercial formulations containing donor and acceptor monomers will be free of additional photoinitiators, inhibitors and other UV absorbing additives.

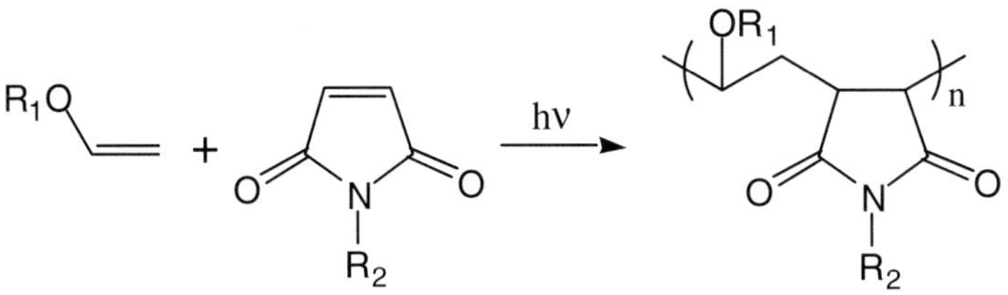

Scheme 2
Photopolymerization of maleimide and an electron rich alkene.

Recently the use of UV spectroscopy for monitoring the concentration of photoinitiators was reported.[13] This technique can only be used in the case of photoinitiators that bleach which means that the photodecomposition products absorb at shorter wavelengths than the photoinitiator itself. The concentration of photoinitiator (PI) is proportional to the intensity of the long wavelength absorption of the photoinitiator. As an example, Scheme 3 presents the decomposition of the photobleachable photoinitiator 2-benzyl-2-dimethylamino-1-(4-morpholinophenyl)-butanone-1, **3**, (Irgacure 369 from Ciba Geigy). The UV spectra of **3** as a function of the irradiation time are presented in Figure 8. In cases where well known photoinitiators that decompose in a unimolecular fashion are used, the amount of reactive species formed can easily be derived from photoinitiator concentration. Due to the complex nature of photopolymerization processes, photoinitiator concentrations do not reveal chemical or physical properties of the photoformed network directly. But, in combination with other techniques that do monitor the cure of a formulation, valuable information concerning photoinitiated polymerization processes can be obtained. This subject will be discussed in more detail section 3.iv. (real time UV monitoring).

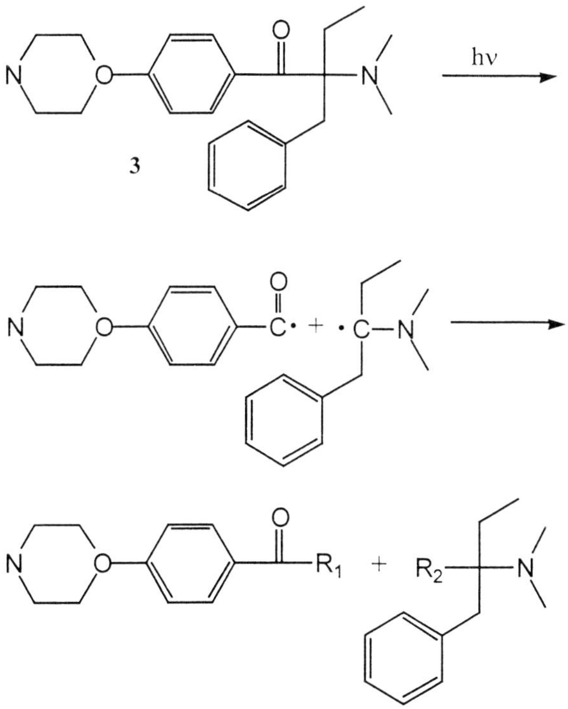

Scheme 3
Photodecomposition of 3 (Irgacure 369)

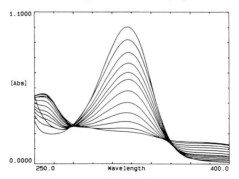

Figure 8
The UV spectra of the photoinitiator 3 (Irgacure 369) as a function of the irradiation time. Taken from reference 13, with permission.

In many cases polymeric materials have been characterized by UV/Vis spectroscopy through absorptions of probe molecules added to the formulation that absorb at long wavelengths.[37] As long as the concentration of the probe molecule is low, one can assume the addition of the probe neither alters the physical properties of the polymer nor interferes with the photopolymerization process. Examples of probes that have been employed for characterization of polymer films are solvatochromic and photochromic probes.

Solvatochromic probes monitor the polarity of the medium in which they are incorporated by means of a shift in their absorption maximum.[38,39] An example of a solvatochromic probe is the pyridinium N-phenolate betaine dye (**4**, Reichardt's Dye) that is widely employed for determining the polarity of organic solvents by means of their $E_T(30)$ values, see Scheme 4 and Equation 4. This molecule has a large dipole moment in the ground state that is strongly reduced in the excited state. Therefore, the CT absorption is blue shifted as the polarity of the medium is increased due to specific ground state stabilization. Other solvatochromic probes that are employed for characterization of the medium by the Kamlet, Abraham and Taft (KAT) method are 4-nitro-N,N-dimethylaniline, 4-nitroaniline and 4-nitroanisole. Solvatochromic dyes have been employed for the characterization of polymeric films made by spin coating.[40] As far as we know, solvatochromic probes have not been used for characterization of photoformed polymers and monitoring photopolymerization processes presumably due to a limited photostability and a limited shift in absorption maxima of the dyes.[41]

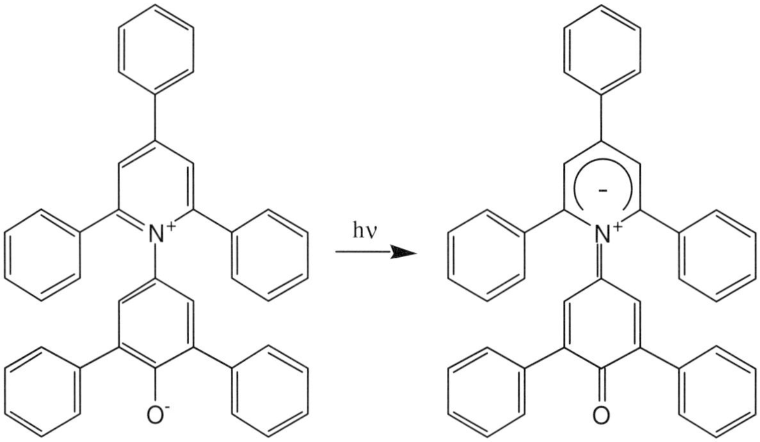

Scheme 4

Reichardt's dye (4).

$$E_T(30) = hcN\nu_{max}(\mathbf{4}) = 2{,}859\nu_{max}(\mathbf{4})$$

$E_T(30)$ = solvent polarity indicator (in kCal mol^{-1})
h = Planck constant (6.626 10^{-34} Js)
c = speed of light (2.998 10^8 ms^{-1})
N = Avogadro constant (6.022 10^{23})
ν_{max} (**4**) = Emission wavenumber of **4** at the emission maximum (in cm^{-1})

Equation 4

In the last decade photochromic probes have been employed to characterize polymeric materials by monitoring the free volume that is available for the probe to undergo a specific photochemical isomerization.[42,43,44,45,46,47,48] In most cases, photochromic probes were used to study the physical ageing of linear polymers, a process by which the free volume in a polymer decreases as a function of time. Examples of such probes are stilbenes and azostilbenes (Scheme 5) which require volume to undergo a *trans-cis* isomerization. A number of photochromic probes have been employed and the free volume that is required for a photochemical *trans-cis* isomerization has been determined, see Table 4. Only probes that have sufficient free volume, so called mobile or free probes, will undergo a photochemical *trans* to *cis* isomerization (Equation 5). Probes that are encapsulated in the polymer, so called immobile or bound probes, will not undergo a photochemical *trans* to *cis* isomerization (Equation 6). By irradiating a sample and monitoring the decrease of the absorption caused by the photochemical *trans* to *cis* isomerization in time, one can easily determine the ratio between free and bounded probe molecules. The ratio between free and bound probe molecules is a measure of the amount of free volume in the polymer matrix. An example is given in Figure 9. By employing photochromic probes of different sizes one can get an impression of the size distribution of the available free volume.

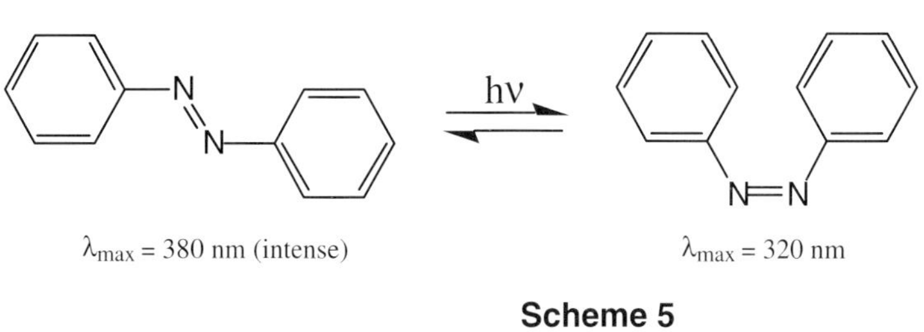

λ_{max} = 380 nm (intense) λ_{max} = 320 nm

Scheme 5
Isomerization scheme for azostilbene.

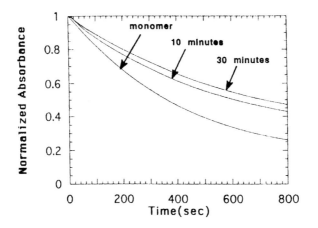

Figure 9
Absorbance of a photochromic probe as a function of the irradiation time in pure monomer and partly polymerized samples obtained after 10 and 30 minute thermal polymerizations. Taken from reference 49, with permission.

Table 4
Isomerization volumes required for the trans-cis isomerization of different stilbene and azobenzene derivatives. Data taken from reference 48.

Probe	Isomerization Volume Å^3
azobenzene	127
p-azotoluene	193
stilbene	224
4,4'-dinitrostilbene	285
4,4'-diphenylazobenzene	356
4,4'-diphenylstilbene	575

Equation 5

Equation 6

t_f, t_b = trans isomer in the free and bound state, respectively.

c_f, c_b = cis isomer in the free and bound state, respectively.

Recently these probes have been employed for monitoring photopolymerization process in real time.[49,50] As a general technique for investigating *photoformed* polymeric films, UV spectroscopy using photochromic probes is not a useful procedure since light needed to induce the trans-cis isomerization, typically 365nm, will also excite the photoinitiator and alter the cure of the material. Using photochromic probes for monitoring photoinitiated polymerizations in real time is a valuable technique that will be discussed in section 3. iv.

As an analytical technique, UV/Vis spectroscopy is straightforward, fast and user friendly. Samples need to be transparent at the wavelength at which one wants to monitor the polymerization process. A direct measurement of cure in which the absorptions of the monomer and the polymer are monitored, can be achieved if the formulation is devoid of photoinitiators and stabilizers which absorb in the same region of the spectrum as the monomer and the polymer. The optimum film thickness which allows for accurate measurements is determined by the extinction coefficients of the monomer and the polymer. In cases where the absorption of the photoinitiator or an added probe is monitored, films of various thicknesses can be examined since the absorption is the product of the extinction coefficient of the probe or the photoinitiator times the film thickness (Equation 2). Specific UV-Vis absorbing probes must absorb at longer wavelengths than the monomer, polymer, photoinitiator, inhibitors and other additives. It should be noted that addition of probes may influence the rate of polymerization, due to competitive absorptions and possible interference with reactive species in the polymerization process. In addition they may alter the final colour of the photoformed material.

UV spectroscopy is suitable for real time monitoring applications. For on line monitoring, UV spectroscopy appears to be less appropriate. Interference by other substrates (laminates) and fluctuations in sample thickness and lamp intensity impose problems.

As for IR measurements, samples need to be of optical quality and preferably of a uniform thickness.

iii. c) Emission Spectroscopy

Emission spectroscopy is widely employed for characterizing polymer films[51,52,53] and monitoring photopolymerization processes. Monomers,[54,55] photoinitiators,[56,57] inhibitors, additives[58] and contaminants may be luminescent. Though monitoring the inherent luminescence of formulations has the advantage that nothing is added to the formulation, it has the disadvantage that the method is neither very sensitive nor selective and may be highly irreproducible, so is not widely employed. In many cases, it is not clear which species are responsible for the luminescence or by what type of mechanism the properties of the formulation are being monitored.

In most applications, luminescent probes are added to formulations and then used to monitor photopolymerization processes by measuring their fluorescence. Since fluorescent probes are highly selective and sensitive, minor amounts are required for most formulations. For 15µm, thin films probe concentrations are generally in the order of 0.01% by mass or 5×10^{-4}M while thicker films require concentrations decreased proportionally. These low concentrations have the distinct advantage that competitive absorption by the fluorescent probe is limited and quenching of excited states and reactive species by the probe unlikely. In this section, the discussion will be limited to those cases where luminescent probes are added to the formulation.

Luminescent probes can be fluorescent, emitting from the singlet excited state, or phosphorescent, emitting from the triplet excited state[59] (see Chapter 1). Fluorescent probes have short lifetimes τ, typically a few nanoseconds (10^{-9}s) high quantum yields of emission (Φ_F approaches unity in many cases),and are generally not sensitive to quenching. Phosphorescent probes have much longer lifetimes τ, generally in the order of milliseconds (10^{-3}s), have low quantum yields of emission (Φ_P usually does not exceed a few percent), and are sensitive to quenching by other species, notably oxygen. Phosphorescent probes have been reported in the literature in a few cases,[60] but are not of general practical value. It should be noted that in specific cases the use of phosphorescent probes may be advantageous. Due to longer lifetimes, phosphorescent probes can be employed for investigating processes within the polymeric matrix that are slower by a few orders of magnitude. Nevertheless, in this section the focus will be on fluorescent probes.

In general, the singlet excited state formed by direct excitation of a probe molecule decays by three different processes: (1) Fluorescence (F), (2) Internal Conversion (IC) and (3) Inter System Crossing (ISC).[61] Each have their own rate constants of k_F, k_{IC} and k_{ISC}, respectively, see Scheme 6 and Equation 7. The lifetime of the excited state τ is defined as the reciprocal of k (Equation 8) and can be measured by monitoring the decay of fluorescence using time resolved spectroscopy according to Equation 9. The quantum yield of fluorescence Φ_F is defined by Equation 10 which states that the non radiative decay processes, internal conversion and intersystem crossing, compete with the fluorescence process. If the non radiative processes are stimulated, which means that the values of the rate constants k_{IC} and k_{ISC} increase, both the lifetime τ and the quantum yield of fluorescence Φ_F will decrease.

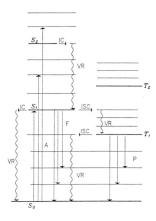

Scheme 6

Jablonski Diagram. Taken from reference 52 with permission

$$k = k_F + k_{IC} + k_{ISC}$$
Equation 7

k = rate constant for the decay of the excited state (in s^{-1})

k_F, k_{IC}, k_{IST} = rate constant for fluorescence (F), internal conversion (IC), and intersystem crossing (ISC), respectively (in s^{-1})

$$\tau = 1/k = 1/(k_F + k_{IC} + k_{ISC})$$
Equation 8

k = rate constant for the decay of the excited state (in s^{-1})

k_F, k_{IC}, k_{ISC} = rate constant for fluorescence (F), internal conversion (IC), and intersystem crossing (ISC), respectively (in s^{-1})

τ = lifetime of the excited state (in s)

$$I_{(t)} = I_{(0)} e^{-t/\tau}$$
Equation 9

$I_{(t)}, I_{(0)}$ Emission intensity at t and 0 seconds after excitation

τ = lifetime of the excited state (in s)

$$\Phi_F = k_F/(k_F + k_{IC} + k_{ISC})$$
Equation 10

Φ_F = quantum yield of fluorescence

k_F, k_{IC}, k_{ISC} = rate constant for fluorescence (F), internal conversion (IC) and intersystem crossing (ISC), respectively (in s^{-1})

In cases where the monomer and the polymer are non-fluorescent and a fluorescent probe has been added, the cure of a formulation cannot be determined directly.[62] One of the most striking changes taking place on a macroscopic scale during a (photo)polymerization process, is the increase of the viscosity and rigidity of the medium; the formulation changes from a liquid to a solid, glassy polymeric network. On a microscopic scale an increase of the microviscosity and a decrease of the available free volume are observed. Most fluorescent probes have been developed to monitor changes in the amount of free volume or in the microviscosity of their immediate environment. Table 5 shows a schematic overview of different methods for monitoring a (photo)polymerization process by using a fluorescent probe. The table summarizes the measured physical quantities, which aspects of the probe's behaviour are monitored, and what the probe requirements are, apart from a high quantum yield of fluorescence Φ_F and good solubility in a resin.

Table 5

Schematic representation of different modes of using fluorescent probes for characterizing polymeric materials

Physical Quantity	Probe Behaviour[a]	Probe Requirements[b]
Anisotropy of emission $I_\parallel/I_\perp$	Rotation (perpendicular to the light beam)	Polarized Absorption and Parallel Polarized Emissions
Excited state lifetime τ	Quenching	Medium Dependent Quenching
Emission Intensity I_{max}	Quenching	Medium Dependent Quenching
Emission Maximum λ_{max}	Stabilization	Medium Dependent Stabilization

a   referring to the behaviour of the probe in the excited state that is monitored
b   general requirements like high fluorescence quantum yield and high solubility are not listed

Most fluorescent probes for monitoring (photo)polymerization processes are small organic molecules that consist of fluorescent chromophores. The probe molecules are dissolved in the monomer and in the polymer. By attaching a reactive functionality to the chromophore, fluorescent probes are made which are dissolved in the monomer and will be covalently attached to the photoformed polymer. Such probes have the advantage that the incorporated probe molecules will not migrate through the polymer matrix. Their position is known and does not change in time. Reactive fluorescent probes that are covalently attached to substrate molecules are widely employed for biomedical applications, notably labelling experiments. Due to their higher price, fluorescent probes bearing a reactive functionality are not expected to play a major role in commercial applications for monitoring photopolymerization processes.

- *Polarization of Emission*

Monitoring the polarization of emission (P_F), sometimes referred to as the anisotropy of emission, is a photophysical technique employed in many fields of (polymer) chemistry.[63,64] In it a probe molecule is excited by linear polarized light and the polarization of the probe's emission is analyzed, see Figure 10. The direction of polarization is (partly) conserved if the probe is immobilized.[65] By rotation of the probe around the propagation axis of the excitation light, the polarization is lost, see Scheme 7 and Equations 11-13. The rotational relaxation time of the probe, ρ, is viscosity dependent. Other factors that determine the value of ρ are the size and shape of the probe. In order to monitor changes in viscosity by measuring the polarization of fluorescence, P_F, accurately, the rotational relaxation time of the probe ρ should be of the same order of magnitude as the lifetime of the excited state τ. When time resolved measurements are employed in which the polarization of emission P_F is determined as a function of time after a short laser pulse, Equation 12 can be used. For measuring P_F using steady state fluorescence, Equation 14 is employed which interrelates P_F, τ and ρ.

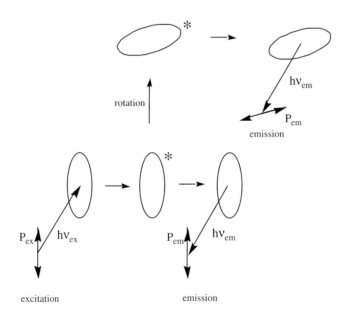

Scheme 7.

Schematic representation of the polarization of the emission of a fluorescent probe. The excited state is denoted by an asterix (*).

$$P_F = (I_\parallel - I_\perp)/(I_\parallel + I_\perp)$$
Equation 11

P_F = degree of polarization

$I_\parallel/I_\perp$ = Emission intensity parallel and perpendicular to the excitation beam.

$P_{F(t)}, P_{F(0)}$ = degree of polarization at t and 0 seconds after excitation

$$P_{F(t)} = P_{F(0)}\, e^{-t/\rho}$$

ρ = rotational relaxation time of the probe
Equation 12

$$P_{F(0)} = 3\cos^2\theta - 1/\cos^2\theta + 3$$

θ = angle between the polarization of the probe's absorption and emission
Equation 13

$$(1/P_F - 1/3) = (1/P_{F(0)} - 1/3)(1 + 3\tau/\rho))$$
Equation 14

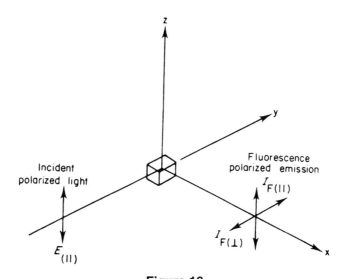

Figure 10
Optical coordinate system for measurements of the degree of fluorescence polarization. Taken from reference 51, with permission.

Monitoring the polarization of emission is not widely employed for monitoring neat polymerization processes. The technique requires additional polarization optics, and since the additional optics severely limit the intensity of both the excitation and emission beam, the method is relatively time consuming. A more severe limitation of this method is its lack of sensitivity at high conversions. In general, the technique is sensitive only at low viscosities usually below the gel point. After gelation occurs, which is almost instantaneously for resins containing multifunctional monomers, no changes in anisotropy of emission are observed for most probes.[66,67]

The requirements for fluorescent probes which are employed for monitoring a photopolymerization process by polarization of emission are not stringent. Apart from a high quantum yield of emission and the presence of well defined strongly polarized absorption and emission bands no other requirements apply. Anthracene is a good example of a probe which is well suited for measuring the viscosity of a medium by the polarization of emission. Due to the modest probe requirements, measuring the polarization of emission can be applied using probes that have been developed for other methods. Therefore determining the anisotropy of emission of a probe is often employed as an additional technique. An example is given in Figure 11. As far as we know this method has not been used for monitoring *photoinitiated* polymerization processes.

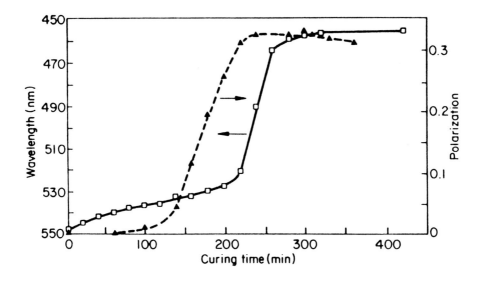

Figure 11
Fluorescence maximum (in nm) of 18 in (P)MMA and fluorescence polarization as a function of the thermal curing time. Curing was performed at 85°C; spectroscopic measurements at room temperature in time-gated mode with zero delay. Taken from reference 66, with permission.

- *Lifetime of Emission*

Monitoring the lifetime of fluorescence of a probe is a technique often employed in photophysical research. In general, the lifetime of the excited state τ increases as the rigidity of the sample increases. Nonradiative decay by Internal Conversion (IC) is often stimulated by collisions with other molecules and conformational changes of the excited state. Both processes are slowed down during a (photo)polymerization process by an increase of the microviscosity and a decrease of free volume in the medium. Lifetimes of phosphorescent species in the order of micro seconds, μs, are easier to determine than are those for fluorescent species which are in the order of nano seconds, ns. However, today emission spectra with nano (ns) and pico second (ps) resolution can be recorded using standard equipment and lifetimes of fluorescent probes can be determined routinely.

Probes for life time measurements must have high quantum yields of emission, and excited state lifetimes that are measurable and dependent on the rigidity of the medium. In many cases, lifetime determinations are performed as an additional technique in combination with other techniques employing probes that were developed especially for those other techniques.[34,66] Figures 12 and 13 show as the emission maxima and the lifetime of the fluorescent probe 2-dimethylamino-7-nitrofluorene (DANF, Scheme 16) as a function of the irradiation dose in the γ-induced polymerization of MMA and TEGDA.

From Figures 12 and 13 it can be concluded that the excited state lifetime of fluorescent probe τ is a very sensitive quantity for monitoring a polymerization process. In both resins MMA and TEGDA, τ increases from 1 to 3-4ns upon polymerization. It should be noted that the increase of τ is accompanied by an increase of the probe's emission intensity during the polymerization process.

As illustrated by Figures 12 and 13, the emission maximum λ_{max} strongly shifts to shorter wavelengths in the course of the polymerization. Qualitatively, the upper curves (emission maxima versus the irradiation time) and the lower curves (life times versus the irradiation time) in the Figures 12 and 13 are very similar though differences such as a higher sensitivity of the lifetime τ during the first stages of the MMA polymerization can be observed.

Figures 12 and 13 also illustrate the different behaviour of a monofunctional and a difunctional monomer during a radiation induced polymerization reaction. The monofunctional monomer does not undergo noticeable changes during the first stages of the polymerization reaction. After a gel is formed, the rate of polymerization dramatically increases due to the Trommsdorff effect[68] and drastic changes are also observed in other physical properties of the sample. In difunctional monomers the polymerization process starts at high rates immediately after the beginning of initiation. Changes in physical properties occur more gradually. Local gel formation through the formation of networks is the most likely explanation for the behaviour of multifunctional resins.

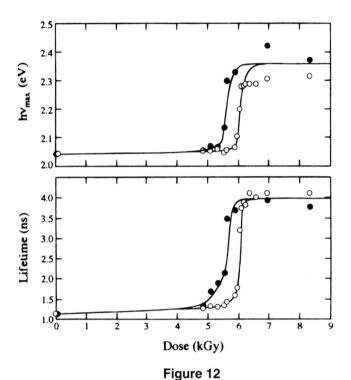

Figure 12
The dependence of the fluorescence emission maximum (upper) and decay time (lower) on the radiation dose for deaerated MMA using DANF (16) concentrations of 0.5 mol m^3 (filled circles) and 2 mol m^3 (open circles). Taken from reference 34, with permission.

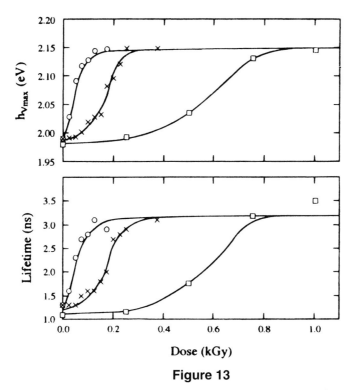

Figure 13

The dependence of the fluorescence emission maximum (upper) and decay time (lower) on the radiation dose for 1 mol m^3 DANF (16) in TEEGDA: (open circles) fully deaerated and sealed sample; (x) air saturated but sealed cuvette; (open squares) air saturated open cuvette. Taken from reference 34, with permission.

Depending on the number of transients required to get an adequate signal to noise ratio, lifetime determinations can be very fast. Assuming that recording one transient takes 50 to 100 ns, and 10 (accumulated) pulses give a reasonable signal to noise ratio, one lifetime determination would require approximately 1 µs. Since in most cases irradiation of the sample excites the probe, luminescent lifetime determinations are not suited for real time monitoring under continuous irradiation conditions. For monitoring (pulsed) laser induced irradiations,[69] however, this technique might be useful.

- Intensity of Emission

As already mentioned Equations 8 and 10, monitoring the intensity of the emission of a probe during a polymerization process basically provides the same information as lifetime measurements. Both techniques differ in the experimental set-up needed to obtain the data. The emission intensity of many probes which have been used for monitoring photoinitiated polymerization processes, increases as the polymerization process proceeds.[34,70,71] Notable exceptions are rigid probes like coumarins or polycyclic aromatics that have quantum yields of fluorescence close to unity regardless of the (micro)viscosity of the medium. During a polymerization process, the microviscosity of the medium increases and the free volume in the medium decreases. Both factors slow down the non radiative decay of the probe (internal conversion) thereby raising the emission quantum yield Φ_F, see Equation 10. (Page 263)

Probes which monitor polymerization by an extremely strong increase in emission intensity are the rotor probes reported by Loutfy and Law, see Scheme 8.[72,73,74,75,76,77,78,79]

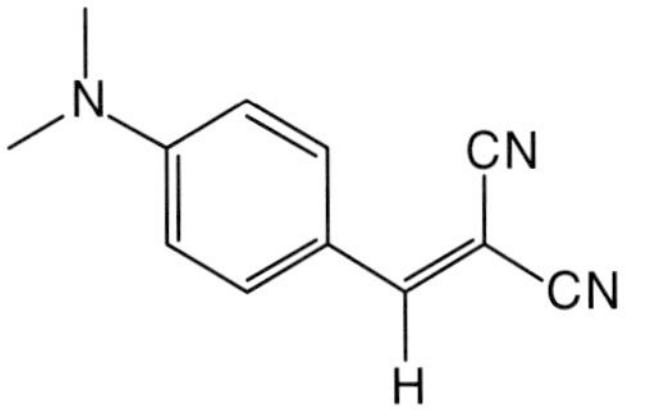

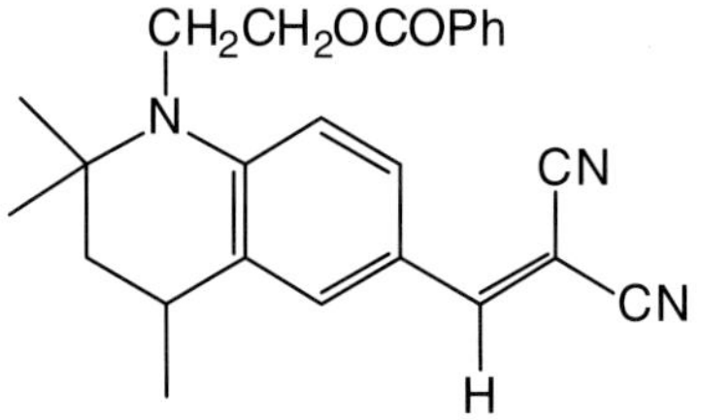

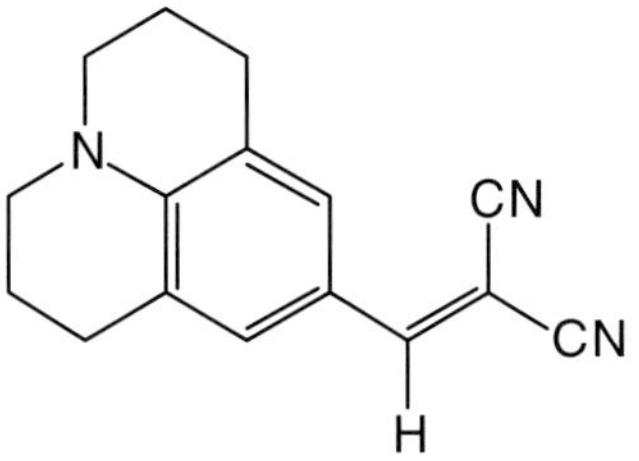

Scheme 8
Intensity sensitive rotor probes developed by Loutfy (5-7).

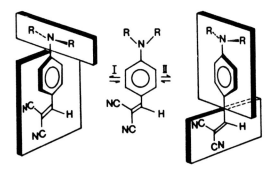

Scheme 9
Proposed rotary decay of the Loutfy probes (5-7). Taken from reference 77, with permission.

Each probe depicted in Scheme 8, contains a 2,2-dicyanovinyl moiety and a dialkylamino group attached to a phenyl group, which is free to rotate (**5**), partially immobilized (**6**), or totally immobilized (**7**). A rotor mechanism is proposed for these probes in which the radiationless decay or internal conversion (IC) is facilitated by rotation of the dialkylamino moiety (**5** only) and rotation of the larger 2,2-dicyanovinyl functionality (**5, 6** and **7**), see Scheme 9. For the small dimethylamino functionality, the free volume required for rotation is much smaller than the free volume required for rotating the 2,2-dicyanovinyl group. The rotational freedom of these moieties strongly depends on the (local) viscosity and the free volume of the medium in which the probe is incorporated. The rate of the rotation of the dimethylamino group in **5** is estimated at $2 \times 10^{11} \mathrm{s}^{-1}$ in solution. The non radiative decay rates of the excited states of **5, 6** and **7** in MMA are 2.5×10^{11}, 1.3×10^{11} and 9.3×10^{10} s^{-1}, respectively. In the same solvent, the singlet excited state lifetimes of these probes are 4.6, 7.5 and 10.7ps, respectively.

During the thermal polymerization of MMA, the emission intensity of **5, 6** and **7** increases by factors of 10, 30 and 40, respectively (Figures 14 and 15). In other monofunctional acrylates, methacrylates and styrene, the observed increases in emission quantum yield are notably smaller. Apparently the available free volume in PMMA is smaller than in any of the other polymers. The fact that the smallest increase in intensity of emission, as well as the smallest quantum yield of fluorescence in both monomer and polymer was observed for **5**, is explained by the availability of relaxation pathway 1 for this probe, Scheme 9. The rotor probes **5-7** have also been used to monitor free volume, physical ageing[80] and the tacticity of PMMA. As we know, rotor probes like **5-7** have not been used for monitoring photoinitiated polymerization processes.

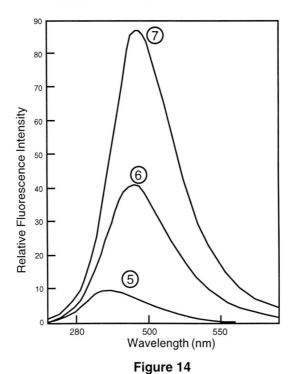

Figure 14

Relative emission spectra of 5, 6 and 7 in PMMA, measured at 23°C. Taken from reference 77, with permission.

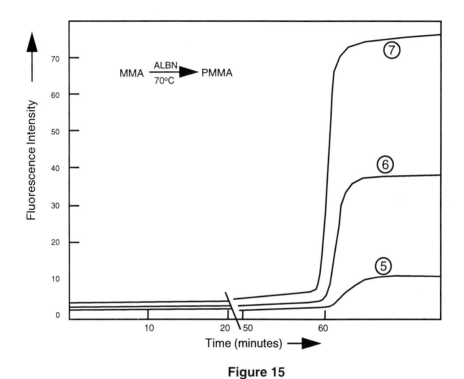

Figure 15
Dependence of the fluorescence intensity of 5, 6 and 7 on MMA polymerization time (polymerization carried out at 70°C. Taken from reference 77, with permission.

In many studies in which the shift in emission wavelength is monitored during a photoinitiated photopolymerization process, the intensity of emission is measured along with other parameters during the reaction.[34,66,70,84,116] In many cases, a distinct increase in emission intensity is observed during the first stages of the photopolymerization process.[81]

A recent development in the field of radiation induced polymerizations is the use of maleimide functionalized chromophores as fluorescent probes, see Scheme 10. Maleimide functionalized chromophores are commercially available for labelling experiments and are often used for biomedical applications.[82] In most cases the probe is virtually non-fluorescent when the maleimide functionality is intact, since the maleimide functionality quenches the luminescence of the attached chromophore. After saturation of the double bond by means of an addition reaction, a succinimide is formed which does not quench the emission of the chromophore.[83,84] When maleimide functionalized probes are added to an acrylate, only those probe molecules incorporated in the polymer backbone emit light. Therefore, the intensity of the chromophore emission is proportional to the cure of the sample. It should be noted that the emission intensity is proportional to the cure of the resin only if the probe molecules:

i.   are incorporated statistically

ii. do not decompose photochemically

iii. are not effected by (physical) changes that occur in the medium,

iv. are not effected by chemical processes other than the acrylate addition.

By a sensible choice of the attached chromophore, multifunctional probes can be made. Such probes can simultaneously monitor the cure (by means of the emission intensity) and other properties of the resin (by means of changes in the emission characteristics of the chromophore). The maleimide functionalized fluoroprobe **8**[90,91] is a good example of a multifunctional probe. Employing **8**, the cure of MMA is accurately monitored at *all* conversions. The cure at low conversions is detected by the increase in fluoroprobe emission intensity, whilst the cure at higher conversions is monitored by means of the blue shift in the emission maximum λ_{max} of the chromophore, see Figure 16. The blue shift of the chromophore is caused by an increase in microviscosity of the medium, see pp 281-284. Since the shift of the probe's emission is only observed at higher double bond conversions, after gelation, both quantities are required to accurately monitor the cure of MMA throughout the reaction.

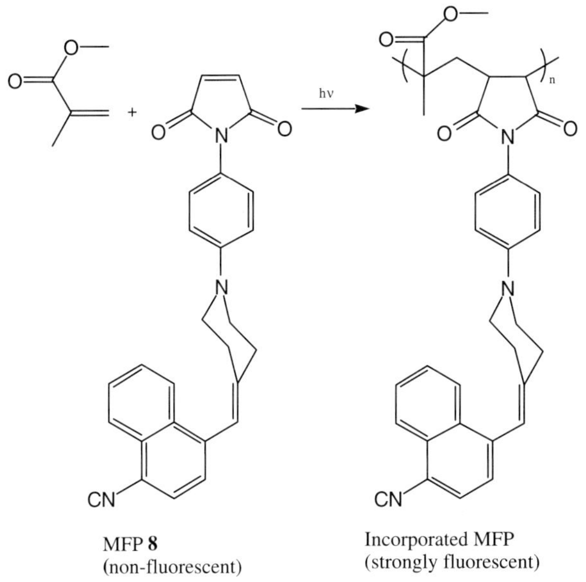

Scheme 10
Maleimide functionalized fluoroprobe MFP (8).

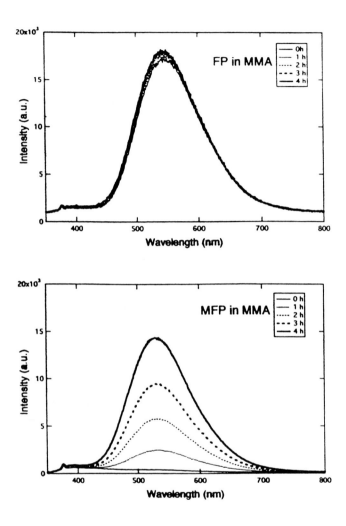

Figure 16

Emission spectra of "maleimido-fluoroprobe" MFP (8) versus "fluoroprobe" FP (18) in the early stages of the MMA polymerization. Taken from reference 84, with permission.

Monitoring the cure of a formulation by means of the intensity of emission is fast and simple. Data can be obtained from "normal" emission spectra and, if faster measurements are required, from the emission at a fixed wavelength. A distinct disadvantage of monitoring the intensity of emission during a photopolymerization process is the possibility of photobleaching of the chromophore. Photobleaching or photodecomposition diminishes the emission intensity regardless of the photopolymerization process and has to be avoided. Other factors that might interfere with the accuracy of the measurements are; fluctuations in excitation intensity, an uneven distribution of the probe throughout the sample and an uneven sample thickness. Fluctuations in excitation intensity are usually not a problem with commercial spectrofluorometers. Uneven sample thicknesses and probe distributions can be dealt with by keeping the sample in a fixed position[85] or by random sampling at many different positions. As long as photobleaching of the probe does not occur, intensity sensitive probes can be used for real time monitoring of a photoinitiated polymerization process. The sensitivity of this methodology the distribution of the probe throughout the sample, the thickness of the sample and to the sample's geometry, make this method less attractive for on line monitoring.

- *Position of the Emission Maximum, (λ_{max})*

Many probes that monitor (photo)polymerization processes via shifts in emission maxima have been developed and a multitude of mechanisms have been proposed to describe this. From a practical point of view, wavelength shifting probes are superior to all others. The technique is very sensitive, fast[86] and measurements are simple, taking "normal" emission spectra using conventional equipment. This is a distinct advantage over measuring the anisotropy of emission or taking time resolved spectra. Secondly this technique, which either monitors the emission maximum of the probe λ_{max} or the ratio of the emission intensity at two fixed wavelengths $I_{\lambda 1}/I_{\lambda 2}$, is applicable to samples of different shapes and geometries as well as those containing fillers and other additives. The technique is not influenced by the concentration of the probe, which implies that photobleaching of the probe is usually not a problem.[87] This is a distinct advantage over monitoring the emission intensity of a probe at a fixed wavelength.[88]

Probes that monitor a polymerization process by means of a shift in emission spectra are of many types and act by different mechanisms. Most probes contain one luminescent chromophore and are employed at such low concentrations that emission spectra of isolated chromophores are monitored. The one exception to this rule are excimer forming probes which form excimers by an *intra*molecular mechanism (bichromophoric probes) or an *inter*molecular mechanism (high probe concentrations). For all fluorescent probes discussed in this section the emission wavelength depends upon the microviscosity of the medium, or the free volume in the polymeric matrix.

All probes respond to a polymerization by a blue shift in emission, meaning that in rigid media the excited state of the probe is stabilized to a lesser extent. A schematic representation is depicted in Scheme 11. After excitation of the probe, the Franck-Condon excited state relaxes to its lowest vibrational level. This relaxation is very fast, and occurs regardless of the medium in which the probe is embedded. Subsequent relaxation processes can occur, which involve conformational changes, rotations and translations of both probe and solvent molecules. The rates at which these processes occur depend on the microviscosity of the medium and the free volume in the medium and can be relatively slow compared to the lifetime of the excited state. The extent to which these relaxations proceed during the lifetime of the excited state determines the wavelength at which the probe emits. A number of the "slow" medium sensitive processes that lead to stabilization of an excited state are summarized in Table 6.

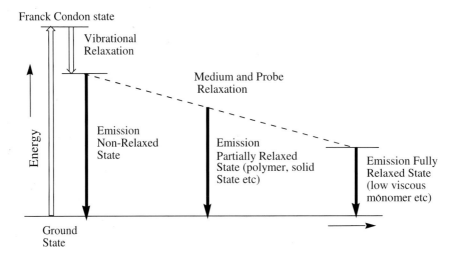

Scheme 11

Schematic representation of the emission of a fluorescent probe in different media.

Table 6
Schematic representation of different types of relaxation processes that can be employed for fluorescent probes that monitor medium rigidity.

Relaxation process	Requirements	Examples
Conformational rearrangements of the probe	free volume in the matrix	Organic salts (p 309)
Rearrangement of the solvent	translational and rotational mobility of solvent molecules	ICT probes (p 291)
Specific probe-solvent interaction	translational and rotational mobility of solvent and probe molecules	Functionalized chromophores, see reference 53
Chromophore-chromophore interactions	probe diffusion	Excimer probes (p 284)

Most of the processes mentioned above and summarized in Table 6, involve translational and rotational motion that are retarded in viscous media. In order that these be monitored by the probe's emission, a significant retardation or inhibition must occur in the ns (10^{-9} s) time domain.

In many cases, it is not a trivial task to separate the different contributing factors and often several processes take place simultaneously. At this point it is worthwhile to observe that probes which require free volume in order to undergo a conformational change, thereby probe the mobility of the first shell of "solvent" molecules *exclusively*. Probes that are stabilized by rearrangement of the solvent might, depending on the mechanism, probe the mobility of a larger section of their direct environment.

Apart from the specific requirements of different types of probes, the major requirement for fluorescent probes that probe the rigidity of their environment by means of a shift in emission, is that the Frank-Condon excited state and the solvent are not in equilibrium. As already mentioned, both rearrangement of the solvent and conformational changes of the Frank-Condon excited state can lead to a decrease of the energy content of the excited state of the probe. In many cases this non-equilibrium condition is achieved by a significant change of the charge distribution between the ground state and the first excited state.

i. Excimer forming probes

A well known technique for monitoring the local viscosity of a medium is to use excimer forming probes. Excimers are complexes formed by the association of an excited chromophore with another chromophore in the ground state. In general, excimers are fluorescent and exhibit a characteristic broad, structureless emission at longer wavelengths than the fluorescence of the isolated chromophore. A schematic representation of this process is given in Scheme 12.[59]

$$P \longrightarrow P^*$$
$$P+P^* \longrightarrow PP^*$$
$$P^* \longrightarrow P+h\nu_M \text{ "normal" emission}$$
$$PP^* \longrightarrow 2P+h\nu_D \text{ excimer emission}$$

Scheme 12.
Schematic representation of *inter*molecular excimer formation. P is the fluorescent probe and excited states are denoted by an asterix (*).

Assuming complexes between chromophores are not formed in the ground state, the formation of excimers takes place exclusively during the lifetime of an excited state. Therefore the extent of excimer formation depends on the microviscosity of the medium. An example of a fluorescent chromophore that readily forms excimers is pyrene. Normal, unimolecular emission of pyrene is observed at 380nm, while the pyrene excimer emits at 490nm. When employing pyrene as an excimer forming probe, one monitors the ratio between the emission of the isolated chromophore and the excimer emission I_m/I_d. Pyrene itself can be used as a fluorescent probe,[89] but due to the high concentrations required to observe excimer formation (usually in the order of 10^{-3}M) it is more convenient to use a bichromophoric probe like 1,3-bis-(1-pyrene)propane (**9**), see Scheme 13[90,91,92,93] Probes like **9** form excimers by an *intra*molecular mechanism and are employed at much lower concentrations, typically 5 x 10^{-5}M.

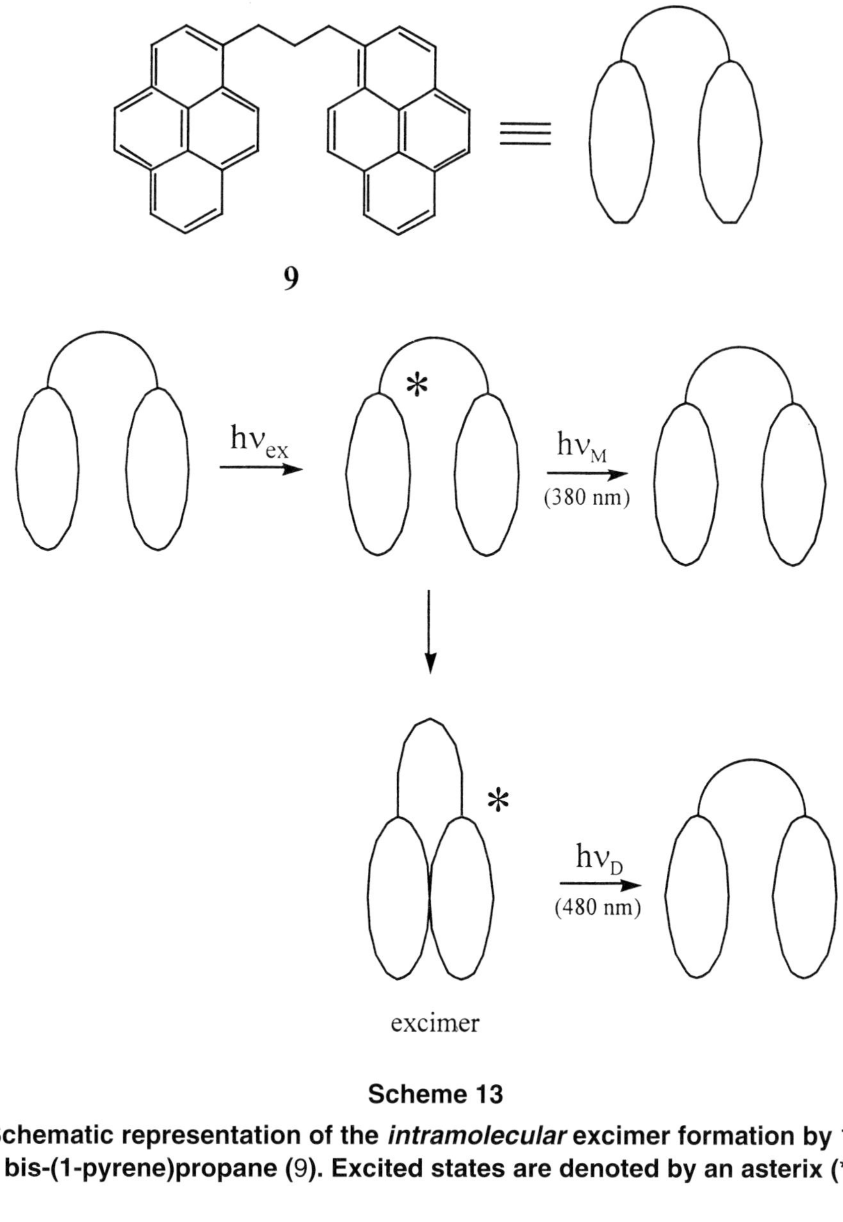

Scheme 13
Schematic representation of the *intramolecular* excimer formation by 1,3-bis-(1-pyrene)propane (9). Excited states are denoted by an asterix (*).

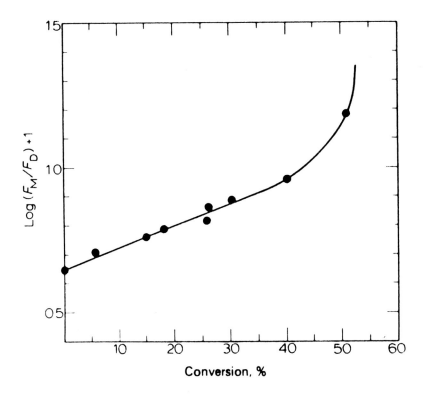

Figure 17
Plot of the intensity ratio log (F_M/F_D) of **9** against the percentage conversion α. F_M and F_D are referred to as I_M and I_D in the text. Taken from reference 92, with permission.

Excimer forming probes, like pyrene and 1,3-bis-(1-pyrene)propane (**9**) have been used to monitor the thermal radical initiated polymerization of MMA and the thermal polymerization of a diepoxy diamine system. A typical characteristic of excimer forming probes is a high sensitivity at low conversions before the gel point and a lack of sensitivity at conversions beyond this point. An example is shown in Figure 17. If pyrene is employed for monitoring the thermal polymerization of MMA, the probe commences monitoring the formation of polymer from the beginning. In the same system other probes, like the rotor probes (**5-7**), and ICT probes like DMANS (**15**), DANF (**16**) and fluoroprobe (**18**), do not monitor changes in the medium below conversions of 40 to 50% (Figure 15). At higher conversions, excimer formation is strongly suppressed and no information can be extracted from the emission of the probe. These characteristics make excimer forming probes, along with maleimide functionalized probes, (Page 271-279) complementary to the rotor probes **5-7** and ICT probes like **15,16** and **18**. Since resins containing multifunctional monomers form a gel at very low conversions, excimer forming probes might not be well suited for monitoring the cure of multifunctional resins.

As far as we know, excimer forming probes have not been employed for monitoring photopolymerization processes. Employing simple probes like pyrene is not an attractive proposition due to the high concentrations that are needed for excimer formation. Though pyrene does not interfere with the thermally induced radical polymerization of MMA,[89] competitive absorption by pyrene itself can retard the polymerization process. When bichromophoric probes are used, interference of the probe will be limited. However, probe stability is expected to be a serious problem. For example, significant photodecomposition was already observed for probe **9** while recording emission spectra!

ii.   Twisted Intramolecular Charge Transfer (TICT) Probes

In recent years the development of fluorescent probes for monitoring photopolymerization processes that form a Twisted Intramolecular Charge Transfer (TICT) state has been reported.[94,95,96,97,98,99] Such molecules are composed of an electron donating group and an electron accepting group linked by an aromatic moiety (D-π-A). Molecules that form a TICT state exhibit two emitting excited states. First, a Franck-Condon (FC) excited state is formed which has the same conformation as the ground state. The degree of charge separation in the Franck-Condon excited state is limited. The other excited state, the TICT state, is formed by conformational changes most likely being rotation of the donor or the acceptor from the plane defined by the aromatic moiety. In most cases, rotations between the rigid aromatic moiety and the electron donor, the electron acceptor or both are possible. In the TICT state, complete charge transfer has taken place giving this molecule a highly dipolar character. In some cases calculations suggest that a twisting process which takes the donor and the acceptor out of conjugation is a prerequisite for complete charge transfer. The formation of TICT was first reported by Grabowski[100] and later reviewed by Rettig.[101] A schematic representation of the process is displayed in Scheme 14.

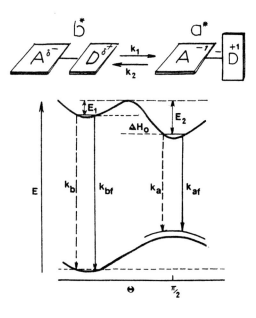

Scheme 14

Schematic representation of formation of the Franck Condon excited state (b*) and the TICT excited state (a*). Taken from reference 94, with permission.

The spectroscopic behaviour of TICT probes is very complex. To the first approximation, one expects an emission spectrum composed of two contributions. The emission of the Franck-Condon state has a relatively small Stokes shift and is relatively close to the absorption spectrum. The TICT state is formed by a spontaneous twisting process, which lowers the energy content of the excited state and hence the TICT emission is red shifted. In order to form the TICT state, free volume is required so the ratio of the Franck-Condon and the TICT emission should be a measure of the amount of free volume in the medium.

In reality, the situation is more complex. While the emission of the Franck-Condon state occurs at a fixed wavelength, irrespective of the medium in which the probe is embedded, the emission maximum of the TICT state is strongly dependent on the medium. Due to the strongly dipolar character of the TICT state, its emission is sensitive to the polarity (solvatochromism)[38,39] and the rigidity of the medium. Therefore, the ratio between the Franck-Condon and the TICT excited states depends on the available free volume, while the position of the TICT emission depends on the microviscosity and the polarity in the medium.

Despite the complexity of the photophysics of TICT probes, the observed emission of most fluorescent probes of the D-π-A architecture during a (photo)polymerization process is not as complex as one might expect. Though many fluorescent probes for monitoring polymerization processes have been classified as TICT probes,[101] the dual emission that is characteristic for such systems is demonstrated in only a limited number of cases. As an example, Figure 18 shows the emission spectra of N-(1-naphthyl)-carbazole (**10**) during the thermal polymerization of MMA. It should be noted that N-(1-naphthyl)-carbazole (**10**) requires a large amount of free volume for the twisting process and PMMA is a polymer that offers little. Dual fluorescence has also been observed for para N,N-dialkylaminobenzoic acid esters (**11**) and para-N,N-dialkylaminobenzonitriles (**12**) in PMMA matrices, see Scheme **15**.[102]

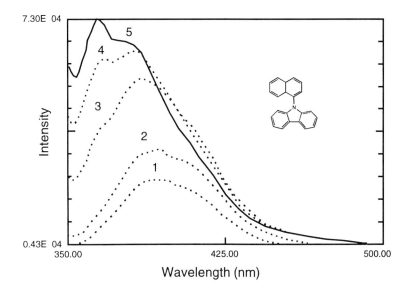

Figure 18

Emission spectra of N-(1-naphthyl)-carbazole (10) during the thermal polymerization of TMPTA-VP (9:1) mixtures. Note the dual fluorescence at high conversions. Taken from reference 98, with permission.

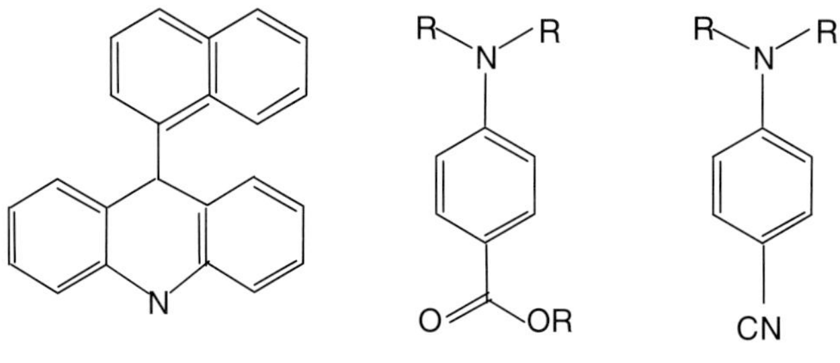

Scheme 15
TICT probes (10-12).

In most other cases where D-π-A probes are employed for monitoring (photo)polymerization processes, no dual emissions have been found, but the emission is highly sensitive to medium rigidity and medium polarity. This implies that the excited state is highly dipolar and intramolecular charge transfer is observed under all conditions. In most cases, the polymers formed in a polymerization are less densely packed than PMMA and the probes contain smaller electron donors (like the dimethylamino group) and smaller acceptors. This might explain the lack of dual fluorescence in such cases. Whether or not the formation of the charge transfer state is preceded by a twisting process does not appear to be relevant. Intermolecular charge transfer takes place under all circumstances from the monomer to the fully cured polymer. Therefore, the D-π-A probes that are employed for monitoring photopolymerization processes will be regarded as Intramolecular Charge Transfer probes and are discussed in the next section.

iii. Intramolecular Charge Transfer (ICT) Probes

The majority of fluorescent probes developed for monitoring (photo)polymerization processes exhibit intramolecular charge transfer. Most probes are of the D-π-A type but charge transfer probes of the D-σ-A type have also been developed. A number of well characterized D-π-A molecules that are employed for monitoring (photo)polymerization processes are listed in Scheme 16.[34,70,103,104,105,106,107]

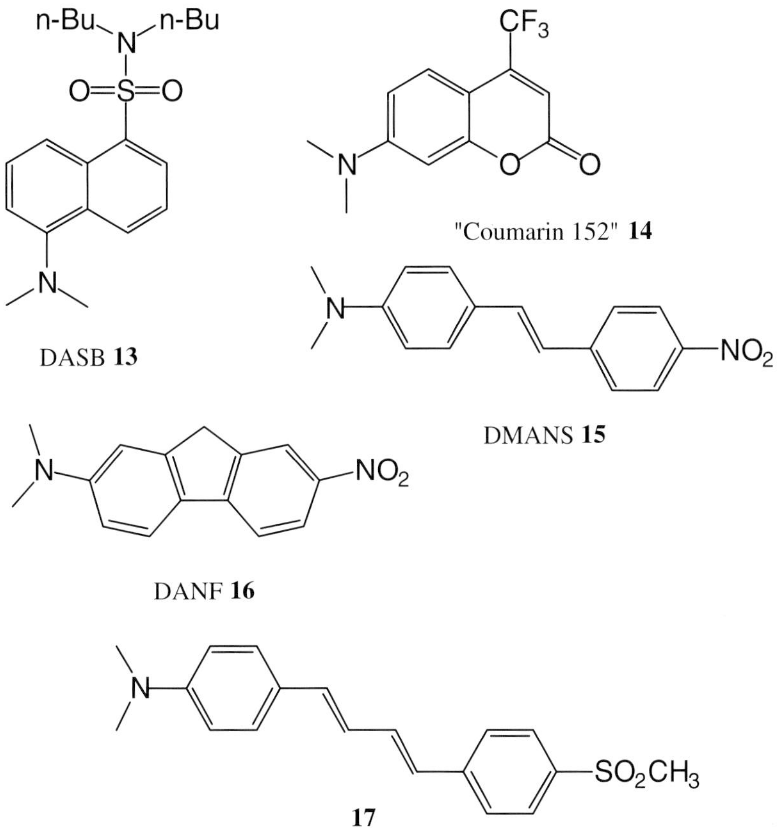

DASB **13**

"Coumarin 152" **14**

DMANS **15**

DANF **16**

17

Scheme 16

Examples of ICT probes of the D-π-A type.

Most ICT probes are of the D-π-A type, and many of these molecules are well known as laser dyes or materials for second order Non Linear Optical (NLO) applications.[108] The architectural variations are manifold. In most cases, a dimethylamino group is used since this group is a very strong electron donor. Also the dimethylamino group is synthetically easily accessible. The electron acceptor is generally a nitro, sulphonyl or cyano moiety. In general, the nitro and cyano group are stronger electron acceptors leading to higher dipole moments in the excited state. The sulphonyl functionality is a weaker electron acceptor, but may be employed in cases where probes absorbing at shorter wavelengths are required.

A large variety of aromatic moieties is available for the π-system interconnecting the donor and acceptor group. Phenyl groups are not generally employed since they absorb in the deep UV and exhibit low quantum yields of emission. Naphthyl, anthryl, fluorenyl, coumarin, stilbene and diphenylbutadiene groups are commonly used. In general those probes that have a rigid aromatic group have higher fluorescence quantum yields, especially in solution. The size of the aromatic moiety influences the position of both the absorption and the emission. Increasing the size of the group results in red shifts. In those cases in which flexible aromatic moieties such as stilbene, 1,4-diphenylbutadiene and 1,6-diphenylhexatriene are used, a strong decrease of the quantum yield of emission upon increasing the size of the aromatic group is observed.[109,110]

The photophysics of D-π-A molecules is well established. In the ground state such molecules have modest dipole moments. Upon excitation a strong increase in dipole moment is observed due to charge separation, (Scheme 16). For DMANS, the most sensitive probe of the D-π-A series, dipole moments of 8 Debye in the ground state and 32 Debye in the excited state have been reported.[111,112,113] A schematic representation of the photophysical behaviour of such ICT probes is depicted in Scheme 17.

As is shown in Scheme 17, a strong dipole moment is generated upon excitation of the fluorescent probe. Subsequently the excited state S_1 is stabilized by the action of solvent molecules to form the stabilized or relaxed excited state S_1', which is fluorescent. Mobile solvent molecules can rearrange into an orientation which solvates the strongly dipolar excited state, thereby reducing its energy. The more polar the solvent is, the stronger the solvation and the probe's emission shifts to the red upon increasing the solvent polarity. Since the lifetime of the excited state τ is limited to a few nanoseconds at most, the mobility of the solvent molecules is of paramount importance for solvation. Reducing the mobility of solvent molecules can lead to an incomplete solvation of the excited state, since rotational and translational movement of solvent molecules is reduced. Therefore a reduction of the mobility of the solvent molecules leads to blue shifts in the probe's emission.

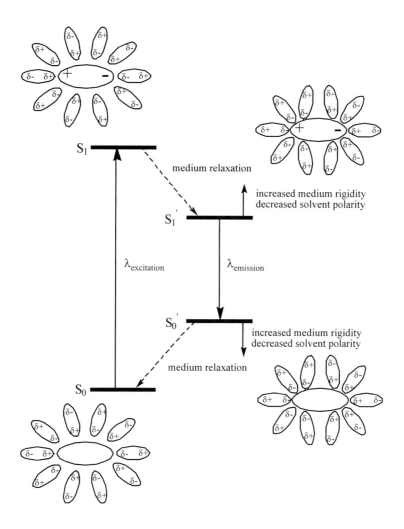

Scheme 17

Schematic representation of the influence of the medium on the emission of an ICT probe. Probe molecules are represented by large elipsoids, while solvent molecules are represented by small elipsoids.

The difference in dipole moment in the ground and excited state is clearly manifested in the emission spectra of D-π-A molecules. The strongly dipolar excited state is far from equilibrium with its direct environment and its emission is very sensitive to the medium

The following relations between the emission of ICT probes and the medium have been reported.

First, these molecules are solvatochromic. As an example the emission spectra of **17** in a number of solvents of different polarity are shown in Figure 19. Equations 15 and 16 describe the emission wavelength of a probe in a specific solvent as a function of the difference of the dipole moments in the ground and the first excited state, the radius of the cavity in which the probe resides and the solvent polarity. By plotting the emission wavenumber v_{em} as a function of the solvent parameter Δf, the sensitivity to solvent polarity or "solvatochromic power" of a probe can easily be determined as the slope of the line interconnecting the data points, see Figure 20.

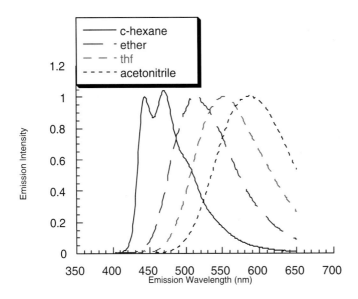

Figure 19

Normalized emission spectra of 17 in solvents of different polarity.

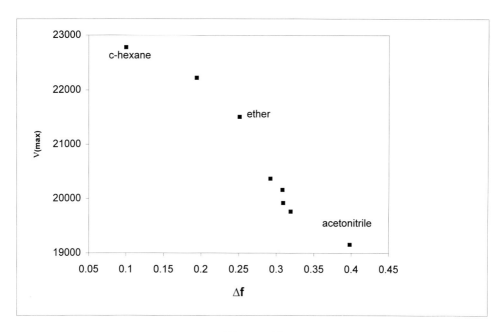

Figure 20

Plot of the emission wavenumber at the maximum emission of 17 (v_{max}) versus the solvent parameter Δf.

$$v_{CT} = v_{CT\ (0)} - (2[\mu_e - \mu_g]^2/r^3 hc)\Delta f$$

Equation 15

$v_{CT}, v_{CT\ (0)}$ = emission wavenumber in solvent and gas phase (in cm^{-1})

μ_e, μ_g = dipole moments in excited and ground state (in Debye)

r = radius of the cavity in which the probe fits (in Å)

c = speed of light

h = Planck constant

Δf = solvent parameter

$$\Delta f = (\varepsilon-1)/(2\varepsilon+1)-(n^2-1)/(4n^2+2)$$
Equation 16

ε = dielectric constant
n = optical refraction index

Secondly, these molecules are sensitive to the rigidity in the medium; an increase of the rigidity of the medium is monitored by means of a blue shift in emission. The ability of ICT probes to monitor the medium rigidity has not only been demonstrated during polymerization reactions. In glasses that are formed by rapidly cooling dimethacrylates, large blue shifts are observed for DMANS (**15**) in DEGDMA and TREGDMA, as is shown in Figure 21.[114] Upon heating these samples, a process in which the viscosity of the medium strongly decreases, the emission wavelengths of **15** return to their previous values.

In addition to being sensitive to the polarity and the rigidity of the medium, ICT probes are also sensitive to morphology changes in their immediate environment.[115] For instance, the crystallization of dimethacrylates is detected in the emission of DMANS. The emission maxima of **15** in TEGDMA upon slowly heating a glass formed by rapidly cooling the solvent are shown in Figure 22. Both the formation and the melting of crystalline TEGDMA are clearly detected.[126]

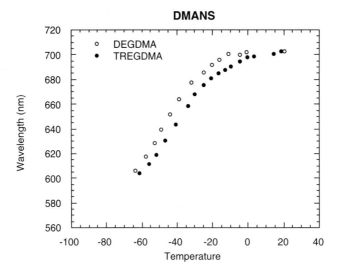

Figure 21
Emission maxima of 15 in DEGDMA and TREGDMA as a function of the temperature.

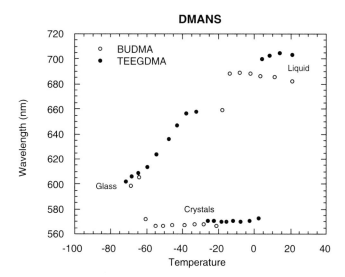

Figure 22

Emission maxima λ_{max} of 15 IN BUDMA and TEEGDMA as a function of the temperature.

Upon photopolymerization of the medium in which they are dissolved, ICT probe emissions exhibit strong blue shifts. Such blue shifts for stilbenes like DMANS (**15**), are accompanied by a strong increase in emission intensity. The emission spectra before and after polymerization and the emission maximum as a function of the double bond conversion of **15** in EGDMA are shown in Figure 23 and 24.

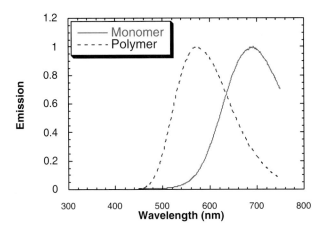

Figure 23
Normalized emission spectra of 15 in EGDMA before and after photoinitiated polymerization. Taken from reference 70, with permission.

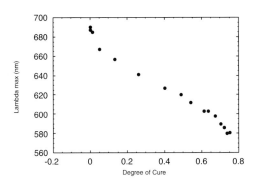

Figure 24
Emission maxima λ_{max} of 15 as a function of the double bond conversion α during the photoinitiated polymerization of EGDMA.

The emission maxima of **15** before and after photopolymerization in a series of diacrylates and dimethacrylates (Scheme 18) are shown in Table 7.[70,116] Clearly the emission of DMANS (**15**) in these di(meth)acrylates is solvatochromic. The emission of the probe shifts to longer wavelengths as the polarity of the formulation increases by adding either ethylene glycol units (in the series EGD(M)A to TEGD(M)A) or by removing ethane units (in the series HEXD(M)A to EGD(M)A). The solvatochromism in the probe's emission does not only manifest itself in the monomers but also in photoformed polymeric networks.

After the photoinitiated polymerization of the formulations, large blue shifts in the emission of **15** are observed. The largest blue shifts upon photopolymerization, up to 114 nm in EGDA, are found in those resins with the shortest spacer connecting both (meth)acrylate units. This means that despite the fact that the lowest double bond conversions is observed for the resins with the shortest spacers the largest increase of the microviscosity of the medium is registered by the probe in such di(meth)acrylates.

Another point worth noting is that after the photopolymerization process is completed and no increase in double bond conversion can be detected by FTIR spectroscopy, the emission of DMANS (**15**) continues to shift to the blue, presumably due to physical ageing.

Finally, it has been demonstrated that the sensitivity to solvent polarity (solvatochromic power) and the magnitude of the blue shift of the probe's emission during a photoinitiated polymerization are proportional for many D-π-A probes,[70] as is shown by Figure 25.

Table 7

Emission maxima of 15 in different resins before and after photopolymerization and the final cure of polymerized samples. (15μm films between glass plates). Data taken from references 70 and 124.

Backbone	Phase	Emissions			
		DMA		DA	
		λ_{em}	(cure %)	λ_{em}	(cure %)
DOD	monomer	639	(91)		
	polymer	555			
	shift	84			
HEX	monomer	664		684	
	polymer	570	(86)	583	(92)
	shift	94		101	
BUD	monomer	677		698	
	polymer	576	(81)	597	(88)
	shift	101		102	
EG	monomer	691		721	
	polymer	585	(78)	607	(84)
	shift	106		114	
DEG	monomer	699		714	
	polymer	595	(82)	610	(91)
	shift	104		104	
TREG	monomer	695		714	
	polymer	601	(86)	615	(94)
	shift	94		99	
TEG	monomer	704		701	
	polymer	613	(92)	629	(96)
	shift	91		72	

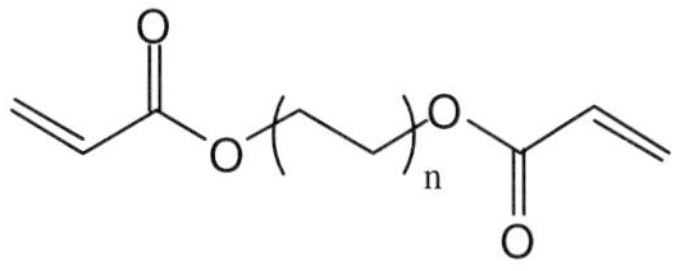

n=1 EGDA
n=2 BUDA
n=3 HEXDA (HDDA)

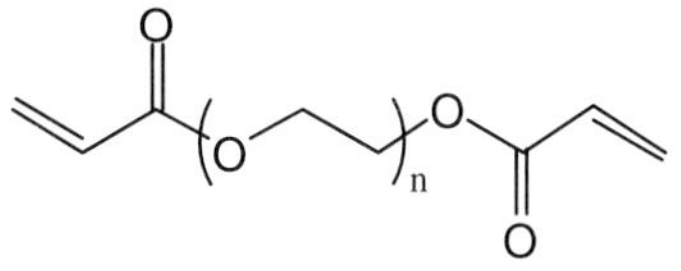

n=1 EGDA
n=2 DEGDA
n=3 TREGDA
n=4 TEEGDA

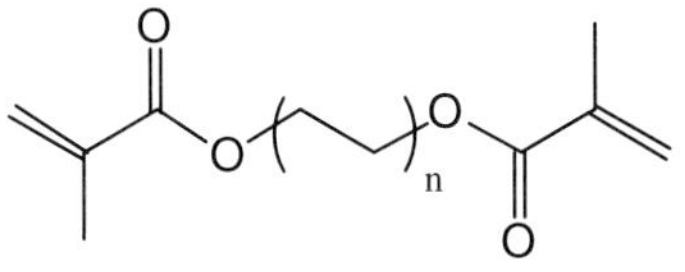

n=1 EGDMA
n=2 BUDMA
n=3 HEXDMA
n=6 DODDMA

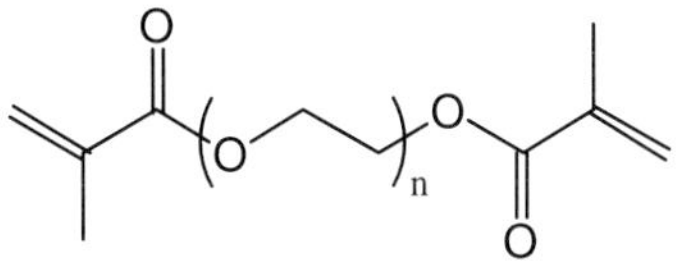

n=1 EGDMA
n=2 DEGDMA
n=3 TREGDMA
n=4 TEEGDMA

Scheme 18
(Meth)acrylates employed for photoinitiated polymerization processes.

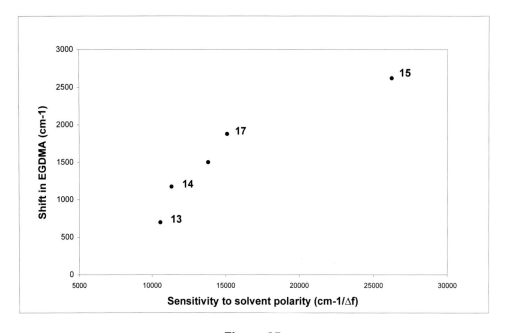

Figure 25
Shift of the emission wavenumber at the maximum emission of various probes v_{max} in EGDMA versus the sensitivity for solvent polarity for different ICT fluorescent probes of the D-π-A type.

Special types of ICT probe are the D-σ-A molecules developed by Verhoeven and coworkers, see Scheme 19.[117,118,119] In the ground state, the donor and the acceptor group behave like isolated units; the absorption spectrum of "fluoroprobe" (**18**) is the sum of a the absorption spectra of a cyanonaphthalene and a dialkylaminoaniline chromophore. The dipole moment in the ground state is also very small. In the excited state, however, complete charge separation occurs and a dipole moment of 27 Debye is reported for **18**. The molecule is highly fluorescent and due to the unprecedented difference in dipole moment between ground and excited state, it is the most solvatochromic molecule known today, see Figure 26. Since solvatochromism and rigidochromism are interconnected, fluoroprobe is also regarded as the most sensitive probe for monitoring (photoinduced) polymerization processes.

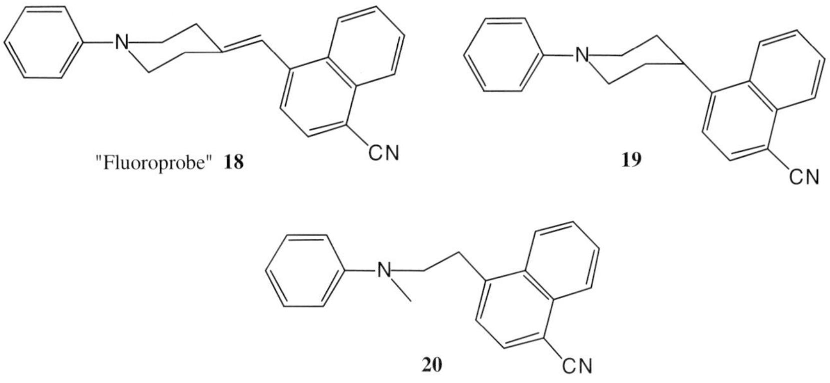

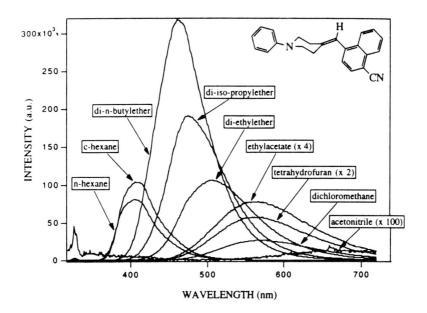

Scheme 19

Examples of ICT probes of the D-σ-A type, developed by Verhoeven and coworkers.

Figure 26

Emission of "fluoroprobe" (18) in solvents of different polarity. Taken from reference 119, with permission.

Several derivatives of **18** with more flexible σ-systems interconnecting the donor and acceptor system have been synthesized, see Scheme 19. Due to the flexibility of the σ-system, the photophysics of these molecules is generally more complex than that of **18**. Scheme 20 summarizes the photophysical behaviour of "fluoroprobe" (**18**) and related D-σ-A probes.

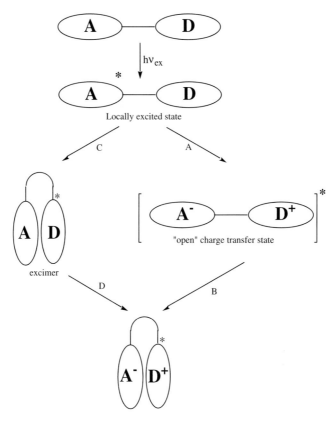

Scheme 20

Schematic representation of the excited state processes for various D-σ-A probes.

Probes with a rigid σ system like "fluoroprobe" (**18**) undergo charge transfer by path A exclusively. For probes with a more flexible σ system like **19**, in which efficient charge transfer takes place, it was found that the highly dipolar "open" excited state collapses in a process called "harpooning" (path B) to form the "closed" charge transfer state. No efficient charge transfer takes place in the open conformation of probes with an even more flexible σ system, like **20.** Instead an exciplex is formed (path C) that subsequently undergoes charge transfer. (path D)

In many cases the locally excited state, the exciplex, the "open" charge transfer state and the "closed" charge transfer state are fluorescent. This can lead to multiple emissions, in which the ratio between emitting species depends on the free volume available for either excimer formation (path C) or the closing of the charge transfer state by the "harpooning" mechanism (path D). Though **18** has proven to be an excellent probe for monitoring polymerization processes, probes like **19** and **20** might be interesting too.

So far fluoroprobe has not been tested in photoinitiated polymerization processes, but the use of this probe for monitoring the thermal and the γ irradiation induced polymerization of MMA has been reported, see Figures 27 and 28. In both processes a 115 nm shift upon polymerization was observed. Also this probe has been employed for characterization of polymers[120,121] for monitoring phase transitions[115] and for monitoring polymer processing[122,123].

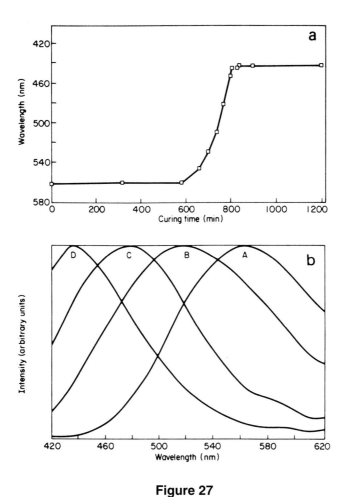

Figure 27

(a) Shift of the fluorescence maximum of 18 in MMA as a result of thermal curing (85°C). (b) Typical fluorescence spectra at four stages of polymerization: curve A, freshly distilled MMA; B, after 720 min; C, after 770 min; and D after 1200 min. All spectra were recorded in continuous mode at 20°C. Taken from reference 66, with permission.

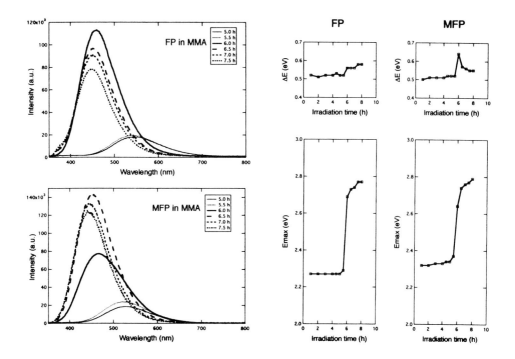

Figure 28
(a) Fluorescence spectra taken after exposure of *ca* 2 x 10-5 M MMA solutions of "fluoroprobe" FP (18, top) and "maleimido fluoroprobe" MFP (8, bottom) to 60Co γ-ray irradiation with exposures from 5-7.5 h. (b) Change with irradiation exposure time of the fluorescence maxima of *ca* 2 x 10^{-5} M MMA solutions of fluoroprobe FP (18, left) and maleimido fluoroprobe MFP (8, right). Taken from reference 84, with permission.

iv.   Organic Salts

Organic salts of the D-π-A+ X- type are another type of fluorescent probes that monitor (photo)polymerization processes by means of a shift in their emission wavelength.[124] Such molecules are used as laser dyes[125] and second order Non Linear Optical (NLO) materials.[126] From a structural point of view organic salts strongly resemble ICT probes of the D-π-A type. Examples of organic salts of the D-π-A type are given in Scheme 21.

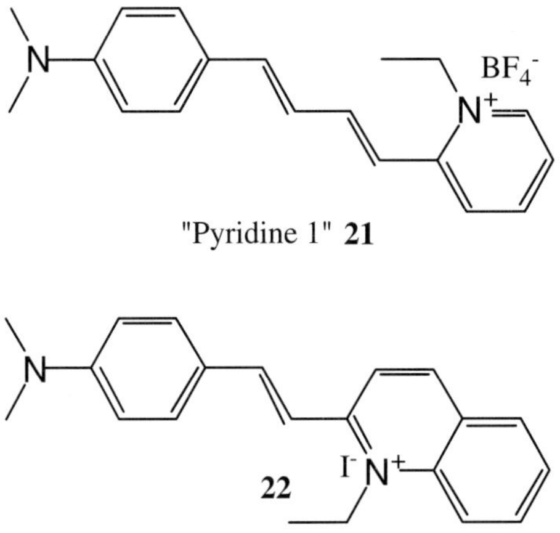

"Pyridine 1" **21**

22

Scheme 21
Examples of organic salts of the D-π-A+ X⁻ type.

The photophysics of the D-π-A+ cation resembles that of D-π-A molecules. After excitation an electron is transferred from the dimethylamino group to the electron acceptor. As a result, a positive charge is located at the dimethylamino group and the rest of the molecule is uncharged. This process does not lead to charge separation, but to a relocation of the positive charge. Hence these molecules are sometimes referred to as charge *resonance*, instead of charge *transfer*, compounds. Semi-empirical calculations of the Highest Occupied Molecular Orbital (HOMO) and Lowest Unoccupied Molecular Orbital (LUMO) for the 4-dimethylamino-4'methylpyridinium stilbene cation support the proposed charge relocalization.[127,128] In order to stabilize the excited state, the negatively charged counterion X⁻ should follow the positive charge and migrate from the pyridinium group to the dimethylamino group. A schematic representation of this process is depicted in Scheme 22.

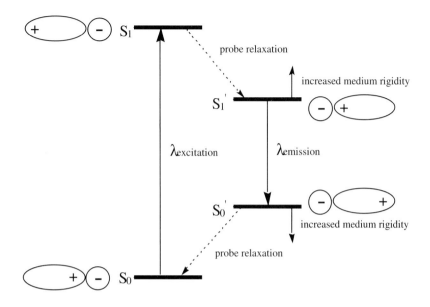

Scheme 22

Schematic representation of the emission of organic salts of the D-π-A⁺ X⁻ type.

The spectroscopic behaviour of organic salts strongly deviates from that of ICT probes. While ICT probes are strongly solvatochromic, organic salts like **21** and **22** emit at virtually identical wavelengths regardless of the solvent polarity.[129] Like ICT probes, organic salts are sensitive to the rigidity of their direct environment. The strong increase in the rigidity of the medium during photopolymerization reactions, results in strong blue shifts in the emission of organic salts. Similar blue shifts in the emission of organic salts can also be observed upon rapidly cooling a dimethacrylate, a process by which a glass is formed.[126] Unlike ICT probes, organic salts do not monitor phase transitions. The crystallization of dimethacrylates, such as TEGDMA, could not be monitored by "pyridine 1" (**21**).

The emission characteristics of the charge resonance probes can be explained by the relaxation mechanism proposed in Scheme 22. The probes are not solvatochromic in solvents of low viscosity since the counterion is mobile and can fully stabilize the excited cation in media of low viscosity. The resulting relaxed excited state has no significant dipole moment and, from an electrostatic point of view, closely resembles the ground state. Conversely the ion is less mobile in highly viscous or rigid media. Hence the probes are rigidochromic since free volume is required for the relaxation of the excited probe.

Table 8

Emission maxima of 21 and 22 in different formulations before and after photopolymerization. (15 μm films between glass plates). Data taken from reference 124.

Monomer Backbone	Phase	Probe 21	22
DODDMA	monomer	659	
	polymer	589	
	shift	70	
HEXDMA	monomer	660	
	polymer	600	
	shift	60	
BUDMA	monomer	664	635
	polymer	607	598
	shift	57	37
EGDMA	monomer	668	635
	polymer	611	602
	shift	57	33
DEGDMA	monomer	672	628
	polymer	614	597
	shift	58	31
TREGDMA	monomer	673	625
	polymer	618	599
	shift	55	26
TEEGDMA	monomer	674	629
	polymer	621	603
	shift	53	26

The emission maxima of the organic salts **21** and **22** before and after the photoinitiated polymerization of a series of dimethacrylates (Figure 29) are presented in Table 8. Table 8 shows that probe **21** is significantly more sensitive (exhibiting a larger blue shift) than probe **22**. Apparently the increase in the size of the π-system that connects the electron donor and acceptor group has increased the probe's sensitivity.[130] Table 8 also reveals that the magnitude of the blue shift (sensitivity) of **21** and **22** is not influenced by the size of the spacer that connects the methacrylate moieties. Instead a modest increase in sensitivity is observed for **21** upon decreasing the polarity of the medium by going from TEGDMA to DODDMA.

After the photopolymerization process was completed, no further shift in probe emission (physical ageing) was observed for **21** or **22**. Neither was the crystallization and the melting of dimethacrylates monitored in the emission spectrum of **21**. A linear relationship between the emission maximum and the double bond conversion was observed for **21**, see Figure 30.

The differences in emission characteristics between the D-π-A$^+$ X$^-$ and the D-π-A probes are indicative of the different mechanisms by which these probes monitor polymerization processes. Based on the proposed mechanisms one can conclude that charge transfer (ICT) probes primarily monitor the microviscosity of a medium, notably the reorganization of solvent molecules, while charge resonance probes (organic salts) predominantly monitor the free volume in a medium. This would imply that complementary information concerning the resin that is under investigation can be obtained by using both types of probe. At the moment it is obvious that more research, especially on the action of the charge resonance probes, is required to get a detailed picture of the exact mechanism by which these probes monitor changes in the "rigidity" of the medium and to establish the scope and limitations of using these probes to monitor photoinitiated polymerization processes.

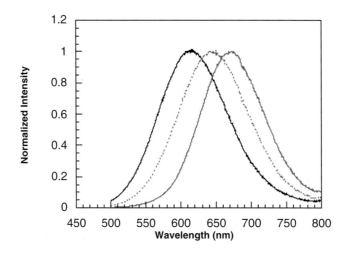

Figure 29
Normalized emission of 21 in EGDMA before irradiation (right spectrum) after 2 min irradiation (38% cure, middle spectrum) and after 20 min irradiation (78% cure, left spectrum). Taken from reference 124, with permission

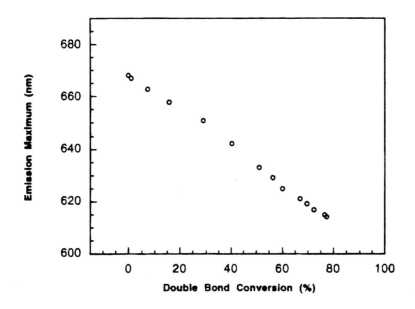

Figure 30
Emission maximum of 21 as a function of the double bond conversion during the photoinitiated polymerization of EGDMA. Taken from reference 124, with permission.

iii. d) NMR Spectroscopy

Photoformed polymeric networks are insoluble and "normal" solution NMR spectra of such polymers can not be obtained.[131] The development of cross-polarization magic-angle spinning (CPMAS) ^{13}C NMR, which allows one to obtain high resolution NMR spectra of solid state samples, has allowed NMR spectroscopy gradually to gain importance as a tool for investigating photoformed polymeric networks. A number of techniques and pulse sequences have been developed to provide information about the detailed molecular structure of polymeric networks and the dynamics of atomic motion in these networks.[132,133,134,135] Resolved isotropic chemical shifts identify chemical environments which may be unique to the solid state. Relaxation experiments exploit the high resolution that can be obtained by magic angle spinning, so ^{13}C spin-lattice relaxation times (T_1) can be used to characterize individually main and side-chain motions. Carbon resolved proton relaxation times $T_{1\rho}$ can be used to determine the homogeneity of photoformed polymeric networks.[116]

Solid state NMR spectroscopy is not a destructive technique in the sense that it does not chemically or physically modify samples. Unlike solution NMR spectroscopy, the sample does not need to be dissolved in a solvent. However, quality spectra are only obtained if very large samples of proper dimensions are made or powders are used. The latter can be made from thin polymer films by crushing the material at low temperatures to particles of 0.5 mm dimensions. NMR spectroscopy is mainly a research tool, and has virtually no use for quality control. Even more importantly, the technique is extremely time consuming; it may take hours to determine accurate $T_{1\rho}$ values.

The most straightforward experiment is to determine the cure of a resin at different stages of the photopolymerization process. This is done by taking solid state NMR spectra and determining the ratio of reacted and unreacted fuctionalities by integration of absorptions that are specific for them. For example, the carbonyl carbons in acrylates and polyacrylates absorb at 165 and 175 ppm, respectively. Even in solid state spectra these absorptions are well resolved, see Figure 31,[2,136] Since the absorptions in solid state ^{13}C spectra are very broad compared to solution spectra carbon absorptions of reacted and unreacted functionalities must be far apart (typically a few ppm) to obtain a good resolution. This point is illustrated by comparing Figure 31, which shows the solid state spectrum of the standard formulation (20% PEGA, 40% TMPTA and 40% DPHPA, Scheme 23)[136] and Figure 32, which shows the ^{13}C spectrum of the same monomers in solution. Figure 32 shows that the cure of the individual components in the standard formulation cannot be monitored directly, using solid state NMR. In principle, it should be possible to investigate polymeric networks by means of ^{19}F, ^{29}Si and ^{31}P solid state NMR spectroscopy.

The first stages of the photopolymerization of multifunctional acrylates can be monitored by solution ^{13}C NMR. For these experiments, 20% solutions in deuterochloroform are used and the polymerization of the individual components in the mixture can be monitored until gelation occurs at 5% conversion. These experiments reveal the relative reactivities of different resins in the first stages of a photopolymerization. It should be noted that the results obtained by these experiments are of limited value, since the experimental conditions in dilute solution strongly deviate from those in neat samples.

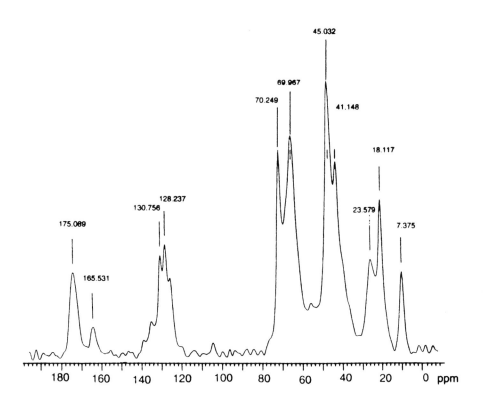

Figure 31
CPMAS ^{13}C Solid state spectrum of the standard formulation "STDF" (20% PEGA, 40% TMPTA and 40% DPHPA) Taken from reference 136, with permission.

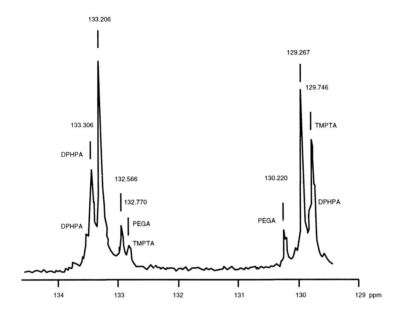

Figure 32
^{13}C Solution spectrum of the monomers PEGA, TMPTA and DPHPA of the standard formulation "STDF" (20% PEGA, 40% TMPTA and 40% DPHPA). A 20% solution in CDCl$_3$ was employed. Taken from reference 136, with permission.

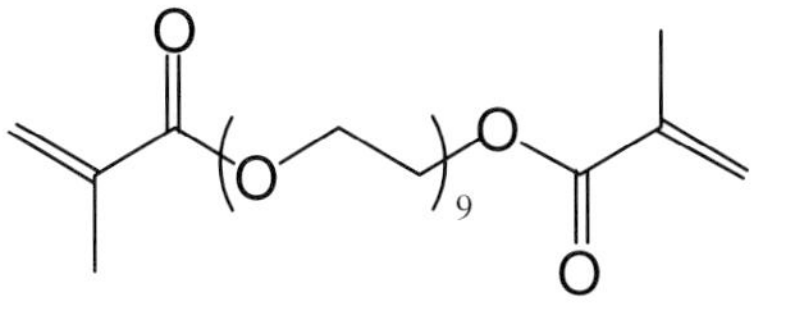

PEGA-400

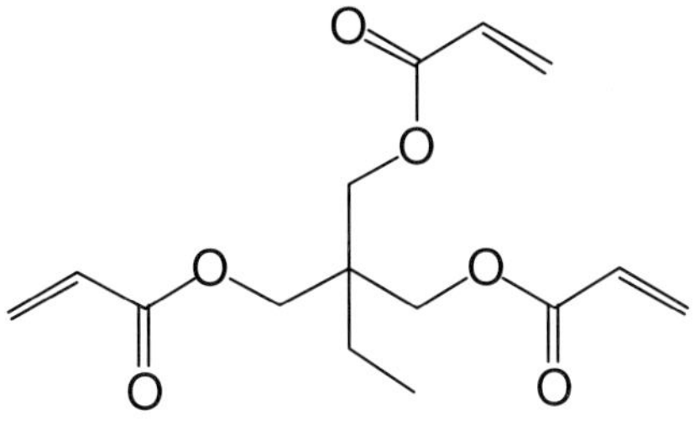

TMPTA

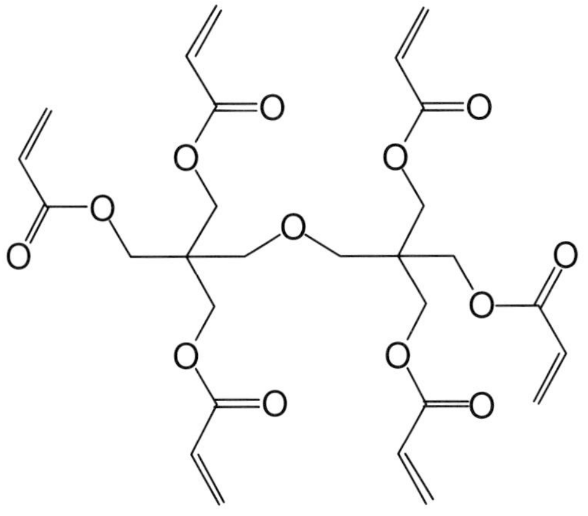

DPHPA

Scheme 23
Components in the Standard Formulation "STDF".

Another technique used to characterize a photoformed network is the determination of carbon spin lattice relaxation times T_1.[2,138] By measuring T_1 values, the mobility of a specific carbon in the network can be determined at different stages of the photopolymerization process. Since the mobility of a specific atom is not the only factor that influences T_1 values, a fair comparison between different networks or different components in a network can only be made by comparing similar carbon atoms such as the methyl of a methacrylate.

An interesting property of polymeric networks is the crosslink density which relates to the molecular architecture of the networks and, in turn, determines their physical properties. During the photopolymerization process of polyfunctional acrylates, ^{13}C T_1 measurements demonstrate that the mobility of main chain carbons *specifically* decreases with an increase in the double bond conversion. By measuring carbon spin lattice relaxation times T_1 of carbon atoms at different positions in the network, cross-link densities can be determined at different stages of the photopolymerization process.[139] For mixtures containing different monomers, crosslink densities can even be attributed to different components provided that the carbon atoms of these components are resolved in the spectra. An example is given in Figure **33**, in which the double bond conversion and the crosslink density in a photocured formulation are given as a function of the laser power. At higher laser powers, the double bond conversion does not increase significantly but a steady increase in crosslink density is observed.

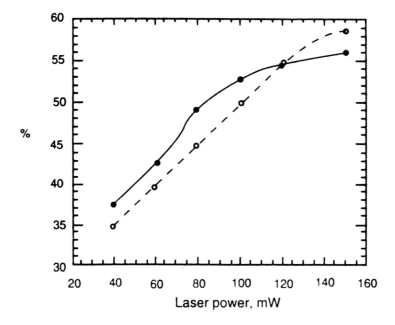

Figure 33

The variation of the degree of double bond conversion (filled dots) and the degree of crosslinking (open circles) for the standard formulation STDF as a function of the laser power. Taken from reference 136, with permission.

The influence of spin diffusion on proton relaxation times can be quite useful for investigating networks composed of different monomers. High-resolution magic-angle techniques used to obtain ^{13}C NMR spectra in solids can be employed to monitor the proton spin diffusion between the fragments originating from the different monomers incorporated in the polymer. For example in a polymer network formed from two components, all the protons may or may not average to a common relaxation time. If the rate of spin diffusion (determined by the domain size and the spin diffusion constant) is large in comparison to the difference in the relaxation rates between the component monomers, only one $T_{1\rho}$ value is obtained for all protons. In such a case the components are all incorporated in *one* homogeneous network (on the scale of spin diffusion, <15-20 nm). If the rate of spin diffusion is small compared to the difference in the relaxation rates (large domains/small diffusion constant), the separate components of the network retain their relaxation behaviour. Different $T_{1\rho}$ values are determined for the components of the network and it is concluded that a phase separation between the components of the network has occurred.[2] So, by determining the rates of spin diffusion of all components that are incorporated in a photoformed polymer, one can determine whether the network is homogeneous, or whether a phase separation has occurred and monomers are incorporated in spatially separated networks. In fact, if the resolution allows one to distinguish ^{13}C absorptions from different monomers, one can determine which monomers are incorporated in which phase.

Table 9
Relaxation rates for PEGA (K_P), TMPTA (K_T) and DPHPA (K_D) homopolymers at different laser powers. Data taken from reference 2.

Laser Power mW	K_P, s^{-1}	K_T, s^{-1}	K_D, s^{-1}
40	12.0	41.2	71.7
60	11.9	42.3	71.7
80	11.9	42.2	72.4
100	12.3	42.4	72.2
125	12.5	41.9	71.0
150	11.8	42.1	72.3

Table 10

Relaxation rates for TMPTA (K_T) and DPHPA (K_D), in a 1:1 TMPTA/DPHPA "TD" mixture at different laser powers. Data taken from reference 2.

Laser Power mW	K_T, s^{-1}	K_D, s^{-1}
60	46.0	68.6
80	48.2	55.1
120	45.1	53.2
150	47.3	49.4

Table 11

Relaxation rates for PEGA (K_P), TMPTA (K_T) and DPHPA (K_D), in a 20/40/40 PEGA/TMPTA/DPHPA mixture (standard formulation, "STDF") at different laser powers. Data taken from reference 2.

Laser Power mW	K_p, s^{-1}	K_T, s^{-1}	K_D, s^{-1}
40	10	43	49
60	7	34	43
100	7	40	45
120	8	41	53
150	10	47	55

A demonstration of the potential of this technique is given in Tables **9-11**.

In Table 9, the spin diffusion proton relaxation rates[140] for the three homopolymers poly-PEGA, poly-TMPTA and poly-DPHPA (Scheme 23) are displayed as a function of the laser power used in the photopolymerization process. For all formulations the cure increases as the laser power is increased and experimental details are given in reference 2. For each homopolymer, a specific relaxation rate is found which is independent of the laser power. In Table 10, the relaxation rates for TMPTA and DPHPA in a 1:1 mixture of TMPTA and DPHPA are displayed as a function of the laser power. This experiment shows that the relaxation times of both resins approach each other at higher laser powers. At high conversions, a homogeneous network is formed in which both monomer residues are intimately mixed and in close contact with each other.

In Table 11, the relaxation rates for PEGA, TMPTA and DPHPA in the standard formulation "STDF" (20% PEGA, 40% TMPTA and 40% DPHPA) are displayed as a function of the laser power. This experiment shows that for the three components in the mixture two different relaxation times are found. Relaxation rates close to 50 s^{-1} are found, for TMPTA and DPHPA, a value closely resembling that found for TMPTA and DPHPA in the TMPTA/DPHPA mixture "TD." Relaxation rates between 7 and 10 s^{-1} are found for PEGA, values close to those in pure poly-PEGA. It is concluded that upon photopolymerization of the standard formulation "STDF" two phases are formed. One phase is composed of TMPTA and DPHPA, and the other phase is composed of PEGA.

Solid state NMR spectroscopy is a very powerful technique for characterizing polymeric network. The number of available techniques and pulse sequences, especially for multidimensional applications, is already impressive. The technique is developing rapidly (hardware and pulse sequences) and promises to play a very prominent role as a tool for characterizing networks at the molecular level. The long times needed to record spectra and the cost of the equipment, however, make this a research technique *par excellence*.

ii. e) ESR Spectroscopy

ESR spectroscopy[141,142,143] is a technique for identifying and determining the concentration of radicals. Therefore ESR can be employed as a research tool for investigating radical initiated polymerizations only. ESR has been extensively used for elucidating mechanisms of photopolymerization reactions in solution, often in combination with other techniques such as CIDNP NMR spectroscopy.[144] The technique is particularly well suited for identifying the radicals that are formed by photochemical decomposition of photoinitiators. ESR spectroscopy can be performed either in solution or on solid samples. The use of ESR spectroscopy in studying photopolymerization processes is limited. The type of information that is obtained, the fact that ESR spectrometers are not widely employed and the fact that mechanistic studies concerning the photochemistry and photophysics of photoinitiators are almost exclusively performed in solution limits their use. ESR studies have been useful in determining the mechanisms by which photopolymerization reactions are terminated. Kloosterboer and coworkers have demonstrated the presence and elucidated the structure of trapped radicals in photoformed polyacrylate films.[145,146] Under anaerobic conditions these radicals persist in polymeric networks for an almost unlimited time. These radicals remain reactive, however, and by elevating the temperature polymerization reaction may be resumed. The general conclusion from these experiments is that vitrification of the polymer matrix stopped the photopolymerization of small multifunctional acrylates. Reactive radicals as well as unreacted acrylate functionalities were still present in the matrix.

iii. f) Dielectric Spectroscopy[147,148,149,150,151,152]

Dielectric spectroscopy is a useful technique for monitoring polymerization processes. A sample is placed between two electrodes that act as a capacitor. Electrodes can be chosen in many shapes and for investigating thin films, two flat parallel metal plates would be appropriate. For monitoring photopolymerization processes, different electrodes which allow the light to reach the sample are required. An example of such a configuration is the comb electrode depicted on Figure 34. Using such a set-up the electrode is immersed in the resin, covered with two quartz plates and irradiated from above.

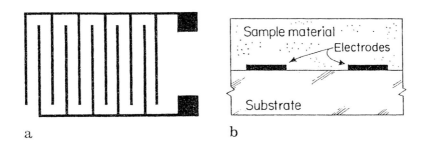

Figure 34
(a) Top view of comb electrode, (b) Cross-section where the insulating substrate is much thicker than the electrode spacing, from reference 153.

In dielectric spectroscopy, current is monitored as a function of an alternating voltage that is applied to the electrodes. The mobility of charge carriers which can be ions (translational freedom) and dipoles (rotational freedom) in the medium, are monitored. Since the frequency and magnitude of the applied alternating voltage can be chosen at will, specific mobility regimes can be selected.

Dielectric spectroscopy has been performed to monitor the cure of thermal,[153] UV and EB induced polymerization processes. In most cases, ion conductivity (not dipole conductivity) has been measured. The ion conductivity is strongly reduced during the curing process, due to an increased viscosity of the medium and the observed current is strongly reduced. By making calibration curves, the ion conductivity of the medium can be related to the cure of the resin. A distinct disadvantage of dielectric spectroscopy is its sensitivity to the presence of contaminants and to temperature. Compared to DSC, dielectric spectroscopy is more sensitive at the later stages of a polymerization reaction.

For a typical experiment samples are mounted between glass or quartz plates using a comb electrode. Dielectric spectra response times can be as short as 55 ms. Since light does not interfere with measurements, dielectric spectroscopy is a suitable technique for real time monitoring of photoinitiated polymerization processes. The possibility of tuning the frequency and the magnitude of the alternating voltage, makes it a versatile and sensitive technique. Since contact between the electrodes and the samples is required, dielectric spectroscopy is not suited for remote control and on line monitoring.

3. Techniques for monitoring photopolymerization processes in real time

i. Introduction

Real time means taking measurements during the photopolymerization process as opposed to interrogating samples after it has occurred. Such techniques have been developed recently and offer the distinct advantage that not only the photoformed polymer, but also the polymerization process by which this polymer is formed can be investigated. The significance of real time monitoring for research in the field of radiation induced polymerization can not be overestimated. For example the mechanism by which a photoinitiator generates reactive species, and the efficiency by which these species are formed, are not necessarily the same in solution and in neat formulations. Large differences for the efficiency of photoinitiators in both systems have been reported, especially for initiating systems that act by bimolecular or trimolecular mechanisms. In addition the polymerization process in neat formulations deviates strongly from the "ideal" behaviour observed in (dilute) solutions.

Data obtained by real time measurements are taken in situ and often differ from those taken from samples after a distinct irradiation period. It is likely that such samples have undergone various kinds of relaxation processes in the time interval between the irradiation and the measurement. This is especially important if one is interested in the exact relationships between sample properties and sample preparations. It has been demonstrated for the fast polymerization processes performed using high intensity irradiation conditions in particular, that relaxation processes, such as volume relaxation, play a significant role in determining the properties of the photoformed material.[145]

As mentioned earlier, real time monitoring techniques are used in the laboratory for testing formulations under different (irradiation) conditions. Data concerning the kinetics of the process and the final cure of the resin can be obtained from such measurements. In general, this methodology is employed for fundamental research and testing novel resins.

The use of real time monitoring techniques for quality control of photopolymerizable resins has also been reported.

ii. Photodifferential Scanning Calorimetry (Photo-DSC)

Photodifferential scanning calorimetry as a technique for real time monitoring of photoinitiated polymerization processes will be discussed in this section.[5] Other calorimetric techniques that monitor the heat evolving from a sample have also been developed and can be employed for monitoring photoinitiated polymerization processes in real time. In most cases, a rise in temperature is measured and by knowing the heat capacity of the total set-up the amount of heat evolved from the reaction can easily be calculated. Since these devices are often home built and differ in many respects, it is generally not easy to compare the results obtained by different research groups with various instruments.[154] These methods will not be discussed in this section.

Photo-DSC is a technique that measures photoinitiated polymerization processes in a controlled atmosphere under isothermal conditions. The technique has been used for 20 years and until recently it has been *the* standard technique for kinetic measurements on photopolymerizable systems. Initially, researchers built their own equipment by mounting quartz windows, through which the sample and the reference cell could be irradiated, on to commercial DSC apparatus. Though differences between individual "home built" machines, in particular the intensity of the irradiation, might exist, results obtained with such instruments are comparable since the irradiation conditions (temperature, atmosphere, pressure) are well controlled. Nowadays, accessories to modify DCS apparatus for use in photo-DSC experiments are commercially available. A typical set-up for photo-DSC experiments is shown in Figure 35.

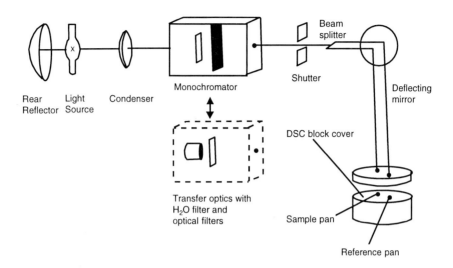

Figure 35

Schematic diagram for the optical set-up of a photo-DSC apparatus. Taken from reference 4, with permission.

Photo-DSC monitors the amount of heat evolved in a photopolymerization reaction. If the heat of polymerization of the reactive groups in the formulation is known, the degree of cure of the resin can be monitored in real time. This technique is universal and has been employed for a large variety of resins such as acrylates, methacrylates, epoxides, vinyl ethers and others, see Table 2.

A typical photo-DSC exotherm curve which displays the amount of heat that evolves per second is shown in Figure 36. From the exotherm curve one can obtain the rate at which the polymerization process proceeds at any time. The maximum value of the rate of the reaction is obtained at t_{max} when the heat evolution dH/dt reaches its maximum. The induction time t_i (T_{ind} in Figure 36), in which stabilizers, oxygen and other contaminants that inhibit the polymerization process are consumed, is also easily determined from the DSC exotherm. By integrating the exotherm curves one can obtain the degree of polymerization as a function of time. A typical example is shown in Figure 37.[155]

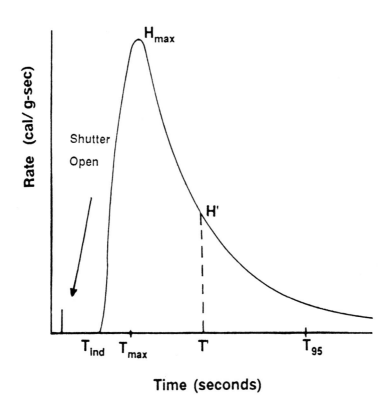

Figure 36
A Typical DSC Exotherm. Taken from reference 5, with permission.

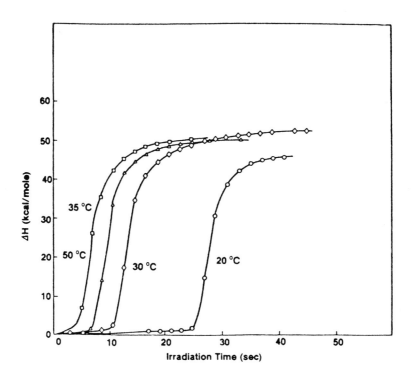

Figure 37
Cumulative heat (integrated area) versus irradiation time for triethylene glycol divinyl ether at several temperatures. Taken from reference 155, with permission.

The kinetics of the polymerization of different formulations can be tested. The DSC exotherms of the photoinitiated polymerization of the diacrylates hexanedioldiacrylate, HDDA and diethyleneglycoldiacrylate, DEGDA and the monoacrylate, n-propylacrylate, n-PA, are shown in Figure 38. From Figure 38 it has been concluded that HDDA is the most reactive resin and, that the reactivity of the monoacrylate n-PA is significantly lower than that of both diacrylates.

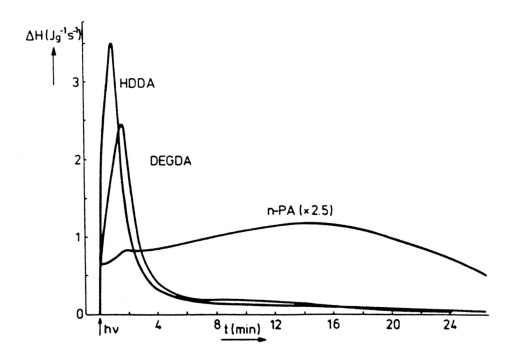

Figure 38

Photo-DSC exotherms from HDDA, DEGDA and n-PA. Taken from reference 1, with permission.

Numerous aspects of the kinetics of photoinitiated polymerization reactions in neat formulations have been investigated by employing photo-DSC. The effect of oxygen, temperature, and the intensity of the light source on photoinitiated polymerizations have been studied.[1,145,156] By combining photo-DSC data with measurements for determining the volume shrinkage of acrylates in real time, the volume shrinkage as a function of the double bond conversion could be determined under various conditions. These experiments demonstrated that at higher intensities, higher rates of polymerization and higher final conversions are obtained. The shrinkage of the resin decreases when higher intensities are used, leading to polymers that contain more free volume, see Figure 39.[145] Merely as an illustration, the DSC curves of photoinitiated polymerizations using a pulsed laser are shown in Figure 40.[69] Figure 40 clearly reveals the pulsed character of this laser initiated polymerization and the direct control of the light source over this reaction.

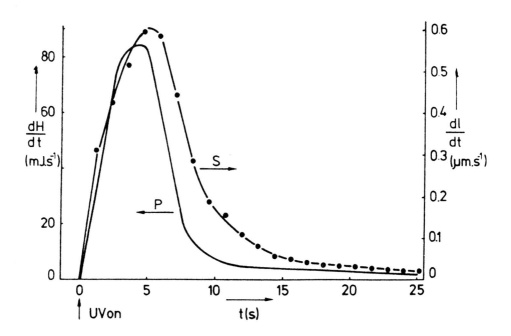

Figure 39

Rate of polymerization and rate of shrinkage. Taken from reference 1, with permission.

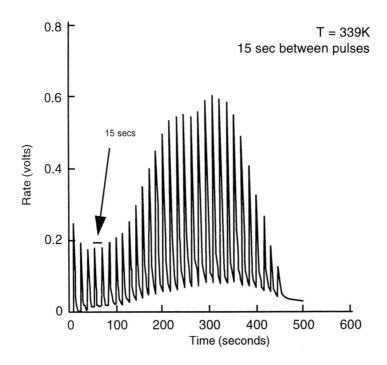

Figure 40

Exotherm curves produced by single laser pulses spaced 15 s apart for HDDA. Taken from reference 5 & 69, with permission.

Using photo-DSC, photoinitiated polymerization reactions can be performed under well controlled conditions (temperature, atmosphere) using different light sources. The size of the samples are typically in the order of 0.5-5 mg depending on the specific heat of the polymerization reaction and the concentration of the reactive moieties in the resin. The technique is sensitive and recording the photo-DSC exotherm of a formulation is a relatively rapid procedure. A distinct disadvantage of photo-DSC as an analytical tool for real time monitoring is the long response time, typically in the order of 2-3 s.[157,158] This means that photo-DSC can only be used to measure relatively slow photoinitiated polymerization processes. In most cases, however, one can tune the rate of the photopolymerization process by using low intensity light sources and low concentrations of photo initiator. Photo-DSC is not useful for real time monitoring of the cure of commercial formulations using high intensity light sources. Such studies can be performed by employing spectroscopic techniques.

Since photo-DSC monitors the heat evolved in a polymerization reaction, the technique is not perturbed by the presence of fillers and pigments which might spoil spectroscopic measurements. Since photo-DSC is a caloric technique and contact with the sample is required, it is not suitable for remote control and on line monitoring of photopolymerization processes.

Photo-DSC is no longer the only technique for real time monitoring of photoinitiated polymerization processes. Spectroscopic techniques such as real time IR, UV and emission spectroscopy have recently been introduced and are developing rapidly. Photo-DSC is not suited for monitoring very fast polymerization processes due to the long response time. In all other cases photo-DSC is a very useful and sensitive technique. Photo-DSC is a standard technique for real time monitoring of photoinitiated polymerization processes which is often employed along with spectroscopic techniques.

iii.   Real time (FT) IR Spectroscopy

One of the most universal and generally employed techniques for monitoring a photopolymerization process is real time IR spectroscopy (RT-IR) developed by Decker and Moussa.[159,160,161] This technique monitors the decrease of an absorption of the monomer that is absent in the photoformed polymer. In most cases the IR absorption is measured in transmission in the normal mode at right angles, Figure 41. For IR measurements in the normal mode the sample is illuminated at an angle of 45°. Different light sources, including lasers, can be employed. Spectra of the relevant IR absorption as a function of the irradiation time are displayed in Figure 42.[162] Real time IR measurements can also be performed in the Single Reflection Absorbance (SRA) mode and in the Attenuated Total Reflectance (ATR) mode. Using these methods, samples lie horizontally and are irradiated at right angles. A schematic representation of a reaction device for real-time ATR-FTIR measurements is shown in Figure 43. Liquid samples can be used and the configurations assembled for SRA and ATR measurements closely resemble those in a production environment. In all cases the temperature, pressure and the composition of the atmosphere surrounding the sample can be controlled.

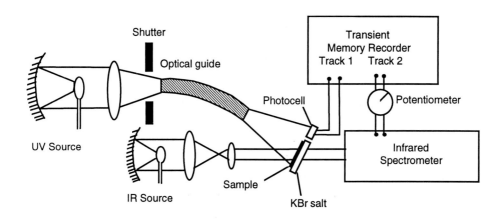

Figure 41
Set-up for recording photopolymerization reactions by RTIR spectroscopy in the transmission mode. Taken from reference 157, with permission.

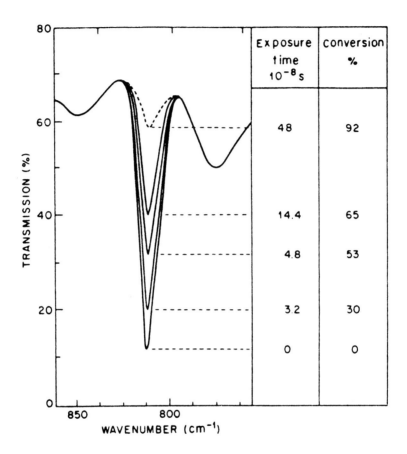

Figure 42
Changes of IR absorption spectra at 810 nm of a photoformulation exposed for various times to nitrogen laser pulses. Taken from reference 162, with permission.

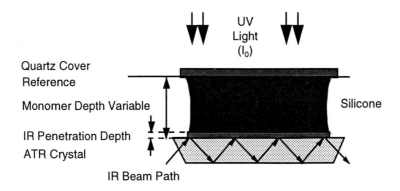

Figure 43

Schematic representation of a real-time ATR-FTIR reaction device. Taken from reference 32, with permission.

Both continuous wave CWIR and Fourier transform FTIR have been employed for real time IR measurements. Employing continuous wave IR spectroscopy is easy; the IR source is kept at a fixed wavelength and the decrease or increase of the IR absorption is measured as a function of time. The method is very fast and response times are in the order of a few milliseconds. The technique is highly reproducible and very suited for comparing the reactivity of different formulations under a variety of conditions. The technique is not necessarily accurate since this method does not correct for shifts in the position of an absorption maximum and the absorption intensity (peak broadening) of a transition. Variations in sample thickness that occur during photopolymerization processes are not compensated for. Corrections can be applied by taking multiple runs of identical samples under exactly the same conditions while recording IR absorptions at different wavenumbers.

More accurate results can be obtained by employing FTIR spectroscopy. By taking full spectra one can accurately measure all relevant absorptions simultaneously. Shifts in the absorption maximum and peak broadening due to temperature effects do not influence these measurements. Variations in sample thickness can be compensated for by comparing the magnitude of the absorption of interest with a reference absorption. Therefore, conversions determined by FTIR spectroscopy should be more accurate and reliable than those obtained by continuous wave IR spectroscopy. In Figure 44, the FTIR spectra of the ambifunctional monomer 1(1-propenoxy)-6-vinyloxyhexane taken at right angles (normal mode) during its photoinitiated polymerization are displayed.

A distinct disadvantage of using FTIR spectroscopy instead of continuous wave IR is the longer response time, typically in the order of 20-50 ms. However, 20-50 spectra can be recorded per second which is enough for most practical cases. Another disadvantage might be the need for data processing which may be time consuming.

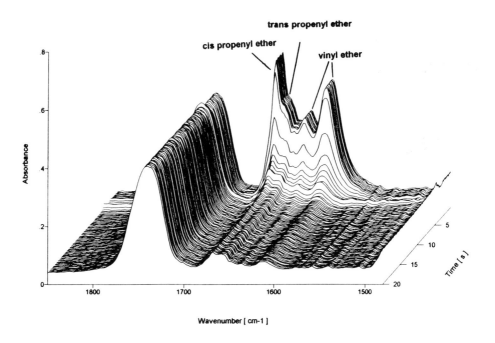

Figure 44
RT FTIR spectra of the photoinduced cationic polymerization of 1(1-propenoxy)-6-vinyloxyhexane. Taken from reference 17, with permission.

For measurements in transmission, the thickness of samples must remain in the 10-50 μm range, a fairly normal thickness for coating applications. A second requirement is that the substrate on which the coating is applied must be transparent at the wavelengths of interest which are usually in the far IR. For research applications salt plates, usually NaCl, are often employed, as are polypropylene, polyethylene and poly(vinylydene chloride) films.

$$R_p = [M]_0 \{A_{mon}(t_1) - A_{mon}(t_2)\} / \{A_{mon}(t_0) * (t_2 - t_1)\}$$
Equation 17

R_p = rate of the polymerization (in mol l^{-1} s^{-1})

$[M]_0$ = concentration of the monomer before irradiation (in mol l^{-1})

$A_{mon}(t_1)$ = specific absorption of the monomer at time t_1

$$\Phi_p = R_p l / 1000 * (1 - e^{-2.3A}) * I_0$$
Equation 18

Φ_p = quantum yield of the photopolymerization process

l = film thickness (in cm)

A = Absorbance at the relevant wavelength

I_0 = Intensity of the incident light (in Einstein cm^{-2} s^{-1})

$$S = t_{0.5} P f$$
Equation 19

S = photosensitivity of a formulation

$t_{0.5}$ = 50% conversion time (in s)

P = radiant power (in mW cm^{-2})

f = fraction of light absorbed by the sample

A typical kinetic profile recorded by employing RT-IR spectroscopy is given in Figure 45. Data that can be extracted from this figure are: the rate of polymerization R_p (Equation 17), the quantum yield of polymerization Φ_p (Equation 18), the induction time t_i, the photosensitivity of a formulation S (Equation 19) and the amount of unreacted reactive groups in the formulation. RT-IR studies can reveal the influence of all kinds of factors on the kinetics of photoinitiated polymerization processes. In Figure 46, the influence of oxygen on the photoinduced polymerization of the highly reactive monoacrylate [(dimethylmethoxy)formamide]ethyl monoacrylate (Acrityl CL 960 from SNPE, Scheme 24) is demonstrated.[159] The kinetic profiles of a number of acrylate monomers are depicted in Figure 47.

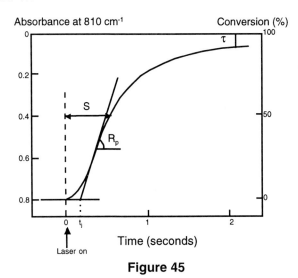

Figure 45

Typical photopolymerization profile recorded by RT-IR. Taken from reference 159, with permission.

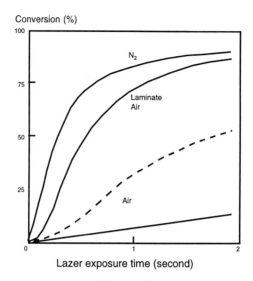

Figure 46

The effect of oxygen and the sample thickness on the kinetics of a typical photoinitiated acrylate polymerization. Taken from reference 159, with permission.

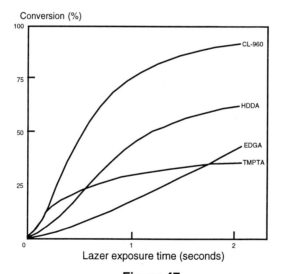

Figure 47

Profiles for the photoinitiated polymerization of various acrylic monomers. Taken from reference 159, with permission.

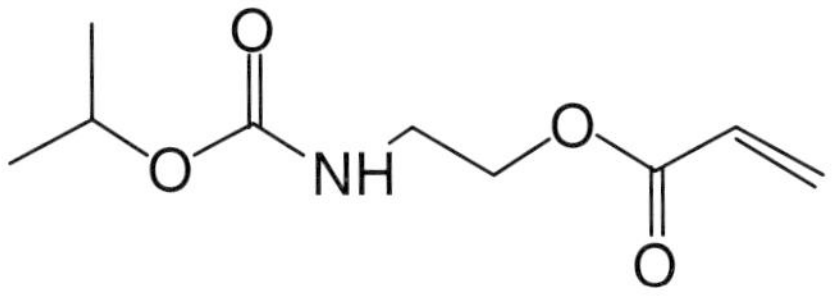

Scheme 24
Structure of [(dimethylmethoxy)formamide]ethyl mono acrylate (Acticryl CL 960).

In other experiments the extent of postcure in laser initiated polymerization reactions of an acrylate formulation has been investigated, see Figures 48 and 49. These experiments show that the postcure, i.e., the cure that has occurred after the light source has been turned off, can be considerable. The postpolymerization is more important during the first stages of the polymerization process and is more pronounced for experiments performed under a nitrogen blanket. The postpolymerization is relatively fast and is completed in 1 (air) to 2 (nitrogen) seconds.[164] The extent of postcure expressed in terms of the total cure can reach values up to 75 to 90% in air and nitrogen, respectively.

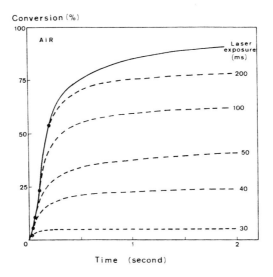

Figure 48

Kinetics of postpolymerization in the dark under air. Taken from reference 159, with permission.

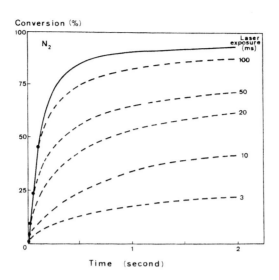

Figure 49

Kinetics of postpolymerization in the dark under nitrogen. Taken from reference 159, with permission.

Recently real time IR spectroscopy was applied in the reflectance mode for monitoring photopolymerization reactions called Real-Time Attenuated Total Reflectance-Fourier Transform Infrared Spectroscopy (ATR-FTIR).[32] The distinct advantage of this technique is that no limits are imposed on the thickness of the sample. Only a very thin slice of the sample attached to the ATR crystal, typically 0.1-0.6μm, is monitored. Using this technique, the sample is irradiated from the top while the cure is measured at the bottom of the sample *exclusively*. By systematically varying the thickness of samples that are irradiated under identical conditions, a spatial (in depth) resolution of the photoinitiated polymerization process can be obtained. Figure 50 shows the kinetic profiles of TMPTA samples of different layer thicknesses. Clearly, the cure at the bottom of the thick samples is strongly reduced due to the reduced light intensity at this position. Inhibition times are strongly increased for thicker samples. The double bond conversion after a 90s irradiation period for different sample thicknesses are displayed in Figure 51.

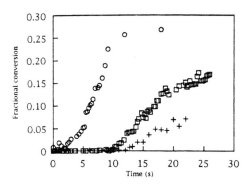

Figure 50

Conversion at the bottom of a sample as a function of the irradiation time as determined by ATR-FTIR. Layer thicknesses are 16 ( ) 130 (O), and 560 μm (+). Taken from reference 32, with permission.

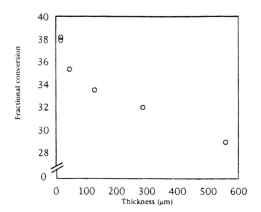

Figure 51

Conversion at the bottom of a sample versus the sample thickness. Taken from reference 32, with permission.

Real time IR spectroscopy has become the major analytical technique for monitoring photopolymerization processes in real time.[165] The technique has some distinct advantages. Firstly, the technique monitors the concentration of reactive groups in the resin under study *directly*. Since reactive groups can be distinguished in the IR spectrum, the concentration of different reactive groups in hybrid systems can be monitored independently. This is particularly the case for real time FTIR measurements. It should be noted that in practical situations, the specificity of IR spectroscopy can be a disadvantage, especially when complex mixtures composed of many components are used. In such mixtures the absorptions of these reactive moieties may be obscured by other absorptions originating from fillers, photoinitiators, pigments and other additives.

The major advantages of RT-IR spectroscopy are speed, reproducibility and accuracy. Due to the fast response times, usually in the order of a few ms in the CW mode, very fast polymerization reactions can be followed. Even in the Fourier Transform mode, which is more accurate, up to 50 spectra can be recorded per second. High intensity light sources, including lasers, can be used when employing this technique. Therefore, commercial formulations can be investigated under realistic circumstances. For the study of photoinitiated polymerization processes in thick(er) samples, complementary curing profiles can be obtained by RT-IR spectroscopy in the transmission mode and RT-IR spectroscopy in the ATR mode. The surface cure is determined in the transmission mode by using thin samples. The cure under the surface is determined in the ATR mode by using samples of varying thicknesses.

iv.   Real Time UV Spectroscopy

Real time monitoring using UV absorption spectroscopy is performed monitoring the absorption of a variety of species during a photoinitiated polymerization process. The concentration of monomers and photobleachable photoinitiators and the isomerization of a photochromic probe added to the formulation have been studied. Using photochromic probes, one can monitor the decrease of free volume in the matrix as the polymerization proceeds. The set-up for most measurements is comparable for that of real time IR monitoring in the transmission mode, depicted in Figure 41. (Page 348) The sample is placed in the sample holder of a spectrophotometer and is irradiated by an external light source under an angle of 45°. Absorptions can be monitored continuously at a fixed wavelength. Absorption at different wavelengths can be monitored by running successive experiments under identical conditions. In many cases real time UV measurements are recorded in combination with other techniques, notably real time FTIR. This does not imply that the techniques are performed simultaneously on the same sample. Instead, successive runs using different techniques are performed with identical samples under identical irradiation conditions.

A direct method for monitoring the cure of a resin in real time by UV spectroscopy was recently reported.[22] The cure of the photoinitiated polymerization of a 1:1 mixture of the electron donor N-vinylpyrolidone (NVP) and the electron acceptor 5-hydroxyphenylmaleimide, which form an alternating copolymer, was examined by monitored the decreasing absorption of the maleimide at 300 nm. The decreases in the

1635cm⁻¹ absorption ascribed to N-vinylpyrolidone and the 829cm⁻¹ maleimide absorption were monitored by real time FTIR spectroscopy for the same formulation. These values were converted to percentage conversion data before being plotted on the same graph as the UV absorption results. Consistent cure versus irradiation time profiles were obtained from the RT-IR and the RT-UV measurements, see Figure 52. Clearly these experiments demonstrate the power and reliability of these real time techniques. It should be noted that formulations without inhibitors, photoinitiators and other species for which the cure can be monitored by UV spectroscopy are exceptional.

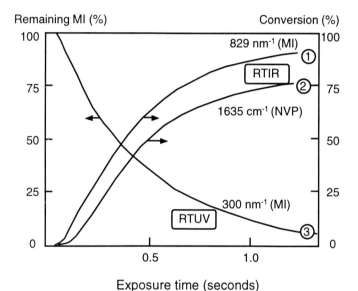

Figure 52

Combined RTUV and RTIR spectra for the maleimide-NVP copolymerization. Taken from reference 22, with permission.

Experiments in which the concentration of photobleachable photoinitiators were monitored in real time, were described by Decker.[13] The photoinitiators under investigation were the radical initiator 2-benzyl-2-dimethylamino-1-(4-morpholino-phenyl)-butanone-1 **3** (Irgacure 369 from Ciba Geigy) and the cationic initiator bis [4-(diphenyl-sulphonio)-phenyl] sulphide-bis-hexafluorophosphate **23** (KI-85 from Degussa), Scheme 25. The practical value of this technique is somewhat limited since only bleachable photoinitiators can be used, but the data obtained from these experiments are expected to be representative for other systems in which similar photoinitiators are used. It should be noted that the experiments described here are performed in neat formulation and present an accurate picture of the behaviour of photoinitiators under realistic circumstances. This is in sharp contrast to many photophysical studies on the decomposition of photoinitiators which are performed in (dilute) solution.

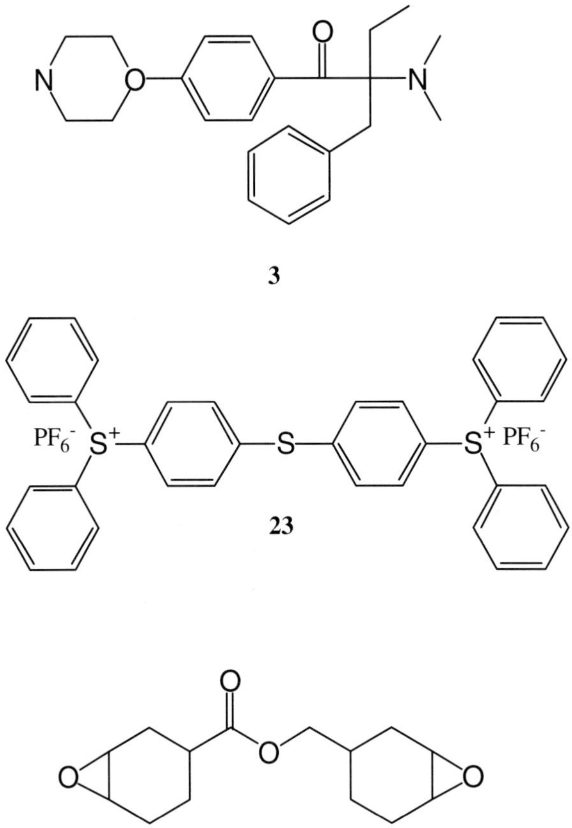

Scheme 25

Molecular structure of photoinitiators 3 and 23 and the epoxy resin 3,4 epoxycyclohexylmethyl-3',4'-epoxycyclohexane carboxylate.

RT-UV measurements at different intensities show that the quantum yields for the decomposition of both photoinitiators are not effected by [PI] and the intensity of the light. A striking observation is that the photodecomposition of **3** is quenched by the presence of acrylates. Apparently energy transfer from the excited state of **3** to acrylates is an efficient process. The presence of oxygen does not effect the photodecomposition of **3**.

By combining RT-IR data which follow the cure of the monomer and RT-UV data which monitor the concentration of the photoinitiator [PI], several conclusions regarding the photoinitiated polymerization of commonly used acrylates and epoxides have been drawn. For acrylate formulations using **3**, the decrease of [PI] is rather slow. Unreacted photoinitiator is abundantly present in all stages of the reaction, and depletion of the photoinitiator is not a limiting factor in the photoinitiated polymerization of acrylates. The presence of oxygen does not effect the decomposition of the photoinitiator but seriously retards the polymerization process by quenching of radicals, see Figure 53.

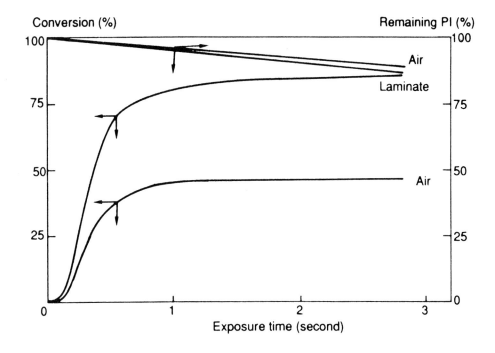

Figure 53

Polymerization and photoinitiator decay profile for an acrylate formulation containing 0.5% 3, upon UV exposure at 9 mW cm-2. Film thickness: 25 μm. Taken from reference 13, with permission.

By combining RT-IR and RT-UV data (taken from identical samples under the same conditions), one can determine the rate of polymerization Rp and the rate of PI decomposition R at any stage of the photoinitiated polymerization process. By taking the ratio of the rate of polymerization over the rate of PI decomposition R_p/R, the kinetic chain length of the polymerization process can be calculated at any stage of the reaction. Using **3** as photoinitiator, in neat acrylate formulations typical kinetic chain lengths between 3000-8000 have been found. Accurate kinetic chain lengths can only be obtained if the number of reactive species formed out of each photoinitiator molecule is constant throughout the reaction. This requirement is fulfilled for **3** which undergoes a unimolecular Norrish 1 photocleavage in which one highly reactive 4-morpholinobenzoyl radical and an α-amino radical of lower reactivity are formed for each photoinitiator molecule that disappears, Scheme 3.

A fast decomposition of the photoinitiator is found in the epoxide system (Scheme 25) which consists of the epoxide 3,4 epoxycyclohexylmethyl-3',4'-epoxycyclohexane carboxylate (Cyracure UVR 6110 from Union Carbide) and the photoinitiator **23** (2.5% by weight). The photoinitiator is consumed in a matter of seconds and therefore it was concluded that depletion of the photoinitiator is an important factor responsible for the low cure of epoxides, typically 25% under normal circumstances. By increasing [PI] up to 10% by weight, high conversions up to 70% can be obtained which proves that the amount of photoinitiator is indeed the limiting factor in this formulation, see Figure 54. The kinetic chain length of the polyepoxides was determined by combining the data collected with RT-UV and RT- IR measurements. Typical values were in the order of 110 which reflects the low reactivity of the epoxide and explains the low conversions obtained under normal circumstances. In order to keep [PI] at acceptable levels, an increase of the reactivity of epoxides and other cationically cured monomers is highly desirable.[14,19,166]

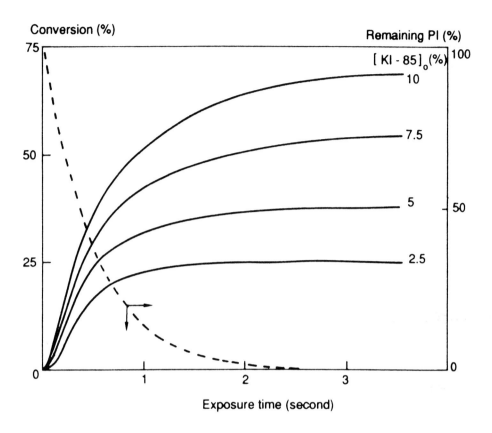

Figure 54

Polymerization profiles recorded by RT-IR spectroscopy upon UV exposure of a cycloaliphatic diepoxy monomer containing 23 at various concentrations. Dashed line: photoinitiator decay profile for [23]= 2.5%. Light Intensity: 55 mW cm^{-2}. Taken from reference 13, with permission.

Photoinitiated polymerization reactions can also be monitored in real time by monitoring the absorption spectrum of a photochromic probe that has been added to the formulation.[49,50] The probe undergoes a photochemical trans-cis isomerization in the mobile or free form only, Equation 20. During the course of the polymerization reaction, both trans and cis isomers are immobilized, see Equations 21 and 22. The absorptions of the photochromic probes in the monomer and in the polymerizing system are given by Equations 23 and 24. A variable z can be defined which is the fraction of probe molecules that are mobile, Equation 25. Taking equal rate constants for the immobilization of the trans and the cis isomers (k_3 and k_4 in Equations 21 and 22, respectively) leads to

Equation 26. Further analysis yields Equation 27, which expresses the fraction of mobile probe molecules z in terms of experimentally available quantities.

$$t_f \underset{k_2}{\overset{k_1}{\rightleftarrows}} c_f \quad (eq\ 20) \qquad t_f \overset{k_3}{\longrightarrow} t_b \quad (eq\ 21) \qquad c_f \overset{k_4}{\longrightarrow} c_b \quad (eq\ 22)$$

$$A_m = \varepsilon_t l[t_f] + \varepsilon_c l[c_f] \quad (eq\ 23) \qquad A_p = \varepsilon_t l([t_f]+[t_b]) + \varepsilon_c l([c_f]+[c_b]) \quad (eq\ 24)$$

$$k_3 = k_4 = \frac{d\ln z}{dt} \quad (eq\ 25) \qquad z = \frac{[P]f}{[P]} = \frac{[c_f]+[t_f]}{[P]} \quad (eq\ 26)$$

$$z = \left. \frac{dA_p/dt}{dA_m/dt} \right|_{t=constant} \quad (eq\ 27)$$

t_f, t_b = trans isomer in the free and bound state, respectively

c_f, c_b = cis isomer in the free and bound state, respectively.

k_1, k_2 = rate constants for the isomerization of free probe molecules (in s^{-1})

k_3, k_4 = rate constants for the immobilization of the probe molecules (in s^{-1})

$\varepsilon_t, \varepsilon_c$ = extinction coefficient of the trans and the cis isomers, respectively (in 1000 $cm^2 mol^{-1}$)

l = thickness of the sample (in cm)

[P] = Total concentration of the probe (in mol l^{-1})

By employing Equation 27, the fraction of mobile probes can be determined at any stage of a photopolymerization process. Required are A_m, the absorption of the probe in an irradiated monomer (without photoinitiator), and A_p, the absorption of the probe in a polymerizing formulation (monomer with photoinitiator), as a function of the irradiation time. Both need to be recorded under identical conditions. In order to obtain the fraction of mobile probe molecules, z as a function of the double bond conversion instead of the irradiation time, the correlation between the cure and the irradiation time, needs to be established by another real time technique. Photo-DSC and real time IR measurements can be employed for this purpose.

The absorption of stilbene as a function of the irradiation time in a monomer and a photopolymerizing formulation are depicted in Figure 55. As the irradiation time increases the concentration trans isomer in the monomer falls as some is converted into the cis isomer. According to equation 20 this will eventually reach an equilibrium value. In the polymerisation experiment less isomerisation occurs. The spectra of stilbene before irradiation (100% trans isomer, a) and after a 10 minute irradiation period in a polymerizing resin (b) and a monomer that does not polymerize (c) are depicted in Figure 56. Both Figures clearly demonstrate that as the polymerization proceeds, a number of the (trans) stilbene molecules are being immobilized, preventing them participating in the trans-cis isomerization. In graph c almost complete isomerization has taken place because the monomer does not polymerise and hence there is no reduction in free volume.

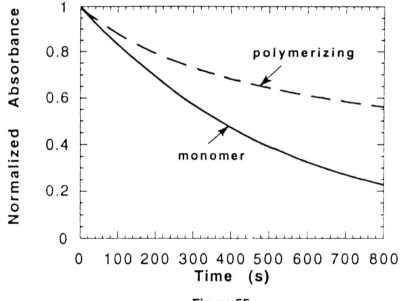

Figure 55
Typical absorbance versus irradiation time curves for a photochromic probe in a monomer system and a photopolymerizing system. Taken from reference 50, with permission.

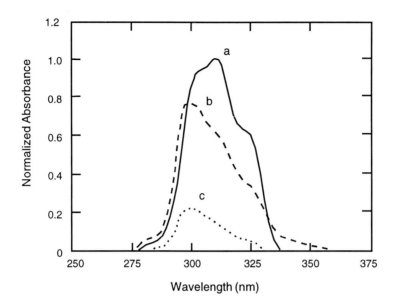

Figure 56

UV spectra of stilbene in (a) an unpolymerized system before isomerization, (b) a system polymerized and isomerized for 10 minutes, and (c) an unpolymerized system (no PI added) after 10 minutes of irradiation. Taken from reference 50, with permission.

In Figure 57, the effect of employing probes of different sizes in DEGDMA is illustrated. Clearly the fraction of mobile probes z, is declining faster and to a further extent if larger probes are used. From plots like Figure 57, one can determine the mean free volume and the distribution of free volume at each stage of the reaction.[167] In Figure 58, the fraction of mobile probes z in a series of dimethacrylates with polyethylene glycol spacers of different sizes is shown. Figure 58 shows that a steeper decrease in free volume is obtained for those resins with shorter spacers. Interestingly, the photoinitiated polymerization for the different resins ceases at various double bond conversions, but at comparable values of z. This result confirms the generally accepted view that the amount of free volume in the polymer matrix is one of the main factors that determine the final degree of cure which a resin can achieve in a (photoinitiated) polymerization reaction.

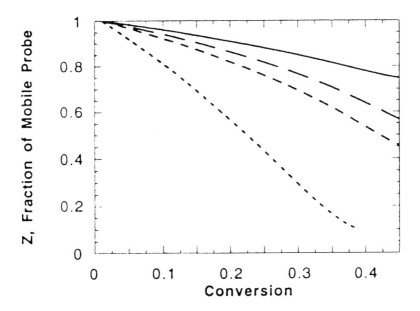

Figure 57

Fraction of probes that are mobile as a function of double bond conversion of DEGDMA polymerized with 0.2 W/cm^2 of UV light and 1 wt% DMPA. (azobenzene; solid line) (p-azotoluene; underbroken line, large) (stilbene; underbroken line, medium) (4,4'-diphenylstilbene; underbroken line, small). For the molecular structure of these probes, see Table 4. Taken from reference 50, with permission.

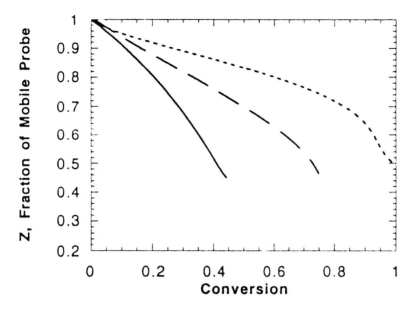

Figure 58

Fraction of mobile probe molecules z, as a function of double bond conversion for DEGDMA (solid line), PEG(200)DMA (underbroken line, large) and PEG(600)DMA (underbroken line, small) polymerized with 0.2 W/cm^2 of UV light and 1 wt% DMPA. Taken from reference 50, with permission.

Real time UV spectroscopy is an important tool for investigating photoinitiated polymerization processes. The technique is rapidly developing and can be used for investigating various aspects of the polymerization process. With the introduction of "initiatorless" formulations consisting of donor and acceptor alkenes, real time UV spectroscopy can monitor the cure of a resin directly. Monitoring the concentration of photobleachable initiators in real time has proven to be a valuable technique for investigating various aspects of photoinitiated polymerization processes. In particular in combination with other techniques which monitor the cure, interesting results are obtained. In particular, the combination of RT-IR and RT-UV is very successful for investigating commercial formulations under realistic conditions. By adding a photochromic probe, the development of free volume in real time can be monitored. In combination with other techniques, the free volume can be monitored as a function of cure. At first glance, this methodology appears to be very attractive for fundamental research.

RT-UV is a very fast and accurate technique. Sample requirements are comparable to those for RT-IR. An additional requirement is that no side reactions occur that produce species absorbing in the wavelength region of interest. This is important in this specific application, since extinction coefficients for UV absorbing species can be very high, and minute amounts of impurities can induce colour in otherwise transparent substrates. The combination of RT-IR and RT-UV seems particularly powerful since IR determines the cure, while UV can provide additional information. Geometrical constraints on the sample, disturbance by substrates to which the sample is connected and the fact that in most cases coloured probes need to be added disqualifies RT-UV as a technique for on line monitoring.

v. Real Time Fluorescence Spectroscopy

Fluorescence spectroscopy has recently been developed as an analytical technique for real time monitoring of photoinitiated polymerization processes. First introduced by Paczkowski and Neckers,[95,96,97,98,99] real time fluorescence spectroscopy was developed for monitoring laser induced polymerization processes employed in stereolithography.[168] One of the major motivations to develop fluorescent probe technology for real time monitoring in this specific application was the fact that spatial resolution can be achieved, for instance by specific irradiation of a section of a photoformed sample. Therefore, the polymerization process can be monitored as a function of time at different positions in a sample (time and spatially resolved). Another advantage of employing fluorescence spectroscopy for monitoring photoinitiated polymerization processes is that samples of a large variety of shapes attached to a wide variety of substrates can be monitored. This makes fluorescence spectroscopy an excellent technique for remote sensing.[103,169,170] The technique is tolerant to many additives such as glass fibres and other fillers that may be added to photopolymerizable formulations.

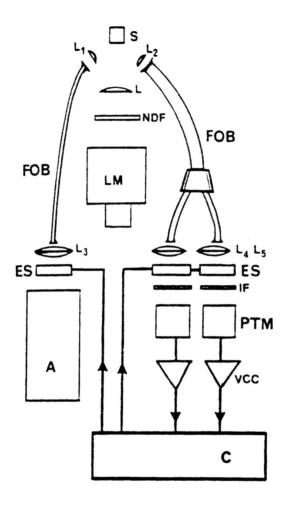

Scheme 26
Optical and electronic arrangement for following the kinetics of laser initiated polymerizations. C, computer; A, argon ion laser; ES, electronic shutter; VCC, voltage-to-current converter; PMT, photomultiplier tube; IF, interference filter; L1,L2,L3,L4,L5, lenses; LM, lamp and excitation monochromator; NDF, neutral density filter; FOB, fibre optic bundle; S, sample. Taken from reference 96, with permission.

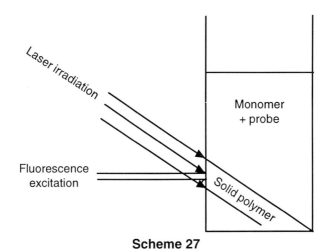

Scheme 27
Geometric scheme for probing the kinetics of photoinitiated polymerizations using the set-up displayed in Scheme 26. Taken from reference 97, with permission.

A fluorescent probe must be added to the resin in order to employ the technology for monitoring photoinitiated polymerization processes. The major requirements for such probes are high sensitivity and non-interference with the polymerization process. The latter is achieved by adding minor quantities of fluorescent probe to the resin, typically 0.02% by weight for thin films. For real time monitoring of photoinitiated polymerization processes, fluorescent probes whose emission shifts during the polymerization process are the most appropriate. Insensitivity for the concentration of the probe (photobleaching), the shape of objects and the high sensitivity of these probes justify this choice. In general the ratio between the emission intensities at two different wavelengths $I_{\lambda_1}/I_{\lambda_2}$ is monitored as a function of time, a process that is easily automated. So far, probes used for real time fluorescence spectroscopy are ICT probes (p 291) and in this section our discussion will be limited to cases where these probes are used.

A conventional fluorimeter equipped with a scanning emission monochromator cannot be used for monitoring photoinitiated polymerization processes in real time since it takes too much time to acquire the emission of a probe at different wavelengths.[171] In order to monitor the probe's emission at two different wavelengths *simultaneously*, the set up shown in Schemes 26 and 27 was developed. Multiple detection heads equipped with photomultiplier tubes, interference filters and shutters which simultaneously detect emitting light at different wavelengths are employed. The system is equipped with bifurcated optical fibres thereby enabling remote sensing and spatial resolution. In the application for which this set-up was developed, a visible photoinitiator, typically a fluorone derivative,[172,173] was excited by an Argon ion laser at 514 nm thereby initiating the polymerization of an acrylate resin. Simultaneously the fluorescent probe, typically a dansylamide like **13** or N-(1-naphthyl)-carbazole **10**, was excited at 345 nm using an additional light source and the intensity of the probe's emission was measured at two wavelengths λ_1 and λ_2. The emission intensities of **10** at 380 and 430 nm and the ratio of these emissions during laser irradiation are displayed in Figures 59 and 60. Since the emission ratio I_{380}/I_{430} as a function of the double bond conversion for **10** in the resin is known from Figure 61, a kinetic profile correlating the double bond conversion and the irradiation time for the photoinitiated polymerizations can be constructed.

For the set-up using two photomultiplier tubes, a response time of 50 ms was reported which is fast enough for monitoring the relatively slow laser initiated polymerizations. It should be noted that response times for photomultiplier tubes can be much shorter and that a similar set-up should be suitable for monitoring faster polymerization processes.

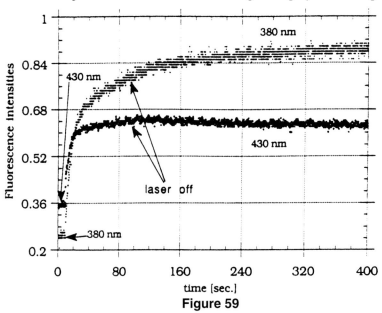

Figure 59

Emission intensity of 10 at 380 and 430 nm during the laser initiated polymerization of a TMPTA/VP (9:1) mixture. Taken from reference 97, with permission.

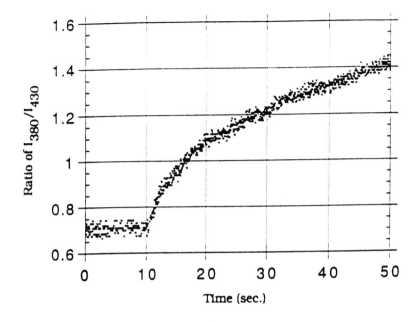

Figure 60

Ratio of emission intensities (I_{380}/I_{430}) for 10 during the laser initiated polymerization of a TMPTA/VP (9:1) mixture. Taken from reference 97, with permission.

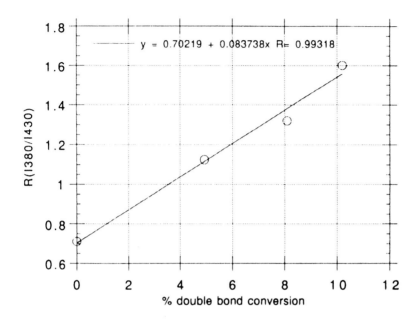

Figure 61

Ratio of emission intensities (I_{380}/I_{430}) for 10 as a function of the double bond conversion in a TMPTA/VP (9:1) mixture. Taken from reference 97, with permission.

In order to obtain a more versatile system for monitoring photoinitiated polymerization processes in real time, an all round detection system featuring a multichannel analyzer and a CCD detector was developed. A schematic representation of this instrument developed in the Neckers group, commonly known as "Cure Monitor",[174] is shown in Scheme 28. Full emission spectra can be obtained with this system at acquisition times down to 25 ms. The instrument thereby couples both speed and versatility. By using a bifurcated optic fibre, the device can be used for off-line (remote sensing) and on-line cure applications as well as for real time cure monitoring.

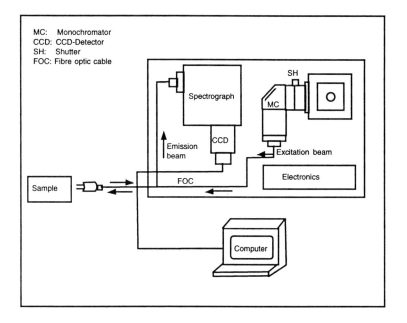

Scheme 28
Block diagram of the Cure Monitor. Taken from reference 175, with permission.

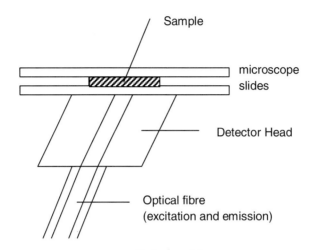

Scheme 29
Geometric scheme for probing the kinetics of photoinitiated polymerizations using the set-up displayed in Scheme 28.

Real-time monitoring of the photoinitiated polymerization of common formulations using the "Cure Monitor" was reported by Popielarz and Neckers.[175] In a typical experiment, a droplet of a photopolymerizable system was squeezed between two glass plates and placed on top of the head of the bifurcated cable of the "Cure Monitor," see Scheme 29. In this set up, the light from the excitation source, typically at 350nm, excites both the fluorescent probe and the photoinitiator. Both the wavelength and the intensity of the excitation beam can be varied systematically and polymerization processes can be investigated under a wide range of conditions. Typical photopolymerization profiles are shown in Figures 62 and 63. In these experiments the irradiation process takes place in a few minutes which is partly due to the relatively weak excitation source. From the kinetic profiles shown in Figures 62 and 63, one can easily determine the induction time t_i, while the initial rate of polymerization R_p and the final degree cure of the formulation can be derived from these curves as well. In order to obtain kinetic data from the kinetic profiles represented in Figures 62 and 63, calibration curves linking the emission ratio recorded for the probe to the cure of the resin need to be taken for each probe-resin combination This can be achieved without difficulty by recording FTIR and emission spectra for the same formulation after fixed irradiation periods. A representative example of a calibration curve that links the cure of a formulation to the emission of the added fluorescent probe is shown in Figure 64.

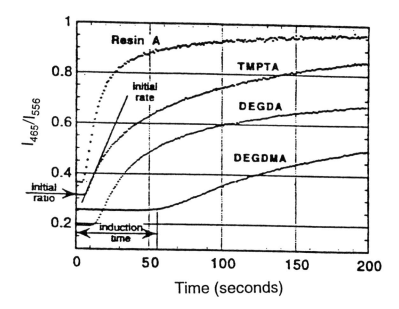

Figure 62

Kinetic profiles of various resins, obtained by curing a 0.1 mm layer of each formulation with the excitation beam of the "Cure Monitor." Taken from reference 175, with permission.

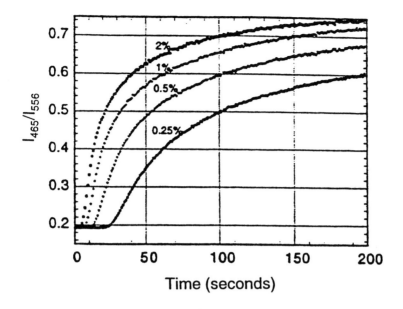

Figure 63
Influence of the initiator concentration on the kinetic profiles of DEGDA.
Taken from reference 175, with permission.

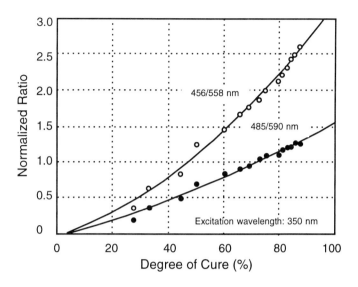

Figure 64
The emission ratio for 13 at two different sets of wavelengths as a function of the double bond conversion of an acrylate formulation. Taken from reference 175, with permission.

Monitoring the cure of a formulation in real time using fluorescence probe technology has some distinct advantages. First, the method is applicable to polymers of a variety of shapes, attached to a variety of substrates. The method is sensitive, universal, easy to apply and can be used for polymerization processes that take place in a few seconds. By applying an external light source that does not emit in the region of the spectrum in which the emission of the fluorescent probe is monitored, higher rates of polymerization can be achieved. If this protocol is applied one can investigate a photoinitiated polymerization process under conditions closely resembling those in a production environment. The temperature, pressure and the composition of the atmosphere can be controlled by placing the sample in a sample holder. Another distinct advantage is that the technique can be applied along with other real time techniques *simultaneously*. Since the light that initiates the polymerization process also excites the fluorescent probe, the only requirement for combining fluorescence with other techniques is the accessibility of the sample to an optical fibre in order to harvest the emission of the probe. Performing two techniques simultaneously on the very same sample is time saving and increases the reliability of the experiments.

A disadvantage of this technique, particularly for research purposes, is that the emission spectra of fluorescent probes are specific for each resin. This means that for kinetic measurements, calibration curves which link the cure of a system to the emission of the probe need to be made for each probe-resin combination. Calibration curves are generally made by irradiating samples for fixed periods of time, after which FTIR and emission spectra are recorded. Due to the time interval between the irradiation and the measurements, data are obtained for samples that might have undergone a dark polymerization (aftercure), or relaxation processes and are cooled down to room temperature.[176] Relatively slow processes which are polymerized under circumstances that are similar to those used for constructing calibration curves, can be monitored both accurately and reliably. Whether the calibration curves obtained from slow polymerization processes can be used for high speed polymerizations, during which large amounts of heat and large excesses of free volume are generated, has not been established. Real time monitoring of laser initiated polymerizations has shown that the emission intensity of a probe is affected by the temperature (estimated at 160°C for laser induced polymerizations) and presumably by the excess of free volume in the sample. By combining real time IR and real time fluorescence spectroscopy, one should be able accurately to determine the correlation between the cure of the resin and the emission of the probe under various conditions.

vi.   Dielectric Spectroscopy

Dielectric spectroscopy has been applied for real time monitoring of photoinitiated polymerization processes.[147] The technique uses special electrodes that are immersed in samples between quartz plates. The ion conductivity in the sample at relatively low frequencies has been measured. The response time of the technique is approximately 55 ms. As an example, the cure of an epoxy forumulation measured by real time dielectric spectroscopy is given in Figure 65. For comparison the accumulated heat of reaction, as determined by DSC, is plotted in the same figure. Real time measurements can be performed by irradiating a static sample with a light source, but also by putting a sample on a conveyor belt traveling through the curing unit.

Dielectric spectroscopy can be used for monitoring fast photoinitiated polymerization processes. As shown in Figure 65, the technique remains sensitive at high conversions which is a distinct advantage over photo-DSC. To what extent the sensitivity of dielectric spectroscopy to temperature and (conducting) contaminants influence these measurements is not clear. Since an electrode is incorporated in the sample, dielectric spectroscopy is not suited for off line and on line monitoring.

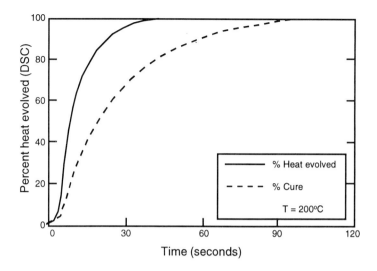

Figure 65
Conversion of an isothermally cured epoxy formulation measured in real time by DSC (solid line) and dielectric measurements (underbroken line). Taken from reference 147, with permission.

4. On line monitoring

i. Introduction.

In this section, analytical techniques that might be suitable for on line monitoring of photocured samples will be discussed. Since on line monitoring must not alter the properties of the samples being examined, only spectroscopic techniques can be employed. Other requirements are that the techniques should be non-contact, have fast response times and must be applicable to objects of differing geometries attached to a variety of other materials. The requirement of non-contact limits useful techniques to those of an optical nature that examine samples by means of electromagnetic radiation. Optical techniques that measure the absorption of electromagnetic radiation are, in general, not suitable for on line monitoring, because the absorption is proportional to the sample thickness, hence geometrical limitations apply. A more serious problem is the fact that most films are attached to substrates that interfere with absorption measurements. Therefore, IR and UV spectroscopy which are excellent techniques for real time monitoring of photoinitiated polymerization processes are not applicable for on line monitoring. Monitoring the amount of heat that evolves during a photoinitiated polymerization by IR techniques has been reported, but this technique is not very accurate. From the techniques discussed in this chapter, fluorescence spectroscopy is the only technique that is generally applicable for on line monitoring.

ii. Fluorescence Spectroscopy for On Line Monitoring

Until now, publications reporting fluorescence spectroscopy for on line monitoring have not appeared in the literature. However, the use of fluorescence spectroscopy for characterization of UV cured materials by means of remote sensing, [103] and real time monitoring[175] have been reported. On the basis of these publications, it was concluded that fluorescence probe technology is sensitive, applicable to objects of different sizes and shapes, accurate and fast. Remote sensing and off line monitoring experiments have focused on two properties of the photocured formulation.[177] The most important property is the cure of a resin derived from the emission ratio of a probe at two different wavelengths $I_{\lambda 1}/I_{\lambda 2}$. In addition, the coat weight of a photoformed polymer can be determined by means of the intensity of the emission of a fluorescent probe.[177] Since the emission intensity is proportional to the number of excited probe molecules, this method can determine the thickness of a photocured coating. It must be noted that for coat weight determinations, many restrictions do apply which are not applicable to cure monitoring. Only flat substrates can be used, the probe should be evenly distributed throughout the sample and may not photobleach. Also the excitation source should be stable and emission from the background is not allowed.

Whether fluorescent probe technology will enter the market place is not a technological question. By employing fibre optics, fast detection systems like a CCD and a computer for data analysis, a system for on line monitoring can be added to any production line, see Scheme 30. In principle one could monitor the photoformed product in the curing unit. In this case, the light that initiates the polymerization process also excites the probe. However, overlap between the emission of the lamps from the curing unit and emission of the probe might be a problem. Also, the samples are still hot and reacting and such measurements would not monitor the final properties of the product that one is interested in. Therefore, on line monitoring is preferred at a position where the polymerization process has ceased and the products are relaxed and cooled down. Optical fibre technology is preferred for harvesting the emission of the probe due to its flexibility and the option to monitor the samples at many positions. By employing a multichannel CCD detector, the spectra from a multitude of fibres can be recorded simultaneously and presented in three dimensional spectra (position and wavelength versus intensity). Excitation of the probe can be done locally by optical fibres or over a broad area by using an external light source. In both cases, overlap between the emission of the light source and the emission of the probe needs to be avoided. Filtering the emission of the light source seems to be the easiest solution. Finally samples must be shielded from background irradiation at the relevant wavelengths.

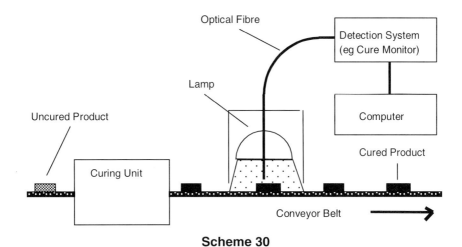

Scheme 30
Schematic representation of the application of fluorescent probe technology for on line monitoring on a curing line

Amongst the factors that determine whether fluorescence probe technology will be applied to on line monitoring is the availability of fluorescent probes that combine a number of desired properties. The features required are probes that are affordable (cheap), sensitive and preferably colourless. The latter property, however, does not seem to be relevant for all applications, but in most cases, colour introduced into products that should be colourless is not appreciated. Other applications, notably those used in food packaging and processing demand that the toxicity of the probe should be considered. Finally, for durable goods, the long term stability of fluorescent probes and their effects on the system must be investigated thoroughly.

On line monitoring of photoformed objects is a desirable technique for a number of applications. In particular, it is essential for those **products produced with strict quality control**. Fluorescent probe technology is the technique that fulfils all the requirements for on line monitoring, such as **speed, sensitivity, non-contact nature and applicable to objects of various sizes and shapes.** From a scientific and technological point of view, using fluorescent probes for on line monitoring is realistic and conceivable. Whether fluorescent probe technology will be successfully introduced in the market place, i.e. whether the addition of fluorescent probes leads to acceptable products and the benefits of on line monitoring outweigh the extra costs, is a question that might be answered in the near future.

References

1. Kloosterboer, J. G. *Adv. Polym. Sci.* **1988**, *84*, 1.
2. Lungu, A.; Neckers, D.C. *Macromolecules* **1995**, *28*, 8147.
3. See for example Rabek, J. F. Experimental Methods in: *Polymer Chemistry; Physical Principles and Applications,* John Wiley: New York, 1980.
4. Rabek, J. F. in: *Radiation Curing in Polymer Science and Technology,* Volume 1, Chapter 7. Fouassier, J. P.; Rabek, J. F. Eds; Elsevier Applied Science: London, 1993.
5. For an extensive review of calorimetric techniques, see: Hoyle, C. E. Chapter 3 in: *Radiation Curing; Science and Technology*, Pappas, S. P., Ed.; Plenum Press: New York 1992.
6. It is well established that the degree of cure at which a photopolymerization process stops, i.e., the final cure, decreases if the length of the spacers between reactive groups is shortened and increases as the temperature at which the photopolymerization process is performed is raised.
7. Bandrup, J.; Immergut, E. H. (Eds), *Polymer Handbook,* 2nd edn., pp II-421.
8. Horie, K.; Otagawa, A.; Muraoka, M.; Mita, I. *J. Polym. Sci., Part A* **1975**, *13*, 18.
9. Liebmann, J. F.; Greenburg, A. *Chem. Rev.* **1989**, 89 1239.
10. Nelson, E. W.; Jacobs, J. L.; Scranton, A. B.; Anseth, K. S,; Bowman, C, N. *Polymer* **1995**, *36*, 4651.
11. For an alternative introduction on FTIR and raman spectroscopy for polymer characterization, see: Griffith, P. R.; Urban, M. W. Chapter 30 in: *Multidimensional Spectroscopy of Polymers; Vibrational, NMR, and Fluorescent Techniques.* Urban, M. W.; Provder, T. Eds; ACS Symposium Series 598, American Chemical Society; Washington DC, 1995.
12. Decker, C.; Moussa, K. *Macromolecules* **1989**, *22*, 4455.
13. Decker, C. *J. Polym. Sci. Polym. Chem.* **1992**, *30*, 913.
14. Crivello, J. V.; Varlemann, U. *J. Polym. Sci. Polym. Chem.* **1995**, *33*, 2473.
15. Crivello, J. V.; Kim, W-G. *J. Polym. Sci. Polym. Chem.* **1994**, *32*, 1639.
16. Crivello, J. V.; Yang, B. *J. Polym. Sci. Polym. Chem.* **1995**, *33*, 1881.
17. Crivello, J. V.; Löhden, G. *J. Polym. Sci. Polym. Chem.* **1996**, *34*, 1015.
18. Andersson, H.; Gedde, U. W.; Hult, A. *Macromolecules* **1996**, *29*, 1649.
19. Crivello, J. V.; Lai, Y.-L.; Malik, R. *J. Polym. Sci. Polym. Chem.* **1996**, *34*, 3103.
20. Crivello, J. V.; Löhden, G. *J. Polym. Sci. Polym. Chem.* **1996**, *34*, 2051.
21. Crivello, J. V.; Yang, B. *J. Polym. Sci. Polym. Chem.* **1995**, *33*, 1381.
22. Decker, C.; Morel, F.; Jönsson, S.; Clark, S. C.; Hoyle, C. E. *PMSE Proceedings* **1996**, *2*, 198.
23. The absorption at the wavelength of interest (A=ecd) should not exceed 2 for accurate measurements.
24. Pemberton, D. R.; Johnson, A. F. *Polymer* **1984**, *25*, 529.
25. Lee, C.; Hall, H. K. *Macromolecules* **1989**, *22*, 21.
26. Hall, H. K.; Padius, A. B. *Acc. Chem. Res.* **1990**, *23*, 3.
27. Broer, D. J.; Boven, J.; Mol, G. N.; Challa, G. *Makromol. Chem.* **1989**, *190*, 2255.

28. Broer, D. J.; Lub, J.; Mol, G. N.; Challa, G. *Macromolecules* **1993**, *26*, 1244.
29. Hikmet, R. A. M.; Lub, J.; Higgins, J.A. *Polymer* **1993**, *34*, 1736.
30. Favre-Nicolin, C. D.; Lub, J. *Macromolecules* **1996**, 29, 6143.
31. Miyashita, K. Chapter 33 in: *Multidimensional Spectroscopy of Polymers; Vibrational, NMR, and Fluorescent Techniques*. Urban, M.W.; Provder, T. Eds; ACS Symposium Series 598, American Chemical Society: Washington DC, **1995**.
32. Dietz, J.E.; Elliot, B. J.; Peppas, N.A. *Macromolecules* **1995**; *28*, 5163.
33. Buehner, R. W. *Radtech Proceedings North America* **1996**, *1*, 407.
34. Schaeken, T. C.; Warman J. M. *J. Phys. Chem.* **1995**, *99*, 6145.
35. Jonsson, S.; Sundell, P-E.; Hultgren, J.; Sheng, D.; Hoyle, C. E. *Prog. Org. Coatings* **1996**, *27*, 107.
36. Hoyle, C. E.; Jonsson, S.; Clark, S. C.; Miller, C.; Shomrose, M. *Radtech Europe; Academic Proceedings*, **1997**, 115.
37. By employing probes that absorb at long(er) wavelengths, the absorption of the probe can be monitored without exciting the photoiniator and initiating polymerization.
38. Reichardt, C. *Chem. Rev.* **1994**, *94*, 2319.
39. Reichardt, C. *Solvents and Solvent Effects in Organic Chemistry*, VCH Weinheim, second edition 1988.
40. Paley, M. S.; McGill, R. A.; Howard, S. C.; Wallace, S. E.; Harris, J. M. *Macromolecules* **1990**, *23*, 4557.
41. Preliminary tests using Reichardt's dye in the photopolymerization of dimethacrylates reveal a strong bleaching of this probe. Typical shifts for **4** are in the order of 20 nm. Jager, W. F.; Neckers, D. C. *Unpublished results.*
42. Lamarre, L.; Sung, C. P. S. *Macromolecules* **1983**, *16*, 1729.
43. Yu, W.-C.; Sung, C. S. P.; Robertson, R. E. *Macromolecules* **1988**, *21*, 355.
44. Yu, W.-C.; Sung, C. S. P.; *Macromolecules* **1988**, *21*, 365.
45. Victor, J. G.; Torkelson, J. M. *Macromolecules* **1987**, *20*, 2241.
46. Victor, J. G.; Torkelson, J. M. *Macromolecules* **1988**, *21*, 3490.
47. Royal, J. S.; Victor, J. G.; Torkelson, J. M. *Macromolecules* **1992**, *25*, 729.
48. Royal, J. S.; Torkelson, J. M. *Macromolecules* **1992**, *25*, 4792.
49. Anseth, K. S.; Rothenberg, M. D.; Bowman, C. N. *Macromolecules* **1994**, *27*, 2890.
50. Anseth, K. S.; Walker, A.; Bowman, C. N. Chapter 10 in: *Multidimensional Spectroscopy of Polymers; Vibrational, NMR, and Fluorescent Techniques*. Urban, M. W.; Provder, T, Eds; ACS Symposium Series 598, American Chemical Society: Washington DC, **1995**.
51. Rabek, J. F. *Mechanisms of Photophysical Processes and Photochemical Reactions in Polymers*; John Wiley & Sons; Chichester, UK, **1987**, Chapter 4.
52. For an alternative introduction see: Soutar, I. Chapter 20 in: *Multidimensional Spectroscopy of Polymers; Vibrational, NMR, and Fluorescent Techniques*. Urban, M. W.; Provder, T, Eds; ACS Symposium Series 598, American Chemical Society: Washington DC, **1995**.

53 For a review of fluorescent probes for application as sensors and switches, see: de Silva, A. P.; Guanaratne, H. Q. N.; Gunnlaugsson, T.; Huxley, A. J. M.; McCoy, C. P.; Rademacher, J. T.; Rice, T. E. *Chem. Rev.* **1997**, *97*, 1515.
54 Wang, S,-K.; Sung, C. S. P. *Polym. Preprints* **1996**, *1*, 467.
55 Grunden, B.; Kim, Y. S.; Sung, C. P. S. *Polym. Preprints* **1996**, *1*, 477.
56 Meijer, E. W.; de Leeuw, D. M.; Geidanus, F. J. A. M.; Zwiers, R. J. M. *Polymer Commun. 1985*, *26*, 45.
57 Meijer, E. W.; Zwiers, R. J. M. *Macromolecules* **1987**, *20*, 332.
58 Optical brighteners are examples of strongly fluorescent additives.
59 Turro, N. J. *Modern Molecular Photochemistry* Benjamin Cummins: Menlo Park, **1978**.
60 Rawlings, K. A.; Lees, A. J.; Fuerniss, S. J.; Papathomas, K. I. *Chem. Mater.* **1996**, *8*, 1540.
61 In cases where quenching of the singlet excited state by a specific agent Q takes place, the term $k_Q[Q]$ should be added to equations 7, 8 and 10.
62 To the best of our knowledge, reactive probes that undergo specific changes in emission intensity upon incorporation in a polymer (see section - page 278) are the only fluorescent probes capable of monitoring the cure of a resin directly.
63 Rutherford, H.; Soutar, I. *J. Polym., Sci. Phys. Ed.* **1980**, *18*, 1021.
64 Nishijama, Y.; Mito, Y. *J. Pol. Sci., Part C* **1970**, *31*, 353.
65 Full conservation of the polarization is possible only in those cases in which probe molecules are aligned and immobilized. The dipole moment of the absorption and the emission should be parallel to the polarization of the excitation beam.
66 van Ramensdonk, H. J.; Vos, M.; Verhoeven, J. W.; Möhlmann, G. R.; Tissink, N. A.; Meesen, A. *Polymer* **1987**, *28*, 951.
67 The sensitivity at higher viscosity might be increased by employing phosphorescent probes and smaller, rounder probes.
68 Trommsdorff, E.; Koehle, H.; Legally, P. *Makromol. Chem.* **1947**, *1*, 169.
69 Hoyle, C. E.; Trapp, M. A. *J. Imag. Sci.* **1989**, *33*, 191.
70 Jager, W. F.; Volkers, A. A.; Neckers, D. C.; *Macromolecules* **1995**, *28*, 8153.
71 Lin, K-F.; Wang, F. W. *Polymer* **1994**, *35*, 687.
72 Loutfy, R. O. *Pure Appl. Chem.* **1986**, *58*, 1239.
73 Law, K. Y.; Loutfy, R. O. *Polymer* **1983**, *24*, 439.
74 Loutfy, R. O.; Teegarden, D. M. *Macromolecules* **1983**, *16*, 452.
75 Loutfy, R. O. *J. Polym. Sci. Phys. Ed.* **1982**, *20*, 825.
76 Law, K. Y.; Loutfy, R. O. *Macromolecules* **1981**, *14*, 587.
77 Loutfy, R. O. . *Macromolecules* **1981**, *14*, 270.
78 Loutfy, R. O.; Law, K. Y. *J. Phys. Chem.* **1980**, *84*, 2803.
79 Law, K. Y. *Chem. Phys. Lett.* **1980**, *75*, 545.
80 Royal, J. S.; Torkelson, J. M. *Macromolecules* **1993**, *26*, 5331.
81 In most cases, a decrease in emission intensity is observed later on. This is due to photodecomposition or photobleaching of the probe.
82 Haugland, R. P. *Handbook of Fluorescent Probes and Research Chemicals*, Larison, K. D. Ed. Molecular Probes Inc: Eugene, OR, 5th edition, **1992**.

83. Verhey, H. J.; Hofstraat, J. W.; Bekker, C. H. W.; Verhoeven, J. W. *New J. Chem.* **1996**, *20*, 809.
84. Warman, J. M.; Abellon, R. D.; Verhey, H. J.; Verhoeven, J. W.; Hofstraat, J. W. *J. Phys Chem. B.* **1997**, 101, 4913.
85. The sample position can be fixed by leaving the sample in the sample compartment of a spectrofluorometer as was performed in physical ageing experiments, see reference 80.
86. Complete emission spectra can be taken in as little as 25 ms, using CCD devices for detection, see section (page 359).
87. As long as the products that are formed by photobleaching do not absorb the emission of the probe and are non-fluorescent photobleaching is not a serious problem. In most cases, the products formed by photodecomposition of fluorescent probes absorb at shorter wavelengths and are non-fluorescent.
88. Using intensity sensitive probes a second fluorescent probe can be used for internal calibration. This makes these probes suitable for samples of different geometries but does not correct for photodecomposition of the probe, see reference 93.
89. Valdez-Aguilera, O.; Pathak, C. P.; Nechers, D. C. *Macromolecules* **1990**, *23*, 689.
90. Wang, F. W.; Lowry, R. E.; Grant, W. H. *Polymer* **1984**, *25*, 690.
91. Wang, F. W.; Lowry, R. E. *Polymer* **1985**, *26*, 1046.
92. Wang, F. W.; Lowry, R. E.; Cavanagh, R. *Polymer* **1985**, *26*, 1657.
93. Wang, F. W.; Lowry, R. E.; Fanconi, B. M. *Polymer* **1986**, *27*, 1529.
94. Paczkowski, J.; Neckers, D. C. *Macromolecules* **1991**, *24*, 3013.
95. Paczkowski, J.; Neckers, D. C. *Chemtracts: Macromol. Chem.*, **1992**, *3*, 75.
96. Paczkowski, J.; Neckers, D. C. *Macromolecules* **1992**, *25*, 548.
97. Zhang, X.; Paczkowski, J.; Kotchetov, I. N.; Neckers, D. C. *J. Imaging Sci.* **1992**, *36*, 322.
98. Paczkowski, J.; Neckers, D. C. *J Polym. Sci. Polym Chem.* **1993**, *31*, 841.
99. Kotchetov, I. N.; Neckers, D. C. *J. Imaging Sci.* **1993**, *37*, 156.
100. Rotkiewicz, K.; Grellmann, K. H.; Grabowski, Z. R. *Chem. Phys. Lett.* **1973**, *19*, 315.
101. Rettig, W. *Angew. Chem., Int. Ed. Engl.* **1986**, *25*, 971.
102. Al-Hassan, K. A. *J. Photochem. Photobiol. A: Chem.* **1994**, *84*, 207.
103. Song, J. C.; Neckers, D. C. Chapter 28 in: *Multidimensional Spectroscopy of Polymers; Vibrational, NMR, and Fluorescent Techniques.* Urban, M. W.; Provder, T. Eds; ACS Symposium Series 598, American Chemical Society: Washington DC, **1995**.
104. Wang, Z. J.; Song, J. C.; Bao, R.; Neckers, D. C. *J Polym. Sci. Polym. Phys.* **1996**, *34*, 325.
105. Jacobson, A.; Petric, A.; Hogenkamp, D.; Sinur, A.; Barrio, J. R. *J Am. Chem. Soc.* **1996**, *118*, 5572.
106. Gvishi, R.; Narang, U.; Bright, F. V.; Prassad, P. N. *Chem. Mater.* **1995**, *7*, 1703.
107. Lin, K-F.; Wang, F. W. *Polymer* **1994**, *35*, 687.

108. For example, see: *Polymers for Second-Order Nonlinear Optics*. Lindsay, G. A.; Singer, K. D. Eds.; ACS Symposium Series 601, American Chemical Society: Washington DC, Plenum Press, **1994**.
109. Slama-Schwok, A.; Blanchard-Descre, M.; Lehn, J-M. *J Phys. Chem.* **1990**, *94*, 3894.
110. Shin, D. M.; Whitten, D. G. *J. Phys. Chem.* **1988**, *92*, 2945.
111. Kanis, P. R.; Ratner, M. A.; Marks, T. J. *Chem. Rev.* **1994**, *94*, 195.
112. Lippert, E. *Z. Electrochem.* **1957**, 61, 962.
113. Liptay, W. *Z. Naturforschung*, **1965**, *20a*, 1441.
114. Jager, W. F.; Neckers, D. C. *Unpublished results*.
115. Jenneskens, L. W.; Verhey, H. J.; van Ramensdonk, H. J.; Witteveen, A. J.; Verhoeven, J. W. *Recl. Trav. Chim. Pays-Bas* **1992**, *111*, 507.
116. Jager, W. F.; Lungu, A.; Chen, D Y.; Neckers, D. C. *Macromolecules* **1997**, *30*, 780.
117. Mes, G. F.; de Jong, B.; van Ramensdonk, H. J.; Verhoeven, J. W.; Warman, J. M.; de Haas, M. P.; Horsman van den Dool, L. E. W. *J. Am. Chem. Soc.* **1984**, *106*, 6524.
118. Scherer, T.; van Stokkum, H. M.; Brouwer, A. M.; Verhoeven, J. W. *J Phys. Chem.* **1994**, *98*, 10549.
119. Scherer, T.; Hielkema, W.; Krijnen, B.; Hermant, R. M.; Eijckelhoff, C.; Kerkhof, F.; Ng, A. K. F.; Verleg, R,; van der Tol, E. B.; Brouwer, A. M.; Verhoeven, J. W. *Recl. Trav. Chim, Pays-Bas* **1993**, *112*, 535.
120. Verheij, H. J. "Fluorescence Probing of Polymers" Ph. D. Thesis, University of Amsterdam, 1997.
121. Jenneskens, L. W.; Verhey, H. J.; van Ramensdonk, H. J.; Witteveen, A. J.; Verhoeven, J. W. *Macromolecules* **1991**, *24*, 3013.
122. Verheij, H. J.; Gebben, B.; Hofstraat, J. W.; Verhoeven, J. W. *J. Polym. Sci. Chem.* **1995**, *30*, 399.
123. Verheij, H. J.; Hofstraat, J. W.; Bekker, C. H. W.; Verhoeven, J. W. *New J. Chem.* **1996**, *20*, 809.
124. Jager, W. F.; Kudasheva, D.; Neckers, D. C. *Macromolecules* **1996**, *29*, 7351.
125. Webster, F. G.; McColgin, W. C., U.S. Patent #3,833,863, December 27, 1971.
126. Marder, S. R.; Perry, J. W.; Yakymyshyn, C. P. *Chem. Mater.* **1994**, *6*, 1137.
127. Rettig, W.; Strehmel, B.; Majenz, W. *Chem. Physics* **1993**, *173*, 525.
128. Di Bella, S.; Fragalà I.; Ratner, M. A.; Marks, T. J. *Chem. Mater.* **1995**. *7*, 400.
129. It should be noted that these organic salts are insoluble in apolar solvents like hexane, benzene or ether.
130. More research is needed to clearly establish the relationship between the structure of these type of probes and their sensitivity for monitoring (photoinitiated) polymerization processes.
131. For a review on NMR studies of solid polymers, see: *Polymer Spectroscopy*, Fawcett, A. H. Ed, John Wiley: Chichester, 1996, Chapters 4 and 5.
132. Schaefer, J.; Stejskal, E. O.; Buchdahl, R. *Macromolecules* **1977**, *10*, 384.

[133] Dejean de la Batie, R.; Laupretre, F.; Monnerie, L. *Macromolecules* **1988**, *21*, 2045.
[134] O'Donnell, J. H.; Whittaker, A. K., *J. Polym. Sci. Polym. Chem Ed.* **1992**, *30*, 185.
[135] Sankar, S. S.; Stejskal, E. O.; Fornes, R. E.; Fleming, W. W.; Russell, T. P.; Wade, C. G. Chapter 11 in: *Polymer and Fiber Science; Recent Advances*, Fornes, R. E.; Gilbert, R. D. Eds; VCH Publishers Inc.: New York, 1992.
[136] Lungu, A.; Neckers, D. C. *J. Coat. Technol.* **1995**, *67*, 29.
[137] Torres-Filho, A.; Neckers, D. C. *Chem. Mater.* **1995**, *7*, 753.
[138] Allen, P. E. M.; Bennett, D. J.; Hagias, S.; Hounslow, A. M.; Ross, G. S.; Simon, G. P.; Williams, D. R. G.; Willimas, E. H. *Eur. Polym. J.* **1989**, *25*, 785.
[139] Rawland, T. J.; Labun, L. C. *Macromolecules* **1978**, *11*, 467.
[140] The spin diffusion rates compiled in Tables 9-11 are the reciprocals of the carbon resolved proton relaxation times $T_{1\rho}$.
[141] Rånby, B.; Rabek, J. F. "ESR Spectroscopy in Polymer Research," Springer Verlag: Berlin, 1977.
[142] Kamachi, M. *Adv. Polym. Sci.* **1987**, *82*, 209.
[143] Kamachi, M. *PMSE Proceedings* **1996**, *2*, 649.
[144] Kolczak, U.; Rist, G.; Dietliker, K.; Wirz, J. *J. Am. Chem. Soc.* **1996**, *118*, 6477.
[145] Kloosterboer, J. G.; van de Hei, G. M. M.; Gossink, R. G.; Dortant, G. C. M. *Polymer Commun.* **1984**, *25*, 322.
[146] Kloosterboer, J. G.; Lijten, G. F. C. M.; Greidanus, J. A. M. *Polymer Commun.* **1986**, *27*, 268.
[147] Smith, N. T.; O'Malley, T. F. *Radtech Proceedings North America* **1996**, *1*, 417.
[148] Gotro J. *Proceedings of the 19th NATAS Conference*, **1990**, 523.
[149] Day, D. R.; Sheoard, D. D. *Proceedings of the 16th NATAS Conference*, **1987**, 52.
[150] Zentel, R.; Strobl, G. R.; Ringsdorff, H. *Macromolecules* **1985**, *18*, 960.
[151] Bormuth, F. J.; Biradar, A. M.; Quotschalla, U.; Haase, W. *Liq. Cryst.* **1989**. *5*, 1549.
[152] Hikmet, R A. M.; Lub, J.; Maassen vd Brink, P. *Macromolecules* **1992**, *25*, 4194.
[153] Senturia, S. D.; Sheppard Jr, N. F.; *Adv. Polym. Sci.* **1986**, *80*, 1.
[154] The lack of temperature control during the polymerization process is the main reason why measurements obtained by such caloric techniques are hard to compare.
[155] Crivello, J. V.; Lee, J. L.; Conlon, D. A. *J. Rad. Curing* **1983**, *10*, 6.
[156] Kloosterboer, J. G.; Lijten, G. F. C. M. *Polymer Commun.* **1990**, *31*, 95.
[157] Decker, C.; Moussa, K. *J. Coatings Technology*, **1990**, *62*, 55.
[158] Anseth, K. S.; Wang, C. M.; Bowman, C. N.; *Polymer* **1994**, *35*, 3243.
[159] Decker, C.; Moussa, K. *Macromolecules* **1989**, *22*, 4455.
[160] Decker, C.; Moussa, K. *Eur. Polym. J.* **1990**, *26*, 393.
[161] Decker, C.; Moussa, K. *Macromol. Chem.* **1990**, *191*, 963.
[162] Decker, C. in: *Materials for Microlithography*, ACS Symposium Series 266, Thomson, L. F.; Wilson, G. G.; Frechet, J. M. J. Eds American Chemical Society, **1984**, 207.
[163] Moussa, K.; Decker, C. *J. Polym. Sci. Polym. Chem., Part A* **1993**, *31*, 2197.

164. When employing thick, deairated samples composed of purified monomers, the postpolymerization process can be much slower, and takes minutes to complete (see reference 34).
165. Anseth, K. S.; Decker, C.; Bowman, C. N. *Macromolecules* **1995**, *28*, 4040.
166. Crivello, J. V.; Bratslawski, S. A. *J. Polym. Sci. Chem. Ed*. **1994**, *32*, 2919.
167. For such calculations, a gaussian distribution of the free volume is assumed.
168. Neckers, D. C. *Polym. Eng. Sci*. **1992**, *32*, 1481.
169. Song, J. C.; Neckers, D. C. *Polym. Eng. Sci*. **1996**, *36*, 395.
170. Song, J. C.; Wang, Z. J.; Bao, R.; Neckers, D. C. *PMSE Proceedings*, **1995**, *72*, 591.
171. It should be noted that a conventional fluorometer can be used for monitoring the emission intensity at a fixed wavelength at reasonable rates.
172. Neckers, D. C.; Valdez Aguilera, O. M. in: *Advances in Photochemistry*; Volman, D. H.; Hammond, G. S.; Neckers, D. C.; Eds.; Wiley New York, 1993; Vol. 18, pp 315-393
173. Tanabe, T.; Torres-Filho, A.; Neckers, D. C. *J. Polym. Sci. Polym. Chem*. **1996**, *33*, 1691.
174. The "Cure Monitor" was developed in cooperation with Oriel Instruments and Spectra Group Limited (SGL) and is now commercially available.
175. Popielarz, R.; Neckers, D. C. *Radtech Proceedings North America* **1996**, *1*, 271.
176. The time scale for dark polymerizations are in the order of a few seconds for thin films, see reference 159. For thick deaereated samples consisting of purified monomers the dark polymerization can take minutes, see reference 34.
177. Neckers, D. C.; Song, J. C.; Torres-Filho, A.; Jager, W. F.; Wang, Z. J. U.S. Patent number #5,606,171, February 25, 1997.

INDEX

A
α-amino radical 143
α-hydroxyketene 106
Absolute rates 6
Absolute rate constants 6
Absorbance probe 258
Absorbance 227, 228, 258
Absorption 12, 13-14, 14-15, 15, 17-19, 23-29, 124, 154, 156, 205-209, 210-216
Absorption spectra 15, 26, 28, 44, 123, 124, 129, 173 204, 206, 208, 212
    benzophenomethyl-tri-n-butylammonium tetrafluoroborate in benzene 28
    benzophenone in ethanol + cyclohexane 15
    cyanine dyes 44
    DIBF 204, 206, 208, 212
    methylphenylglyoxylate in benzene 26
    p-benzoylphenyldiphenyldimethyl (RAD) 123, 124
    RAD 123, 124
    TBR (transient) 129
Absorptions and degree of cure (IR) - various resins 243
Acceptor 17, 29, 97-98, 154-166, 166-168, 170, 189
Acetone 32-33
Acetonitrile 48, 173-180, 188
Acetyl initiator 197
Acridines 196
Acrylates 3, 6-7, 55, 117, 161-164, 191-195
Actinometer 34
Actinometrical measurement 34
Acyl radical 95-98, 105
Added probe molecules 269
AFM (Atomic Force Microscopy) 63-68, 80, 111-112, 113-114
Afterglow 29
AIBN 10-11
Alcohols 102-103, 140
Aldehydes 92-95, 101
Aliphatic ketones 32-36
Alkynes 100-101
Alkoxide 57-58
Amine precursors 60-63, 169-171
Amines 50, 51, 52, 53, 57-68, 97, 98-100, 140-141, 142-144, 182
Amplifying images 55
Analytic techniques - monitoring polymerization processes 239-321
    calorimetric: differential scanning (DSC) 239-241

classification of techniques 236
        on line/off line monitoring 237-238
                spectroscopic 241-261 (See Spectroscopic techniques + specific technique)
Analysis of polymeric films (research techniques) 241-321
        (See Spectroscopic techniques + specific technique
Argon 77-81, 191
Argon ion lasers 77-81, 138
Aromatic hydrocarbons 103-105
Aromatic carbonyl compounds 91-95, 103-105, 105-109
Aromatic ketones 105-109
Arsines 46
Aryl iodonium salts 46, 47-50
Asymmetric iodonium bromides 185-186, 187-189
ATR (Attenuated Total Reflectance) mode 244, 251-252
Attenuated Total Reflectance-Fourier Transform Infrared (ATR-FTIR) 251-252
Avagadro's number of photons 33-34
Azo photoinitiators
        AIBN 10-11
        azobenzene/stilbene derivatives 259-260
        azoquinomethine 196-197
        historical use 8, 10-11
        long wavelength UV 8, 9
        visible radiation 10
        solvent cage 10

B
β-naphthylmethyl radical 180
Back electron transfer 46, 181-182
Benzaldehyde 155
Benzene 109-110, 139
Benzoin and benzoin ethers 12, 51-52
Benzophenone 9, 13, 15, 16, 24-25, 27-29, 30-32, 51-55, 91, 98-100, 101-105, 107, 116-119, 127, 134-135, 142-144, 169-170, 230
Benzoyl formic acid 106
Benzoyl radical 95, 96, 118-119
Benzoyloxy radical 119
Benzyl radical 118-119
Bichromic probes (intramolecular) 279
Bimolecular reaction of radicals 6
Bimolecular electron transfer 135
Bimolecular reactions 75, 106-107, 144
Biphenyl 172-173
Biradical state/attack 30, 100-101, 105-106, 108, 166-168
Bleaching 196
Blue shifts 295-297
Bond conversion 187-188
Bond dissociation 41-47, 91-109, 179-180

Bond homolysis 91-109, 181
Borate salts as coinitiators 45, 154-155
Borate iodonium complexes (DPIBB, DMPIBB, DMPITB, BPPIBB, OPPIBB, OPPITB) 185-186, 187-191, 191-195
Borates 99, 169-170, 170-183, 183-184, 184-186, 187-191, 191-195
  as electron donors 183-184
    iodonium borates 184-186, 187-191, 191-195
Butyltriphenyl borate salts 188-189

C
2 + 2π cycloaddition 100-101
Calorimetric techniques (DSC) 239-241
  degree of cure: acrylate/methacrylate/epoxide/vinyl ether 241
  DMA (Dynamic Mechanical Analysis) measurements 241
Camphor quinone 98, 140-142
Carbenium ion (acylium ion) 96
Carbocation
  from ammonium ion 50
  entrapment by ester carbonyl group 218
  triphenylmethyl ('trityl cation') 144
Carbon spin lattice relaxation times 315-319
Carbonyl region of spectrum 62
Carbonyl group 92-98, 98-100, 102-101, 101-105, 116-140, 140-144, 144-153
  π-π* transition 92-94
  π-π* transition 94-95, 95-96
  2 + 2• cycloaddition 100-101
  acidity of protons alpha 92
  alkyne reactions 101
  as targets using carbonyl group 116-140
  benzophenone carbonyl group 117-119
  benzophenone triplets 142-144
  benzoyl radical 96
  bonding π orbital 92-93
  carbonyl compound initiators hydrogen abstraction route to crosslinks 101-105
  electron density 94
  extinction coefficients 94
  intramolecular oxidational reduction 144-153
  of acyl radical 95-98
  Paterno-Büchi reaction 100-101
  photochemistry 116-140
  reactivity route 116
  reactivity 92-95, 95-98, 116
  reduction 116, 142-144
  reduction with amines 142-144
  resonance structure 92
  sigma 92, 95-98
  target 119-121

            transition state formation 97-98
Cationic chain reactions and polymerization 47-50, 217-219
Chain length - kinetics 6-8
Chain propagation 57-61
Charge transfer complex photoinitiators 154-195
        "initiator free" 155
        acrylate replacement 163
        acrylates 162, 163, 191-195
        alkyl group from borate to olefin 169-170
        biradical formation 167-168
        borates as electron donors 182-184
        copolymers 164-166
        donor/acceptor complex 154, 155-166, 166-168
        exciplex reaction 166-168
        formulation of radicals via CT 167-168
        intramolecular electron transfer 170-182
        iodonium/borates 184-186, 187-191, 191-195
        maleimide (MI)/vinyl ether (VE) 161, 163-166
        maleimide/vinyl ether pairs 156-160
        mechanism 166-168
        monomer structures 161-162
        n-substituted maleimides 155-163
        photo-DSC 156, 159-160
        polymerization profiles/data MI-VE combination 163-166
        polymerization of tripropylene glycol diacrylate 166
        principles of charge/transfer photoinitiation 155-166
        reactivity of monomer/maleimide pairs 162-163
        styrene 161-163
        synthesis of amine precursors 171
        tertiary amine tetraphenyl borate 171-182
        vinyl maleimide charge transfer 156
        vinyl ethers 156-162, 163
Chromophores 35, 105, 109-116, 135-140, 170, 172-182, 252
CIS/Trans isomer of pent-zenyl phenol glyoxylates 151-152
Cleavage rates 35
C-N bond homolysis 181
Concentration dependence RAD 124
Counterion 44-46
Coupling constant/product 10, 41
Crosslinking
        epoxides 51
        carbonyl compound initiators 101-105
        crosslinked polymer 7-8
        matrix entanglement 7
        thermal reactions 56-63
        photocrosslinking process 113-114, 115-116
Cumene 104-105
Curing 240-241, 243-246

Cyanine borates 183
Coinitiators 195
Cyanine singlet lifetimes 42-45
Cyanines 42-47
Cyanoethyl acetate 133-134
Cyclobutanediones, racemic and distereoisomeric 36
Cyclohexane oxide 50
Cyclopropyl carbinyl esters 106-109

D
3D systems SLA 500 78-81
D-π-A molecules 306-307
D-π-A+X⁻ type organic salts as fluorescent probes 306-311
Dark reaction 7, 128
Decay parameters 137-138, 146, 177-178
Deprotonation 46
Depth of cure measurement 72-73
Diaryl iodonium ion 213
DIBF (2,4-diiodo-3-butoxy-6-fluorone) 199, 204, 205-209, 210-216, 217
    DIBF absorption spectra in MeCN 206
    DIBF absorption decay and absorption spectra 210-216
    DIBF anion reaction with diaryl iodonium ion 213
    DIBF/borate mechanism 214
    DIBF + DIDMA mechanism 215-216
    DIBF:borate:OPPI absorption spectra 212
    photo reduction mechanism 205-209
    photopolymerization 204
    radicals 210-216
    reaction with borate salt 207-209
    reaction with tertiary amines 207-209
    transient spectra 206, 208
    triplets lifetimes/quenching 207, 209
Dielectric spectroscopy 320-321
    alternating potential/mobility regimes 320
    comb electrode 320-321
    ion conductivity 320-321
    monitoring thermal cure 320-321
    polymeric monitoring 320-321
Dimethoxybenzoinyloxycarbonyl masked amines 51
Dimethoxyphenyl carbamates 51
Dimethylketene 35
Disposition of absorbed energy 14-15
DMPIBB $\rightarrow$ iodotoluene and biphenol 185-186, 187-189, 191-195
DMPITB 185-186, 187-189, 192-195
Donor/acceptor compounds 154, 155-166, 166-168, 245-247, 300
Donors 17, 103-104, 158, 159-160, 253, 300
DPIBB 191-195

DSC (differential scanning calorimetry) 56, 156, 159-160
Dye sensitised polymerisation of cycloalphartic epoxide 219
Dye/onium salts 217-219
Dye properties 195-199, 200-203, 205-206
Dye structures 211, 216

E
"Einstein" and Avagadro 33
EGDMA (ethyleneglycol dimethacrylate) 246-247, 308-311
Electron activity
    absorption spectrum 23
    acceptor/donor complexes of iodonium borates 184-186
    back electron transfer 181-182
    distribution 96-98, 142-144
    electrochemistry of TBR 134-135, 135-140
    electron acceptors 29, 42, 155-158, 159-160, 187-202
    electron donors 29, 42, 155, 156, 158-160
    electron paramagnetic/states/resonance (EPR) 23, 31, 32, 37, 92, 122, 142
    electron richness 95151-153
    electron spin 17-19, 21, 30, 95-98
    electron transfer/electrostatic interaction 42
    electron transfer from organoborates 169-170
    electron transfer from borates 170
    electron transfer reactions 32-67, 91-109, 116-140, 140-144, 154-195, 195-217
    electron transfers from chromophores 17, 19
    electron transfers from tertiary amine 98-100
    electron withdrawers 99
    electrons and bonding orbitals 19-21
    flash photolysis 23-25, 37, 39, 41
    intramolecular electron transfer 170-183
    transfer to benzophenone triplets 142-144
    withdrawal 99
Electron beam curing 252
Electrostatic interaction 42
Emission lifetime 270-279
Emission maxima (λ_{max}) 279-311
Emission spectra 16, 23, 31, 75,
    benzophenone in Freon 23
    emission maximum pyridine 1 during polymerization of EGDMA 311
    maleimide-fluoroprobe MA FP: fluoroprobe FP 294
    N-(1-naphthyl)-carbazole:polymerization TMPTA-VP 303
    normalised emission of ICT probes of differing polarity 308
    normalised emission pyridine 1 in EGDMA 310
    Rose Bengal derivatives 16
    rotor probes in PMMA 274-276
    singlet oxygen at $\approx$ 1270nm 31
    xenon lamps 75

Emission spectroscopy 123, 261-311
    anthracene probe 267-268
    characterization of polymeric materials 263-264
    classification of analytical techniques 235
    emission lifetime 268-271
    fluorescent chromophores 263-264
    fluorescent probes 261-264, 264-268, 268-271, 272-279, 285-288, 288-305, 306-311
    fluorescent polarization as a function of time 265-268
    luminescence 261, 291
    microviscosity 282, 297, 309
    optical coordinate system measurement 267
    phosphorescent probes 261
    polarisation of emission 265-268
    polarisation of emission of fluorescent probe 265-268
        fluorescence - influence of radiation dose for emission:decay time 268-271
        internal conversion (IC) (nonradiative decay) 268
        quantum yields of emission 269-271
        Tromsdorff effect 269
    emission intensity 271-279
        maleimide functionalised chromophores fluorescent probes 276-279
        rigid probes: coumarins/polycyclic aromatics 271
        rotor probes -Loutfy and Law 272-276, 284
    excimer forming probes 282-285
        excimer forming - intramolecular 283
        excimer formation-intermolecular 282
        percentage conversion 284
    ICT probes 281, 288-305, 306-311
        D-π-A probes: structure:dipolar photophysics 304-292, 297
        D-σ-A probes (Verhoeven et al) 300-305
        D-σ-A excited state processes:fluorescent activity 300-305
        dipole moment 290, 291-294, 300
        emission maxima 293, 294-298, 300
        fluoroprobe emission 301
        fluoroprobe-solvatochromism/rigidochromism interconnection 300-301
        Franck-Condon state 281, 285-287
        ICT (intramolecular charge transfer) probes 288-305
        influence of medium on emission - ICT probe 290-298
        methacrylates used in photopolymerization 299
        nonlinear optical (NLO) applications 289, 306
        solvatochromic emission 292-294, 296, 298, 300
        translational and rotational motion 281
    organic salts-D-π-A+X$^-$ type 306-311
        photophysics of D-π-A+ cation 306-311
        charge resonance 306, 307
        relaxation mechanism (rigidochromic nature) 307
        blue shift 307-310
        emission maxima: DODDMA: HEXDMA: BUDMA: EGDMA: DEGDMA: TREGDMA: TEEGDMA backbone 307-311

position of emission maximum (λ_{max}) 279-311
    blue shift 280, 294, 295-298
    relaxation processes 280-281
    relaxation-organic salts/ICT probes/chromophores/excimer 281
    wavelength shifts 279-281
TICT (twisted intramolecular charge transfer) probes 285-290
    Franck-Condon/Stokes shift state 288
    dual emission - TICT probes 286, 287, 288
    D-π-A architecture/probes 287-289

Energy changes and electronics 14, 29
Energy gap law (Kasha Rule) 15
Energy transfer 17-19, 20-23 28-32
Enolates 49??
Eosin (xanthene) 196, 197
EPR (electron paramagnetic resonance) 23, 37, 41
Epoxides 51, 56-64
Epoxy resins tests 59
Epoxy ring opening 63
Equimolar mixture 154
Erythrosin (xanthene) 196
ESR 120, 122-123, 319
ESR spectroscopy 319
    determination of polymerization termination 319
    identification radical concentration 319
    RAD 122-123
Excimer forming probes 279, 282-285
Exciplexes 108, 152, 166-168
Excited radiation and fluorescence measurement 15-23
Excited state 14-15, 15-23, 28-32, 32-82, 303
    reactions from 32-82
    formation of free radical by bond dissociation 32-41
    manifold - oxidation/reduction reactions 41-47
    quenching 28-32
Extinction coefficient 139, 260

F
Flash photolysis 23, 25, 37, 39, 41, 146
Film
    monitoring 242-252
    sensitivity parameters 61-70, 65-68, 112, 113-114
    solvent resistance 73-75
Film thickness 54-55, 65, 112, 260
Fluorescein dye 29
Fluorescence 15-23, 44-45, 54, 265, 304, 305
    efficiencies + lifetimes - cyanines 45
    p-benzoylphenyldimethyl radical (RAD) 123, 125, 126-127
    TBR 130-132, 132-134

TBR transient decay 135-138
TBR transient fluorescence 135, 137, 138
Fluorescent probes 261-264, 265-268, 268-271, 272-279, 287-288, 302-305
Fluorescent spectroscopy (real time) 351-362
    kinetic profiles of DEGDA (influence of initiator concentration) 360
    emission ratio - double bond conversion of acrylate formulation 361
    kinetic profiles of various resins 359
    emission intensity-laser initiated polymerization:TMPTA/VP 354-356
    optical+ electronic monitoring-kinetics of laser initiation 352
    geometric scheme for probing kinetics of photopolymerization 353, 358
    real time monitoring laser induced polymerization 351
Fluorine substituted aromatics 184
Fluorones as initiators 198-203, 204, 205-209, 210-216
    spectroscopic properties 199
Fourier Transform mode 254, 264
Free radical 3, 4, 8, 32-36, 91, 101, 127
    formed by bond dissociation 32-36
    free radical and azo 10
    mechanism in free radical polymerisation 4
    photochemical reactions and free radicals 8-12
FTIR spectra of donor acceptor 245-246
Fusion Systems D, H, Q and V bulbs 72

G
γ induced polymerization 252
Gegen ion 62, 189
Gel effect 7, 112
General research techniques for Analysis 239-321
    dielectric spectroscopy 362-363
    photodifferential scanning calorimetry (Photo-DSC) 322-329
    real time fluorescence spectroscopy 351-362
    real time (FT) IR spectroscopy 329-340
    real time UV spectroscopy 340-351
Glycerol 60, 67
Glyoxylates 144-153
    cyclic transition state -thiol esters 149
    cyclisation mechanism -thiol esters 148
    kinetic data - laser flash 148
    photoreaction products and yields 147
    structure/products 149-150
    triplet states/lifetimes 150-153
Gomberg 119, 120, 123
Ground state 14, 18, 19-23
    ICT probe 300
    fluorescent probe 296
    RAD 125-126
    TBR 130-131, 135

H
Heat transfer 14-15
Heats of formation 241
Heteroatoms 265
Hexaphenylethane 119-120
Homolytic decomposition - aliphatic azo 10
Hund's rule 20
Hydrocarbon residues 104
Hydrocarbons 103-105
Hydrogen abstraction 101-105, 106-107, 108-109, 142-144, 148-150, 152-153, 180, 193-195, 213-215
Hydrogen donors 103-105, 180
Hydrogen extraction 180
Hydroperoxide 104-105

I
ICT probes 281, 288-305
'Initiator free' 155
Imaging systems 51-68, 109-112, 112, 113-114, 170
Inhibitors 184-186
Initiation - photocuring 7-12, 12-23, 32-67, 69-81, 154-195, 195-217
    cation 47-50
    charge transfer 155-195
    lamps and lasers 69-81
    organic dyes 195-217
Initiator decomposition 8-12 49, 52-53, 174-182, 189-191, 217-219
Initiators (photoinitiator) 61-69
    bond dissociation:bond homolysis after photon absorption 91-109
    carbon compound 101-111
    charge transfer 154-195
    organic dyes 195-217
    oxidation reduction 144-159
    phenylglyoxylate 105-109, 144-153
    tertiary amines 98-100
Initiator solubility 187
Initiator systems
    benzophenone triplets/amines 142-144
    benzophenone/Michler's ketone 91
    camphor quinone/tertiary amine 91, 140-141
    carbonyl compounds 106-111
Intensity of emission 271-279
Internal conversion 14-15, 18-19, 19, 261-264
Intersystem crossing 19-23, 261-264
Intramolecular electron transfer 144-153
Intramolecular charge transfer probes (ICT) 288-305

Intramolecular cycloaddition reactions 144
Iodonium borates
    BPPIBB 185-187
    DMPIBB 185-187, 192-195
    DMPITP 185-186, 191-195
    DPIBB 185-187, 191-195
    OPPIBB 185-186, 187-191
    OPPITB 187-191
    photochemical reactions with methyl methacrylate 191-195
    photoinitiating activities 187-191
    photolysis iodonium borate salts 191-195
    with Kesheng Feng 184-186
Iodonium bromide 185
Iodonium butyltriphenyl borate salts 185, 187, 195
Iodonium cation 195
Iodonium salts 46, 47-50, 154-155, 185-186, 187-191, 191-195, 217-219
IR (Infrared) spectra 39, 40, 60-61, 62
    (2,4,6-trimethylbenzoyl)dimethylphosphine oxide 37, 39, 40
    FT/IR of donor acceptor 246
    transmission EPON + DMBA 60
    transmission PH_4B gegen ion + EPON 838 62
IR spectroscopy 36-41, 243-252
    absorptions and degree of cure - various resins 243-252
    analysis of polymeric networks with varying IR modes 244-247
    Attenuated Total Reflectance (ATR) mode 244, 245, 251
    donor acceptor compound polymerization monitoring 245-247
    film irradiation 244-245
    Fourier Transform Mode (FT) 243
    FT/IR spectroscopy 243
    IR microscopy 251
    Langmuir-Blodgett films 248-250
    Single Reflectance Absorbance mode (SRA) 247-249
    specific absorptions 243
IR transient absorption 37
Isobestic points 205
Isomerization scheme for azostilbene 257
Isomerization volumes for trans-cis isomerization 259-260

J
Jablonski diagram 18, 20, 22, 31, 78
    lasing action 88-89

K
Kamlet Abraham and Taft (KAT) method-solvatochromic probes 256
Kasha rule 15
Ketone chromophores 144-153

Ketones 9, 34-36, 91, 94, 101-102, 142-144, 144
Ketyl biradical 148
Kinetics 7, 10, 27-29, 36-41, 48-50, 75-77, 163-166, 198-203
 azo induced decomposition 10
 chain length 7, 49, 75
 characteristics of formed radical 34-41
 characteristics of triple state and absorption 28-32
 decay and acceleration 38
 pseudo first order kinetics 38
 steady state 198-203
 thermal reactions 3
 trace 27
 zero order 48-49
Kinetics of real time cure measurement 322, 325

L
Lamps 69-81
 CW lasers 81
 dental lamp 69-70
 depth of cure 73
 Fusion System 71, 72, 74
 pulsed sources 76-77
 rare gas 78, 79, 81
 solvent resistance 74
 tungsten/halogen 71, 76
 UV bath 81
 xenon 75, 76-77
Lasers 24-26, 69, 77-81
 emitted wavelengths 79
 He-CD laser 81
 pulsed laser 24-26, 77-81
 rare gas 77-81
 UV Bath 81
Laser beam deflection 64
Lasing action 78
Lifetimes
 DIBF 210
 fluorescent probes - emissions 268-271
 glyoxylate 147-148
 TBR 135-138
Light absorption 12-23, 13-14, 17-19, 20, 22, 27
 neodynium/YAG laser 246
Light emission 15-19, 106 (See also Chromophores, Emission spectra, Emission Spectroscopy, Fluorescence, Fluorescent probes)
 fluorescence 15-19
 phosphorescence 19-23
Light quanta 13-14

Luminescent probes 261, 279

M
7-methoxycoumarin 4-methyl radicals 180
(meth)acrylates for photopolymerisation 299
Matrix entanglement 7
Mead initiators 42
Mechanism in free radical polymerization 3-4
Mercury lamps 71
Methoxy group(vibrational character) 182
Methyl cyclohexame 130, 132
Methyl groups 193-195
Methyl methacrylate 4-6, 118, 191-195
Methyl phenylglyoxylate 26, 105
Methylene blue 196, 216
Mineral oil glass 132-133
Molar extinction coefficient 13-14, 138-140
Molecular dioxygen 21
Molecular rotors 132-134
Monitoring techniques real time (See also Real time) 321-365
Monochromator 13
Monomers/Prepolymers 3-8, 155-156 (For further information See Volume II)
    naphthalene 34
    rate constants 4-6
    reactivities 162-166
Monomer structures 161-162
Morphology in polymeric films 242

N
2-[N,N-dimethyl,N-(p-benzoyl)benzylammonium(triphenyl-n-butylborate)] ethyl methacrylate 54
4-nitro-N,N-dimethylaniline 256
4-nitroanaline and 4-nitroanisole 256
n-π* transition 94-95, 95-96, 142-144, 144, 173
N-phenylglycines 99-100
N-phenylmaleimide acrylate 154-155, 163
N,N,N-trialkyl-N(p-benzoyl)benzylammonium triphenylbutylborate 52
Naphthalene/ground state + triplet 32
Naphthalene ring 143-144
Negative tone imaging systems 57, 68
Neodynium/YAG laser 246
NMR 120
NMR spectroscopy 311-319
    cross-polarization magic-angle spinning (CPMAS) 13°NMR 311, 317
    carbon spin lattice relaxation/times/standard formulation STDF 312, 314, 315-319

    crosslinking:double bond conversion 316
    relaxation rates for homopolymers 317-319
    solid state NMR 312-313
    solution spectrum of PEGA: TMPTA: and DPHPA 314
    spin diffusion:proton relaxation 317-318
N,N-dimethyl-n-butylamine (DMBA) 58-65
Nonradiation conversion of electrons 19
Norrish I 12, 32-36, 41, 96
Norrish II reactions/products 105-107, 110, 115-116, 144-146, 149, 150
Norrish II single electron transfer (SET) 91

O
Octanediolacrylamide (ODA) 243-250
Olefins 36, 46, 107, 150-153, 169-170
Onium salts 46, 217-219
On line monitoring 363-365
   fluorescence spectroscopy 364-365
OPPI (p-octyloxy)phenyliodonium) 207-213
Organic dyes 195-216
   DIBF photoreduction mechanisms 201,213, 214, 215, 216
   DIBF structures 204, 208, 209
   DIBF singlet oxygen decay and quenching 207-208
   oxidation/reduction reactions - visible initiator 195-216
Organic salts -D-π-A+X- fluorescent probes 306-311
Organoborates 169-170
Ortho-nitrobenzyl carbamates 51
Oxetanes 101
Oxidation of excited states 46-47
Oxidation of triphenylmethyl radical 134-135
Oxidation/reduction 140-144, 144-153, 195-216, 217-219
   camphorquinone 140-141
   chains - cationic polymerization by dye/onium salts 217-219
   electron transfer to benzophenone triplets 142-144
   organic dyes 195-216
   reactions with visible initiator systems - organic dyes 195-216
Oxidised borate and dissociation 45
Oxygen 148-150, 175, 190-191, 206-210
   singlet oxygen decay + quenching 207

P
π-π^* transition 92-97, 173
π-π^* triplet 143
π-π^* transition/state 94-95, 95-96, 142, 173
2-pentanone 33
p-acetylbenzyl radical 172
p-acetylbenzyl radical 180

p-benzoylbenzoyloxy radical 117
p-benzoylbenzyl radical 180
p-benzoylphenyldiphenylmethyl radical dimerisation (RAD) 122-128
    absorption spectrum 123, 124
    chemistry in solution 127-128
    chemistry in absence of dimer 128-134
    dimerisation 124, 125
    fluorescence 123, 125, 126
    resonance structures 122
    spectra 123, 124, 126, 129, 130, 131
p-benzoylphenyldiphenylmethyl radical reactions without dimer - (TBR) 128-132
p-benzoyl tertbutyl perbenzoate (PBE) 116-117
p-dimethylaminobenzophenone 142-143
Paramagnetic quenchers 30
Paterno-Büchi reaction 100, 107-109, 150
Pattern transfer 56-61
Pendant dimethylamino functionalities 53, 55
Peresters 116-119
Peroxides 8-10, 117
Peroxy ester chemistry 116-119
Phenylazoisobutyronitrile 48
Phenyl chromophore imaging system 109-111, 111-112, 112-116
    photomer analysis 115-116
    predevelopment images 113-114
    resolution of imaging system 111-112
    sensitivity and contrast 112-113
Phenylfluorenyl radical 121
Phenylglyoxylate chromophore 109-111
Phenylglyoxylate photochemistry (see also Glyoxylates) 105-109, 144-153
    electron transfer 148
    glyoxylate triplet states/reactions 144-153
    light initiated dissociation 106
    photolysis 146, 153
    reactions 106-108
Phosphene oxides 36, 91
Phosphines 36-40
Phosphonyl radical 36-40
Phosphorescence 19-23, 28-32
Photo multiplier tubes 12??
Photodecarbonylation 35-36
Photodecomposition of tetraphenylborates 174-182
Photodissociable benzoyl peroxide 9
Photoinitiation - principles
    anionic polymerization of amines 58
    bond dissociation reactions:bond homolysis photon absorption 91-98
    camphor quinone 140-141
    carbonyl group as target using carbonyl group 116-140
    carbonyl group character 95-98

carbonyl group structures/mechanisms 92-95
cation generation 47-50
charge/transfer complex photoinitiators 154-195
chemical amplification 67
CT complex mechanism 166-167
CT systems reactions and reactivity 154-195
disposition of absorbed energy 12, 14-15
dye/onium salts 217-219
electron transfer to benzophenone triplets 142-144
electron transfer from tertiary amines 98-100
electron transfer/organoborates (Polykarpov, Sarker, Kaneko, Feng) 169-170
excited carbonyl group reactions:$2+2\pi$ 100-101
formation of free radicals by bond dissociation 32-36
formation of phenylfluornene 121
generation of reactive bases 51-68
hydrogen abstraction route → linking with carbonyl compounds 101-105
initiator systems 12-23
intramolecular electron transfer/oxidational reduction (carbonyl chemistry) 144-153
lamp/laser curing 69-81
laser emissions 77-81
light absorption 13-14
light emission (phosphorescence) 19-23
light emission (fluorescence) 15-19
maleimide/vinyl ether pairs 156-166
organic dyes 195-216
oxidation reactions with visible initiator systems 195-216
oxidational reduction 41-47, 144-153, 217-219
peroxy ester chemistry 116-119
phenylborates in acetonitrile 174-182
phenylglyoxylate 105-109, 144-153
phenylglyoxylate chromophore 109-116
phenylglyoxylate reaction mechanisms (Shengkui Hu) 105-109, 144-153
photoreduction p-dimethylaminobenzophenone 143
principles of charge/transfer photoinitiators 154-166
principles of light absorption 12-23
RAD reaction D_2O + CH_3OD 127-128
RAD in solution 127-128
RAD (p-benzoylphenyldimethyl radical) 122-132
reactions of iodoniumborate salts with methyl methacrylate 191-195
reduction with amines 142-144
single electron transfer in carbonyl group compounds 140-144
synthesis of 2[N,N-dimethyl,N-(p-benzoyl)ammonium(triphenyl-n-butylborate)]ethyl methacrylate 59
TBR electrochemistry 134-135
TBR fluorescence (in rigid medium) 132-134
TBR synthesis/characteristics 128-132
TBR transient studies 135-140
thiol ester (glyoxylates) reactions 144-153

transient spectroscopy 23-32
triphenyl radical (Alexander Nikolaitchik) 119-140
Photoinitiators 8-12,
concentration monitoring 254
p-acetylbenzyl radical 172
photochemistry and photophysics 12-23
Type I reactivity + mechanisms 36-41
Photolysis 25, 139, 168, 172, 175, 177, 178-180
2-(bromomethyl)naphthalene 177-178
apparatus 25
of phenylglyoxylates 153
of RAD in PhH + visible light 127
of tertiary amine tetraphenylborate 156, 177-176, 175
of tetrafluoroborate 177
Photolysis of TBR in PhH 127
Photons 14, 18-19, 32, 33-34, 91, 139-140
density and reversal 78-79
Photopolymerization 3-8, 12, 17, 47-51, 51-68, 75-76, 91-219 (See also Photochemistry)
amines 58-68
anionic polymerization of epoxides 58
cation initiated 47-50, 217-219
chain transfer 7
copolymerization of difunctional maleimide vinyl ether monomers 165
copolymerisation of maleimide/vinyl pyrrolidine mixture 165
crosslinking 7
dark reaction 7
decomposition of peroxides 8-12
dye sensitised - cycloalphartic ester 219
epoxide curing 65-68
epoxy materials 56, 57-68
fluorones 198-203, 204, 205-216
free radical and photochemical reactions 8, ???
γ induced 252
induced by xanthene dye/tertiary amine + iodium salt 218
influence of bismaleimide on tripropyleneglycol diacrylate + hexandeioldiacrylate 166
initiation 2, 6
iodonium borates 187-191, 191-195
kinetic chain length 7, 49-50, 75-77
kinetics of monoacrylate: monovinyl ether: monomaleimide 164
laser initiated 78, 79-81
light absorption 12, 13-14, 18-19, 20-21, 26-30 (see also Absorption spectra)
maleimide (MI)vinyl ether(VE) 162-163, 164
monoacrylate and monovinyl ether 164
oxygen/reduction chains cationic-initiated by dye/onium salts 217-219
photocuring 7, 8, 93, 97-98, 98-100
propagation 3-4, 7, 49-50
quantum yield 8-9, 33-36

        rate constants 4-8, 174-177, 179-180
        reactive bases 51-68
        reduction mechanism-fluorones 204-216
        steady state 4, 7, 117-118, 173
        termination 4, 7, 49-50
Photoreactions of tetraphenylborate in acetonitrile 175-182
Photoresists 67-67, 111, 113-114, 189
Photoresists 74, 119
Pinacol products 143
Polarization of emission 265-268, 292
Poly(methylmethacrylate) 132-133
Polyborates 184
Polybrominated biphenyl (PBB) 11-12
Polychlorinated biphenyl (PCB) 12
Polymeric amines 52-68
Polyols 57-58, 67
Polystyrene 104-105
Polyurethane oligomers 56
Positive-tone photoimageable polymers 53
Principles of charge transfer complex photoinitiators 155-166
Probes/types/ absorption, emission 261-264, 265-267, 268-269, 271-279, 279-282, 282-285, 285-288, 288-305, 306-311
Protons 47, 50
Pseudo first order kinetics 38
Pulsed laser 24-25, 77, 79
"Pump and probe" 24

Q
Quantum yields 9, 33-35, 44, 47-50, 51-52, 67-68, 102-103, 118, 132-134, 138-140, 174, 264
Quarternary ammonium borates 52-55
Quenching\quenchers 17, 28-29, 30-32, 39, 43, 144-146, 170-172, 173-180, 206-209, 210-211, 264
        bimolecular quenching of tetrafluoroborate 174-175, 179
        DIBF 206-209
        fluorescent 17
        rate constants (k_q) 174, 176
Quinoid dimer 120, 139
Quinomethine 197
QUV weatherability test 180 ???

R
RAD 122-127, 127-128, 128
RAD in solution 127-128
        benzophenone 127, 128
        deuterated methanol 128

       dimer dissociation 128
       p-benzoylphenyldiphenylmethane 127-128
       p-benzoylphenyldiphenylmethyl ether 127-128
       photolysis 127
       photoreaction 128
       synthesis and characteristics of TBR 129-132
       tetraphenylethylene/1,4-dibenzoylbenzene 127
Radiation filtering 34
Radical anions 142
Radical traps and chain transfer agents 10, 11
       iodine; butanethiol; galvinoxyl; 11
Radicaloid electron distribution 142
Rare gas lasers 77-81
Rate constants 4-8
       bimolecular 40, 179-180
       decay of excited state 174-177, 262-263
       dielectric 294
       dissociation - aliphatic ketones 35
       fluorescence decay 262
       quenching 39
       reduction potentials 174
       SET 48-49, 52, 53-54, 91, 207
       substituent 99-100
Rate of polymerization + shrinkage 327
RBAX (decarboxylated Rose Bengal) 98-100, 197-203
Real Time monitoring techniques 321-363
       dielectric spectroscopy 362-363
       fluorescence spectroscopy 351-362
       (FT)IR spectroscopy 329-340
       photodifferential scanning calorimetry (Photo-DSC) 322-329
       UV spectroscopy 340-351
Reactivity of maleimide/p-methoxy styrene systems 162
Reactivity of N-ethylmaleimide/styrene systems 163
Recombination rate 75
Red shift 123-124, 126-127, 130-132, 188-189
Rehm-Weller equation 42, 91
Regional bands of energies 15
Rehm-Weller calculation 42, 135
Reichardt's dye 256
Relaxation process 280-281, 307, 312-314, 315-319
Resist film 111, 113-124
Resonance 92, 122
Ring opening 63, 101
Ring closure 108
Rose Bengal (RBAX) 16, 98-99, 196-198
       as initiator 198-199
Rotational levels 14, 20
RTIR spectroscopy 164, 165

S

Scanning probe microscopies (SPM) 64
Selenium dioxide oxidation of camphor 140
SEM (scanning electron microscopy) 54-57, 111
SET (Single Electron Transfer) 48-50, 52, 53-54, 91, 207-209
Shuttering 79
Single Reflectance Absorbance Fourier Transform Spectroscopy (SRA-FTIR) 247-249, 329
Singlet oxygen 30-32, 207
Singlet state 17-19, 19-21, 29, 30, 31, 32, 261
Sinussoidal beam profile 80
Slow fluorescence 29
Small molecule organic reactions and azo 10
Sodium tetraphenylborate 189
Solid polymeric networks 238
Solution spectrum (^{13}C) of PEGA, TMPTA, DPHPA 314
Solvatochromic probes/dyes 256-258
Solvent resistance 73
Solvent cage 10, 11-12, 42
Spectral analysis 26
Spectroscopic techniques 241-321 (See also specific technique)
    dielectric spectroscopy 320-321, 362-363
    emission 261-311
    EPR 37
    ESR 319
    fluorescence 351-362
    IR 37, 242, 243-252
    NMR 311-319
    UV/visible light 242, 252-261
Spin coating 256
Spin pairings and singlet state 17
Spin density 41
Spin coated film 60-68, 112
Spin configuration 15, 17
STDF - standard formulation 312-315
Steady state 4, 48, 75, 198
Stereoisomerization 93
Steric effects 58
Stilbene and azostilbenes 257-260
Structures of some dyes/TIHF;TIHCF;TBHF;TBHCF 210-211
Styrene 4-5, 118, 158, 160, 161-162
Sulphur 144-146, 149
Symmetrical iodonium salts 184, 187

T

TBR (see below)

Tert-butyl substituted triaryl methyl radical (TBR) 129-140
    carbocation 134-135
    electrochemistry 134-135
    extinction coefficients 130
    fluorescence in rigid medium 132-134
    fluorescence spectra 136, 137
    fluorescence probes/spectra 132-134, 278
    photolysis 138-139
    transient studies 135-140
    transient absorption spectra 138
    transient decays 136, 137
    triphenylmethyl radical 134
(2,4,6-trimethylbenzoyl)diphenylphosphine oxide 36
1-3 dioxolane quantum yield 48
4-tert-butyliodobenzene quantum yield 48
t-butyl alcohol 9
TEM mask 113-114
TEM grids - G200-Cu; T2000-Cu 55, 111-112
Termination - reaction of radicals 4, 3, 7, 49
    chain termination 49
    gel effect-reduced rate 7
Tert-butoxy radical 94, 103-104, 129-132, 134-135, 138-140
Tertiary amines 46, 57-68, 98-100, 141, 142-144, 170-182, 207-209, 217-218
Tetra substituted tin 46
Tetra-4-t-butylehtylene 139
Tetrabutylammonium perchlorate (TBAP) 173
Tetraphenylboranyl radical 181
    back electron transfer 181
Tetrafluorophenyl borate systems 183-184
Tetramethylsuccinonitrile coupling product 10
Tetraphenyl borates 170-182, 183-184, 187, 190
    decomposition 174
    photocleavage 181, 179
    photolysis 175
    quenching 174, 179-180
    singlet state 179, 180
    spectra 175-179
Thermal reactions 3, 9
Thermal repopulation 29
THF (polymerizable ethers tetrahydrofuran) 47-50
Thiazenes 196
Thiol esters 144-148
    phenylglyoxylic acid 144-149
TICT (twisted intramolecular charge transfer probes) 285-290
Titanium dioxide 76
Toluene 180
Trans-cis isomerization 257-260
Transient absorbance 26-28, 43-44, 135-136, 138-139, 207, 208-209

Transient absorption 29, 138, 175, 176, 177, 178, 179, 205, 207
Transient spectra 41, 146, 175, 176, 177, 178, 179, 196, 197, 198, 199, 205, 208, 212, 233
    4-(bromethyl)benzophenone 195
    absorption tetraphenylborate in acetonitrile 194
    absorption TBR 152
    benzene solution (flash photolysis) 146
    cyanine dyes 44
    decay: (2,4,6-trimethylbenzoyl)diphenylphosphine oxide 41
    DIBF 206
    DIBF; DIDMA 208, 212
    laser flash photolysis tetraphenylborate 177
    tetraphenylborate 175, 176, 178, 179
    time profile tetraphenylborate (normalised) 176
    triplet absorption 28
Transient spectroscopy 25-34, 102
    DIBF 206-209, 212
    DIBF 212
Transitions 15-16, 19, 93-95, 97-98, 100-101
Transparent substrate 244
Triaryl methyl radical 121
Tribenzylamine 217??
Triethylamine 142
Trimethyl borane 183*
Triphenylmethyl radicals 119-121 (See also TBR)
    oxidation 134-135
Triplet quencher 29
Triplet state 19-23, 23, 27-32, 33, 35-36, 142-144, 144, 147-153, 169-170, 170, 173-176, 179-181, 205-209, 210-211, 261
Trityl cation 134-135

U
Unpaired electron spins 30
Unsaturated molecules and photochemical reaction 17
UV (351 and 355nm) and Ar-ion laser 78
UV spectra 41
    (2,4,6-trimethylbenzoyl)diphenylphosphine oxide in deoxygenated heptane 41
    Irgacure 369 as function of irradiation time 255
    TEGDA, pure poly-TEGDA p 266
UV/Vis Spectroscopy 252-261
    absorbance photochromic probe/thermally polymerised sample 258
    isomerization scheme for azostilbene 257
    Kamlet Abraham and Taft (KAT) method 256
    photochromic probes 257, 258-260
    photodecomposition of 2-benzyl-2-dimethylamino-1-(4-morpholinophenyl)-butatone-1 (Irgacure 369) 254-255
    photoinitiator bleach 254
    photopolymerization of maleimide and electron rich alkene 253-254

        probes/types/absorption 256-258, 260
        Reichardt's dye 256-257
        solvatochromic probes 256-257
        stilbene and azostilbene reactions 251-260
        trans-cis isomerization - stilbenes and azobenzene derivative 257-26074
        transition parameters 252
        UV-Vis absorbing probes parameters 260-261

V
Vibrational levels of excited states 15-16, 18-19, 20-21, 78
Vinyl acetate 6
Vinyl polymerization 3
Vinyl/maleimide charge transfer systems 155-166, 166-169
        photo-DSC measurements 159-160
        polymerization data 164-166
        reactivities of systems 162-163
Vinyl initiator systems 196
Visible initiator systems - organic dyes 195-217
        DIBF 205-209, 210-216
        fluorones 199-203, 204
        methylene blue/xanthenes 196
        quinomethine group 197
        RBAX 198
Visible regions of the spectrum 10, 41-42, 48, 48, 78

W
Wavelengths emitted by common "Photochemistry lasers" 79
Wurtz-Fittig reaction 119

X
Xanthene/xanthones 142, 196-198, 205, 217-218
Xenon 74, 75, 76-77

Y
Yellowing, longwave, UV and visible radiation 10, 119, 140, 141

Z
Zero order 49
Zero point vibrational level 15